Australian Island Arks

CONSERVATION, MANAGEMENT AND OPPORTUNITIES

EDITORS: DORIAN MORO, DEREK BALL AND SALLY BRYANT

CSIRO PUBLISHING

National Library of Australia Cataloguing-in-Publication entry

Australian island arks : conservation, management and opportunities / Dorian Moro, Derek Ball and Sally Bryant (editors).

9781486306602 (paperback)
9781486306619 (epdf)
9781486306626 (epub)

Includes bibliographical references and index.

Island ecology – Australia
Islands – Australia – Management.
Islands – Australia – Government policy.

Moro, Dorian, editor.
Ball, Derek, editor.
Bryant, Sally, editor.

Published by

CSIRO Publishing
36 Gardiner Road, Clayton VIC 3168
Private Bag 10, Clayton South VIC 3169
Australia

Telephone: [+613] 9545 8555
Email: csiropublishing@csiro.au
Website: www.publishing.csiro.au

Front cover: (top) View from Mount Gower on World Heritage Lord Howe Island (photo: Jack C Shick); (middle left) Flatback turtles (*Natator depressus*) nest on islands off Australia (photo: Dorian Moro); (middle centre) King Penguins on Macquarie Island (photo: Krudler/Wikimedia Commons, CC BY-SA 4.0); (middle right) Lord Howe Island Stick Insect (photo: Granitethighs/Wikimedia Commons, CC BY-SA 3.0); (bottom) Montebello Islands (WA), selected for the re-introductions of threatened mammals and birds (photo: Judy Dunlop).

Back cover: (left to right) Cocos (Keeling) Islands, situated in the Indian Ocean, site of a local black rat eradication program (photo: Judy Dunlop); Legendre Island, Dampier Archipelago (WA), site of a black rat eradication program (photo: Judy Dunlop); Brampton and Carlisle Islands, Central Great Barrier Reef Region (Qld) (photo: Derek Ball).

Set in 10.5/14 Palatino LT Std and Optima
Edited by Adrienne de Kretser, Righting Writing
Cover design by Andrew Weatherill
Typeset by Desktop Concepts Pty Ltd, Melbourne
Index by Indexiçana
Printed by Ingram Lightning Source

Mar26_RP_ILS

CONTENTS

List of contributors v

1 **Introduction: Australia's island arks** 1
Dorian Moro, Derek Ball and Sally L. Bryant

2 **Values of islands across Australia's states and territories** 11
Keith Morris, Dave Algar, David Armstrong, Derek Ball, Sally L. Bryant, Peter Canty, Peter B. Copley, Chris Dickman, Alaric Fisher, Graeme R. Gillespie, Michael Johnston and Dave Kelly

3 **Enhancing island conservation outcomes: the policy and legal context, need, and options** 45
John C.Z. Woinarski, Andrew A. Burbidge and April E. Reside

4 **The 4Cs approach to island management** 61
Derek Ball, Jodie Williams and Margaret Christian

5 **What's next on the list? Prioritising management actions on Australia's islands** 71
Cheryl A. Lohr, Christopher M. Baker, Michael Bode and Robert L. Pressey

6 **Island pest management** 85
Keith Springer

7 **Australian islands as 'arks' for biodiversity** 99
Andrew A. Burbidge, Sarah Legge and John C.Z. Woinarski

8 **Indigenous Protected Areas and islands** 115
Ellie Bock, Philip Rist, Tom Vigilante and Neil Waina

9 **Integrating resource development with island conservation: Barrow Island as a model for conservation and development** 131
Dorian Moro, Johann van der Merwe, Melissa Thomas, Andrew Smith and Russell Lagdon

10 **Australia's World Heritage islands** 147
Jennie Whinam and Justine D. Shaw

11 **The interplay between tourism and conservation on islands** 165
Derek Ball, Sally L. Bryant and Roland Mau

12 **Partnerships for island conservation: it's all about people** 177
Sally L. Bryant and Peter B. Copley

13 **Value of islands for the marine environment** 191
Mark Read, Graham Hemson and John Olds

14 **Ecological restoration on New Zealand islands: a history of shifting scales and paradigms** 205
David Towns, Keith Broome and Alan Saunders

15 **Managing islands in the context of climate change** 221
Stephen T. Garnett and April E. Reside

16 **Australian islands: current thinking, emerging themes and a way forward to manage these offshore assets** 235
Derek Ball, Dorian Moro and Sally L. Bryant

Index 249

LIST OF CONTRIBUTORS

Dave Algar
Science and Conservation Division, Department of Parks and Wildlife, Woodvale, WA 6946, Australia, Dave.Algar@dpaw.wa.gov.au

David Armstrong
Department of Environment, Water and Natural Resources, PO Box 1047, Adelaide, SA 5001, Australia, david.armstrong@sa.gov.au

Christopher M. Baker
Centre for Biodiversity and Conservation Science, School of Biological Sciences, University of Queensland, Brisbane, QLD 4072, Australia; and CSIRO Ecosystem Sciences, 41 Boggo Rd, Dutton Park, QLD 4102, Australia, c.baker7@uq.edu.au

Derek Ball
Wild Mob Pty Ltd, PO BOX 1742, Milton, QLD 4064, Australia, Derek@wildmob.org

Ellie Bock
Regional Advisory and Innovation Network (RAIN) Pty Ltd, PO Box 104, Mena Creek, QLD 4871, Australia, rainqld@bigpond.com

Michael Bode
Australian Research Council Centre of Excellence for Coral Reef Studies, James Cook University, Townsville, QLD 4811, Australia, bodem@unimelb.edu.au

Keith Broome
Department of Conservation, Private Bag 3072, Hamilton 3240, New Zealand, kbroome@doc.govt.nz

Andrew A. Burbidge
87 Rosedale St, Floreat, WA 6014, Australia, amburbidge@westnet.com.au

Sally L. Bryant
Tasmanian Land Conservancy, PO Box 2112, Lower Sandy Bay, TAS 7005, Australia, sbryant@tasland.org.au

Peter Canty
Department of Environment, Water and Natural Resources, PO Box 1047, Adelaide, SA 5001, Australia, peter.canty@sa.gov.au

Margaret Christian
Norfolk Island Flora and Fauna Society, PO Box 999, Norfolk Island, NSW 2289, Australia, Justmarg28@hotmail.com

Peter B. Copley
Department of Environment, Water and Natural Resources, GPO Box 1047, Adelaide, SA 5001, Australia, Peter.Copley@sa.gov.au

Chris Dickman
School of Life and Environmental Sciences, University of Sydney, NSW 2006, Australia, chris.dickman@sydney.edu.au

Alaric Fisher
Department of Environment and Natural Resources, Palmerston, NT 0831, Australia, alaric.fisher@nt.gov.au

Stephen T. Garnett
Research Institute for the Environment and Livelihoods, Charles Darwin University, NT 0909, Australia, stephen.garnett@cdu.edu.au

Graeme R. Gillespie
Flora and Fauna Division, Department of Environment and Natural Resources, CSIRO Complex, 564 Vanderlin Dve, NT 0828, Australia, Graeme.Gillespie@nt.gov.au

Graham Hemson
Queensland Department of National Parks, Sport and Racing, 61 Yeppoon Rd, Parkhurst, QLD 4702, Australia, graham.hemson@npsr.qld.gov.au

Chantelle Jackson
Australian Wildlife Conservancy, Subiaco East, WA 6008, Australia, Chantelle.Jackson@australianwildlife.org

Michael Johnston
Science and Conservation Division, Department of Parks and Wildlife, Woodvale, WA 6946, Australia, Michael.Johnston@dpaw.wa.gov.au

Menna Jones
Department of Zoology, University of Tasmania, Private Bag 5, Hobart, TAS 7001, Australia, menna.jones@utas.edu.au

Dave Kelly
Lord Howe Island Board, Lord Howe Island, NSW 2898, Australia, dave.kelly@lhib.nsw.gov.au

Russell Lagdon
Australian Business Unit, Chevron Australia Pty Ltd, 250 St Georges Tce, Perth, WA 6000, Australia, rdla@chevron.com

Sarah Legge
National Environmental Science Programme Threatened Species Recovery Hub, Fenner School of Environment and Society, Australian National University, ACT 0200; and Centre for Biodiversity Conservation, University of Queensland, St Lucia, QLD 4072, Australia, sarah.legge@anu.edu.au

Cheryl A. Lohr
Science and Conservation Division, Department of Parks and Wildlife, Woodvale, WA 6946, Australia, Cheryl.lohr@dpaw.wa.gov.au

Roland Mau
Department of Parks and Wildlife, 17 Dick Perry Ave Technology Park, Kensington, WA 6151, Australia, Roland.Mau@dpaw.wa.gov.au

Dorian Moro
School of Veterinary and Life Sciences, Murdoch University, Murdoch, WA 6150, and Science and Conservation Division, Department of Parks and Wildlife, Woodvale, WA 6946, Australia, d.moro@murdoch.edu.au

Keith Morris
Science and Conservation Division, Department of Parks and Wildlife, Woodvale, WA 6946, Australia, Keith.Morris@dpaw.wa.gov.au

John Olds
Queensland Department of National Parks, Sport and Racing, 28 Fraser Rd, Gympie, QLD 4570, Australia, john.olds@npsr.qld.gov.au

Robert L. Pressey
Australian Research Council Centre of Excellence for Coral Reef Studies, James Cook University, Townsville, QLD 4810, Australia, bob.pressey@jcu.edu.au

Mark Read
Great Barrier Reef Marine Park Authority, 2-68 Flinders St, Townsville, QLD 4810, Australia, mark.read@gbrmpa.gov.au

April E. Reside
National Environmental Science Programme Threatened Species Recovery Hub, Centre for Biodiversity and Conservation Science, University of Queensland, St Lucia, QLD 4072, Australia, april.reside@gmail.com

Philip Rist
Girringun Aboriginal Corporation, PO Box 303, Cardwell, QLD 4849, Australia, eo@girringun.com.au

Alan Saunders
Waikato Regional Council, Private Bag 3038, Waikato Mail Centre, Hamilton 3240, New Zealand, alan.saunders@waikatoregion.govt.nz

Vince Scoleri
Department of Zoology, University of Tasmania, Private Bag 5, Hobart, TAS 7001, Australia, vince.scoleri@utas.edu.au

Justine D. Shaw
Centre for Biodiversity and Conservation Science, School of Biological Sciences, University of Queensland, St Lucia, QLD 4072, Australia, j.shaw6@uq.edu.au

Andrew Smith
Australian Business Unit, Chevron Australia Pty Ltd, 250 St Georges Tce, Perth, WA 6000, Australia, andys@chevron.com

Michael Smith
Australian Wildlife Conservancy, Subiaco East, WA 6008, Australia, Michael.Smith@australianwildlife.org

Keith Springer
Private Consultant, keith.springer@gmail.com

Melissa Thomas
Australian Business Unit, Chevron Australia Pty Ltd, 250 St Georges Tce, Perth, WA 6000, Australia, MelissaThomas@chevron.com

David Towns
Department of Conservation, Private Bag 68908, Newton, Auckland 1145, New Zealand; and Institute for Applied Ecology, Auckland University of Technology, Private Bag 92006, Auckland 1142, New Zealand, dtowns@doc.govt.nz

Johann van der Merwe
Australian Business Unit, Chevron Australia Pty Ltd, 250 St Georges Tce, Perth, WA 6000, Australia, johannvdm@chevron.com

Tom Vigilante
Bush Heritage Australia/Wunambal Gaambera Aboriginal Corporation, Level 1, 395 Collins St, Melbourne, VIC 3000, Australia, infohc@wunambalgaambera.org.au

Neil Waina
Wunambal Gaambera Aboriginal Corporation, PMB 16 Kalumburu via Wyndham, WA 6740, Australia, neil.waina@klc.org.au

Jennie Whinam
Geography and Spatial Sciences, University of Tasmania, Churchill Ave, Sandy Bay, TAS 7005, Australia, j.whinam@utas.edu.au

Jodie Williams
Traditions Magazine, 42 Taylors Rd, PO Box 534, Norfolk Island, NSW 2289, Australia, Design@jodiewilliams.com

John C.Z. Woinarski
National Environmental Science Programme Threatened Species Recovery Hub, Charles Darwin University, Casuarina, NT 0909, Australia, John.Woinarski@cdu.edu.au

1

Introduction: Australia's island arks

Dorian Moro, Derek Ball and Sally L. Bryant

Introduction

Australia holds custodianship of one of the most diverse sets of island landscapes of any country. These islands include tropical atolls (e.g. Cocos-Keeling) and coral cays, tropical, subtropical and temperate continental islands, sea stacks such as those off Tasmania, the world's largest sand island (K'gari-Fraser Island), oceanic islands (such as the exposed sea-mounts that form Lord Howe and Norfolk), and finally sub-Antarctic islands at the higher latitudes (Heard-McDonald and Macquarie Islands). Additionally, our islands vary in topography, reaching altitudes as high as 2745 m on Heard Island. As a consequence of this geographical and altitudinal spread, species and ecosystem diversity is understandably high.

Offshore islands under Australian stewardship number some 8411 (Woinarski *et al.* 2014). These islands contribute 40% of Australia's coastline (Geoscience Australia 2004), and thus a significant proportion of Australia's coastal landscape despite representing only 1.26% of its landmass (Woinarski *et al.* 2014). Many islands are repositories or in some cases essential locations for a staggering amount of Australia's biological diversity (Gibson and McKenzie 2012). They support over 35% of threatened species listed under the Commonwealth's *Environment Protection and Biodiversity Conservation Act 1999*.

Importantly for conservation purposes, islands offer relatively enclosed areas where key threatening processes, such as feral mammals, may be cost-effectively eliminated. Furthermore, incursions by feral animals, weeds and diseases that continue to impact biological diversity on mainland Australia can be managed through robust biosecurity arrangements on islands. Finally, islands provide important ecosystem services to human well-being, such as fisheries, energy, freshwater, agriculture and traditional ecological knowledge (Wong *et al.* 2005; Vigilante *et al.* 2013). They therefore play a critical role in the local economy, for the people who reside there, and more widely for the nation. For example, Phillip Island (Victoria), Kangaroo Island and the Whitsunday Islands all rank highly in terms of tourism economy relative to other tourism areas across mainland Australia (Tourism Research Australia 2011).

Australia's islands form the foundation of an estimated A$13 billion/annum nature-based tourism industry. For example, in 2012 the Great Barrier Reef alone contributed A$5.7 billion to the Australian economy and sustained 69 000 full-time jobs (Deloitte Access Economics 2013). In a similar vein, in 2013–14, internationally renowned Phillip Island – primarily, Phillip Island Nature Parks – in Victoria was estimated to contribute A$1.3 billion in total tourism output and, directly and indirectly, to employ 6400 people (Tourism Victoria 2015). On Kangaroo Island in South Australia, tourism contributed A$134 million (c.15%) to the regional economy in 2015, employed 500 people directly and helped to support a further 900 employees indirectly.

At a global scale, islands are recognised as important areas that support biodiversity in terrestrial, freshwater and thus connected marine environments. Yet this biodiversity can be highly vulnerable to intense human visitation and use (Secretariat of the Convention on Biological Diversity 2014). Surprisingly, the critical importance of islands to conservation of Australia's biological diversity and regional economies has not received a commensurate amount of management attention or investment by this nation. Co-operative research centres and aligned management efforts have been formed for other landscapes – coral reefs, rainforests, deserts, the Antarctic and savannahs – but not islands. Islands are also overlooked within protected area systems. For example, nationally, less than 30% of islands are protected within Nature Reserves (GeoSciences Australia, unpublished data).

Historical context of Australia's use of its islands

Aboriginal and Torres Strait Islander peoples have maintained continuous connection with most of Australia's islands since before they were created by the 100+ m rise in global sea levels that stabilised ~6000 years ago. Tasmanian Aborigines, for example, harvested short-tailed shearwaters or mutton birds (*Ardenna tenuirostris*) and their eggs from offshore Tasmanian islands for generations. This farming continues today for many families as an important cultural practice and commercial industry (Skira 1997). Similarly, the Traditional Owners of the coast of the Kimberley region have had a close association with the coastal landscape, including the offshore islands which hold great significance and value for them (Vigilante *et al.* 2013). Many islands, such as these Kimberley islands, are components of larger interlinked Indigenous cultural landscapes and seascapes that are of enormous significance to Traditional Owners all around Australia.

Historical use of islands around Australia by Europeans dates as early as the 1600s, when some were used as staging posts for mariners to record their visit: Dutch mariner Dirk Hartog placed a flattened pewter plate on what is now Dirk Hartog Island off Western Australia as a record of his visit there in 1616 (Playford 1998). Further north, with the establishment of towns such as Roebourne (1866) and Cossack (1972) on the Pilbara mainland, nearby offshore islands in the Dampier archipelago became bases for fishing, pearling and whaling operations particularly because of their proximity to sheltered bays in which to moor vessels (Morris 1989).

During the late 18th and the 19th centuries, Australian islands were also used as important penal colonies because of their remoteness and ocean barriers. Australia itself was viewed as an island on which to maroon convicts, by the British Empire. Many of Australia's offshore islands, especially Tasmania, became *in situ* jails for convicts seen as in need of additional security: Norfolk Island (~1500 km east of Australia) was reinstated in 1824 as a place to send the worst convict reoffenders. Queensland's North Stradbroke Island became the site of Australia's first Catholic mission in 1843 (*Carter et al.* 1994). Other islands such as Barrow, Bernier, Dorre and Rottnest Islands (Western Australia) or Tasmania's Flinders Island were used in the late 1800s/early 1900s as quarantine areas to isolate Aboriginal people or those suffering illnesses such as tuberculosis, measles or venereal disease (Jebb 1984).

Islands were viewed as reliable sources of marine wildlife and many offered important resting sites for whalers and sealers. Due to their remoteness, men were often left onshore with provisions for months, to establish bases, exploit the resources and await returning vessels. Macquarie Island was exploited for fur and oil from elephant seals and four species of penguins in the early 1800s (Cumpston 1968). Other islands (e.g. Norfolk Island; St Peter's Island, Tasmania; Moreton Island, Queensland) were used to establish whaling stations or turtle canneries. Green sea turtles (*Chelonia mydas*) were exploited for the European and Asian markets in canneries such as those on Heron Island in the Great Barrier Reef; female turtles were preferred for their superior flesh, and were killed and rendered into tins of soup (Limpus 2008). Commercial exploitation of Western Australian turtles was operational from at least the 1930s, with green turtles harvested from the Montebello Islands and at reefs around Flying Foam Islands (Western Australia: Limpus 2008). Other island wildlife exploited by Europeans included mutton birds and beche de mer, among other fisheries (Hourston 2010).

Islands have long been used as navigational aids and lighthouse stations. During the 1950s, due to their remoteness from European and eastern Australian cities, islands in the Montebello archipelago were used by the British for nuclear tests, with no recognition of the island wildlife that occurred there. The European mindset that islands were waste grounds and there only to be exploited was typical at that time. However, historical visitations on islands have had consequences, particularly the introduction of non-indigenous species that have had major impacts on local insular and typically predator-naïve fauna and succulent flora.

A significant number – and geographical area – of Australian islands have little or no human visitation, and continue to protect biodiversity values regardless of the islands' formal protected area status. Unlike other parts of the world, the majority of Australia's islands are not populated; less than 1% of Australia's total population lives on islands and those that are occupied comprise only ~30 of the larger island groups (ABS 1999). Many have one or more non-exclusive or complementary uses such as mining and resource extraction industries, tourism, farming and as urban or rural residences. The impacts of these uses may be significant and contrary to biodiversity conservation unless carefully mitigated. Poor or absent biosecurity arrangements leave islands open to further ingress of non-indigenous species that might have negative impacts at the ecosystem level. Overuse of resources such as freshwater, or pollution of water by waste management landfills, may have high and chronic impacts. Poorly managed access to the coastline by vehicles, particularly in erosion- prone areas, causes degradation of habitats and can lead to progressive loss of soil stability and subsequent vegetation loss. Not all island vegetation communities are fire-tolerant, and fires have the potential to remove all vegetation cover, leading to local extinctions for those species unable to find refuge. Finally, artificial lighting can result in severe disruptions to breeding for some species, such as marine turtles (Kamrowski *et al.* 2014) and seabirds (Montevecchi 2006).

The importance of islands in the Australian context

Together with their surrounding marine environments, islands have intrinsic values as functioning and interlinked ecosystems supporting high levels of biodiversity, which in turn provides economic benefits for tourism. Many islands have been vested as nature reserves, national parks or World Heritage Areas due to their significance for nature conservation, geomorphology, as examples of ongoing ecological and biophysical processes or the role they played in human cultural history.

Islands are logical places to safeguard Australia's biological diversity. They are important breeding grounds for seabirds, marine turtles and seals, with many species breeding primarily on islands (e.g. short-tailed shearwater, little penguin (*Eudyptula minor*) and lesser frigatebird (*Fregata ariel*). Islands offer essential refugia for common as well as for rare and endangered species. These either occur only on islands, or have their most secure populations on islands. Unfortunately, Australia has a poor mammal extinction record on islands: of 28 endemic land mammal species in Australia extirpated since 1788, seven (25%) were island endemics (Woinarski *et al.* 2015). A further two species have recently become extinct: the Christmas Island pipistrelle (*Pipistrellus murrayi*) in 2009 (Lunney *et al.* 2011) and Bramble Cay melomys (*Melomys rubicola*) in 2016 (Gynther *et al.* 2016).

Islands, because of their isolation and size, often present the opportunity to holistically restore the landscape in a way that is not possible on the mainland. Eradication (rather than just control) of non-indigenous species is more feasible on islands where suitable biosecurity is in place to reduce the risks of reinvasions. Isolation (and remoteness) also limits the risks from other ecosystem impacts such as disease or further land clearing.

Islands offer value for people. Many islands have deep and abiding spiritual connections for Aboriginal and Torres Strait Islander communities. Many islands are places where people live and gain an income based on recreational and tourism industries. They supply food resources from the marine environment or support opportunities for commercial development and resource extraction. Finally, many islands offer scientific and educational destinations for scientists and land managers to survey the biota and genetic resources of these relatively undisturbed areas, study island biogeography, monitor the health of marine ecosystems and trial management initiatives.

Development of the consciousness of island importance

There is a change occurring in people's values of islands (Secretariat of the Convention on Biological Diversity 2014). This changed paradigm is particularly relevant due to the altering climate and the associated sea level rises accepted by the IPCC (Mimura 1999; Mimura *et al.* 2007), and the recent impacts of tidal inundation on biodiversity of some low-lying islands (Gynther *et al.* 2016).

In addition to these global challenges are salient challenges faced by islands around Australia: border protection and biosecurity, resource management, invasive species management, and whole-island restoration to rebuild or maintain functional ecosystems. In particular, Australia as a nation has undertaken vast changes in its assessment of offshore islands and required conservation-driven actions. This assists prioritisation of island conservation interventions to guide future government and other investments (Ecosure 2009). These changes have been commensurate with the urbanisation of many islands, which has further increased the pressures on their ecosystems through land clearing for agriculture and mining, and infrastructure associated with those industries. The challenge today is to develop priorities among islands to allocate scarce resources and to change the community mindset that islands are not just remote outposts – they are important as arks for biodiversity conservation, and this can support existing and new economic opportunities, i.e. for tourism. Investments to return island ecosystem services through whole-island restoration programs are becoming more broadly acceptable. However, island restoration (and other – narrower – conservation interventions) also has socio-economic implications and requires exemplary community engagement where people have ownership or traditional links to the island. Conservation ecology and its practice is now considering these human elements in island conservation initiatives (Chapter 4).

In recent times, the historical and cultural significance of many islands and archipelagos to Indigenous groups is being recognised more formally and managed through Indigenous Protected Areas (Smyth *et al.* 2016; Chapter 8). Indigenous Protected Areas are areas of land and/or sea that have been voluntarily dedicated as protected areas by their Traditional Owners, recognised and supported by governments, and managed by Indigenous people through a combination of legal and other effective means, often in collaboration with conservation agencies and other partners, consistent with international protected areas guidelines. Coastal and island Indigenous Protected Areas are planned and managed holistically in accordance with the Indigenous concept of 'sea country', in which all the natural and cultural values of land, sea, islands, animal, plants and people are acknowledged as interlinked and interdependent.

It is also becoming apparent that island restoration needs to be conducted holistically due to a high probability that actions directed at one part of an ecosystem may have undesired consequences elsewhere in space and time. For example, taking a holistic approach to the management of invasive species in natural environments was a lesson learned from a pest management program on Macquarie Island: eradicating feral cats (*Felis catus*) led to a trophic cascade effect in the island ecosystem due to increases in rabbit abundance that substantially altered large areas of island vegetation (Bergstrom *et al.* 2009). Lessons from New Zealand ecological restoration programs on islands demonstrate that a multi-species approach to island eradication programs is achievable, particularly when it is driven by support from local community and cultural partnerships (Chapter 14).

National strategic planning

Legislation in Australia relating to the management of islands either remains undeveloped or (most often) poorly considers specific challenges and opportunities of islands and their cultural values and human communities. While inhabited islands offer electoral reasons to support their infrastructure, uninhabited islands (i.e. most) remain unrecognised politically and so threats that degrade them continue largely unabated and with little current prospect of resource investment. There are several shortcomings of current planning that, if reversed, could improve the management of Australia's islands in the future. However, a shift in thinking is now clear: islands hold a special identity as unique ecosystems for Australia and many hold spiritual significance as part of Indigenous sea country. They deserve to be recognised as such.

As a result of the adoption of the Programme of Work on Island Biodiversity in 2006 (COP 8 Decision VIII/1), Australia (together with Italy and Mexico) has developed island strategies that recognise the importance and significance of island biodiversity. Australia's efforts focused on an independent national assessment of the conservation value, and subsequent prioritisation of high conservation status offshore islands and their specific vertebrate pest management issues (Ecosure 2009). The focus of the report was to develop a priority list of 100 islands of high conservation status to help guide future government investment on offshore islands. However, the process was largely flawed in that it only considered islands of 100 ha or larger. Consequently, many smaller yet extremely important islands – such as those used for nesting by seabird and marine turtles – are not considered. This information is now being collated at a state level (Chapter 5).

In *Ten Commitments: Reshaping the Lucky Country's Environment*, Lindenmayer *et al.* (2008) highlighted the ecosystems around Australia where special political and socio-economic attention is required to stem their further degradation; islands as ecosystems are not recognised as warranting special attention.

However, in a follow-up edition six years later – *Ten Commitments Revisited* (Lindenmayer *et al.* 2014) – islands were afforded their own chapter in recognition of their value for biodiversity based on endemism, refuge and breeding grounds. Woinarski *et al.* (2014) identified 10 key issues that will move island conservation and management forward over the next decade in Australia. These focus areas are worth repeating as they underpin the advocacy we require to invest and commit to current and future island conservation initiatives:

1 Implement a comprehensive stocktake of island life.
2 Design and implement a comprehensive biosecurity program for islands, prioritised by risk and biodiversity significance.
3 Design and develop a national program that systematically prioritises, implements, monitors and improves island conservation management interventions.
4 Ensure that Australian islands are included in national conservation management programs.
5 Act to conserve island threatened species, and amend the *Environment Protection and Biodiversity Conservation Act 1999* to better safeguard islands of biodiversity significance.
6 Develop strategic risk-based programs to translocate threatened species among islands and mainland areas.

7 Develop broad-based sustainability programs for island human communities, incorporating social, economic and environmental factors.
8 Improve collaboration among island managers, residents and visitors across Australia, including the ongoing support for and coordination of the island arks program.
9 Increase resourcing to, and longer-term security of, island Indigenous Protected Areas, and enhance their management capacity.
10 Integrate island-based ecotourism industries into holistic management arrangements.

While overarching in motive, these focus areas should guide a national strategic approach to conserving offshore island ecosystems. This strategy is required if the full suite of extant island values is to be conserved into the future.

National-level prioritisation and coordination of island restoration programs, notably including non-indigenous species eradication programs, and translocations of insurance populations of threatened species, is required to maximise return on investment. This should be a sensible framework that recognises islands as arks for Australia's biological diversity, but must also consider the imperatives of maintaining community, cultural and commercial values and future opportunities. National planning needs to be practical to implement, dares to adopt novel approaches to landscape-scale management, addresses evidence-based management in the decision-making process, actively includes local citizens to support on-ground initiatives, and needs to include adaptive learning (especially that made from mistakes). The question is: are we as a nation actively achieving and investing in this level of planning and implementation for conservation of Australia's offshore islands?

Structure and content of this book

Our intent has been to produce an invaluable yet easy-to-understand book for politicians, policy-makers, scientists, practitioners and visitors with interests in Australia's offshore islands. This book highlights the interplay of conservation and land development in or associated with island ecosystems across the Australian region. The book aims to provide the knowledge and shared experience required to effectively plan and implement island restoration projects to maximise their value to biodiversity conservation into the future. The authors are a broad mix of island management practitioners, researchers, planners and policy-makers, ranging from managers with 20+ years of experience to postgraduate students offering fresh perspectives.

Included are a selection of case studies of island management practices and approaches, shared by Australian managers and scientists, about their work on island protection and continued management. These include Aboriginal and Torres Strait Islander values, visitor management, invasive species management, endangered species recovery, partnerships and conservation planning. The rise in community-driven island conservation programs (Chapter 12) demonstrates the volunteer effort that has contributed to restoring populated and unoccupied island ecosystems. The intent of these case studies is to highlight what is possible when well-designed management interventions are undertaken in partnership with investors, stakeholder organisations and island communities.

This book has been developed to focus on key themes of relevance to island conservation and management around Australia. Importantly, we note that these themes also have relevance and application to islands globally. We have structured the book to address 11 focus areas (Table 1.1), such as the political and policy values of islands around Australia including tenure and jurisdictional arrangements (Chapters 2, 3, 8, 10), the biodiversity and associated biosecurity values to protect island ecosystems (Chapters 5, 6, 7, 9), the importance of islands to Aboriginal and Torres Strait Islander peoples (Chapter 8), the interplay between islands and the marine ecosystem (Chapter 13), the vulnerability of islands to a changing climate (Chapter 15), community engagement (Chapters 4, 9, 12) and the socio-economic values of islands to the resource sector and tourism (Chapters 4, 9, 11, 12). In the final chapter, we summarise the advancements in

Table 1.1. The organisation of chapters in this book based on a framework of themes related to island policy, conservation and management around Australia

Island theme and values	Chapters
Policy and status of islands, statutory processes and legislation	2, 3, 8, 10, 16
Community-driven conservation and engagement	4, 9, 12
Tourism	11,12
Island prioritisation	5
Biosecurity and quarantine	6, 9
Biodiversity	7
Indigenous Protected Areas	8
Resource development	9
Marine connections	13
Whole-island management	14
Climate change	15

relation to Australia's offshore islands in terms of conservation, policy and management, and identify emerging priorities and challenges. This will lead to a re-evaluation of the themes identified in Chapter 1, and an attempt to reflect on the current commitments *sensu* Woinarski *et al.* (2014) for the next 20 years for Australia, integrating research, management, and policy.

Concluding comments

The Australian nation has a gift: continental and oceanic islands that collectively are rich in cultural heritage, biodiversity and ecosystem services. Ultimately, these unique ecosystems need institutional support and cross-organisational partnerships to maintain their integrity and to avoid increasing their risk of further degradation. We hope this book will stimulate thought and further action to manage these iconic landscapes sustainably.

References

ABS (Australian Bureau of Statistics) (1999) *Australian Social Trends.* <http://www.abs.gov.au/ausstats/abs@.nsf/2f762f95845417aeca25706c00834efa/65f1ae55ef772f18ca2570ec001117a4!OpenDocument>

Bergstrom DM, Lucieer A, Kiefer K, Wasley J, Belbin L, Pedersen TK, Chown SL (2009) Indirect effects of invasive species removal devastate World Heritage Island. *Journal of Applied Ecology* **46**, 73–81. doi:10.1111/j.1365-2664.2008.01601.x

Carter P, Durbidge E, Cooke-Bramley J (1994) *Historic North Stradbroke Island.* North Stradbroke Island Historical Museum Association Inc. and Watson Ferguson & Co., Brisbane.

Cumpston JS (1968) *Macquarie Island.* Publication no. 93. ANARE Scientific Reports, Series A. Department of External Affairs, Antarctic Division, Melbourne.

Deloitte Access Economics (2013) *Economic Contribution of the Great Barrier Reef.* Great Barrier Reef Marine Park Authority, Townsville.

Ecosure (2009) *Prioritisation of High Conservation Status of Offshore Islands.* Report to the Federal Department of the Environment, Water, Heritage and the Arts. Ecosure, Queensland.

Geoscience Australia (2004) *Geodata topo 100K.* Google Earth, 1:100,000 scale vector map data (DVD). Geoscience Australia, Canberra.

Gibson LA, McKenzie NL (2012) Identification of biodiversity assets on selected Kimberley islands: background and implementation. *Records of the Western Australian Museum* **81**(Supplement), 1–14. doi:10.18195/issn.0313-122x.81.2012.001-014

Gynther I, Waller N, Leung LK (2016) *Confirmation of the Extinction of the Bramble Cay Melomys Melomys rubicola on Bramble Cay, Torres Strait: Results and Conclusions from a Comprehensive Survey in August–September 2014.* Department of Environment and Heritage Protection, Brisbane.

Hourston M (2010) Review of the exploitation of marine resources of the Australian Indian Ocean Territories: the implications of biogeographic isolation for tropical island fisheries. Fisheries Research Report no. 208. Department of Fisheries, Western Australia.

Jebb MA (1984) The Lock hospitals experiment: Europeans, aborigines and venereal disease. *Studies in Western Australian History* **8**, 68–87.

Kamrowski RL, Limpus C, Pendoley K, Hamann M (2014) Influence of industrial light pollution on the sea-finding behaviour of flatback turtle hatchlings. *Wildlife Research* **41**, 421–434. doi:10.1071/WR14155

Limpus C (2008) *A Biological Review of Australian Marine Turtle Species. 2. Green Turtle, Chelonia mydas. (Linnaeus).* Environmental Protection Agency, Queensland.

Lindenmayer DB, Dovers S, Harriss Olson M, Morton S (Eds) (2008) *Ten Commitments: Reshaping the Lucky Country's Environment.* CSIRO Publishing, Melbourne.

Lindenmayer DB, Dovers S, Morton S (Eds) (2014) *Ten Commitments Revisited: Securing Australia's Future Environment.* CSIRO Publishing, Melbourne.

Lunney D, Law B, Schulz M, Pennay M (2011) Turning the spotlight onto the conservation of Australian bats and the extinction of the Christmas Island pipistrelle. In *The Biology and Conservation of Australasian Bats* (Eds B Law, P Eby, D Lunney and L Lumsden) pp. 485–498. Royal Zoological Society of NSW, Sydney.

Mimura N (1999) Vulnerability of island countries in the South Pacific to sea level rise and climate change. *Climate Research* **12**, 137–143. doi:10.3354/cr012137

Mimura N, Nurse L, McLean RF, Agard J, Briguglio L, Lefale P, Payet R, Sem G (2007). Small islands. In *Climate Change 2007: Impacts, Adaptation and Vulnerability. Contribution of Working Group II to the Fourth Assessment Report of the Intergovernmental Panel on Climate Change* (Eds ML Parry, OF Canziani, JP Palutikof, PJ van der Linden and CE Hanson) pp. 687–716. Cambridge University Press, Cambridge.

Montevecchi W (2006) Influences of artificial light on marine birds. In *Ecological Consequences of Artificial Night Lighting* (Eds C Rich and T Longcore) pp. 94–113. Island Press, Washington DC.

Morris KD (1989) The Dampier archipelago: managing people in a nature reserve. In *Australian and New Zealand Islands: Nature Conservation Values and Management.* (Ed. A Burbidge) pp. 183–192. Department of Conservation and Land Management, Perth.

Playford PE (1998) *Voyage of the Discovery to Terra Australis by Willem de Vlamingh in 1696–97.* Western Australian Museum, Perth.

Secretariat of the Convention on Biological Diversity (2014) *Island Bright Spots in Conservation and Sustainability Investing, Scaling and Replicating in Order to Achieve the Aichi Biodiversity Targets.* Secretariat of the Convention on Biological Diversity, Canada.

Skira I (1997) 'I hope you will be my friend': Tasmanian Aborigines in the Furneaux Group in the nineteenth century – population and land tenure. *Aboriginal History* **21**, 30–45.

Smyth D, Gould J, Ayre M, Bock E, Dulfer-Hyams M (2016) Indigenous-driven co-governance of sea country through collaborative planning and indigenous protected areas. *Indigenous Law Bulletin* **8**, 15–20.

Tourism Research Australia (2011) *The Economic Importance of Tourism in Australia's Regions.* Tourism Research Australia, Canberra.

Tourism Victoria (2015) *Value of Tourism to Phillip Island 2013–14.* Research factsheet. Tourism Victoria, Melbourne.

Vigilante T, Toohey J, Gorring A, Blundell V, Saunders T, Mangolamara S, George K, Oobagooma J, Waina M, Morgan K, Doohan K (2013) Island country: Aboriginal connections, values and knowledge of the Western Australian Kimberley islands in the context of an island biological survey. *Records of the Western Australian Museum* **81**(Supplement), 45–81.

Woinarski J, Ball D, Burbidge AA (2014) Islands. In *Ten Commitments Revisited: Securing Australia's Future Environment* (Eds DB Lindenmayer, S Dovers and S Morton) pp. 117-128. CSIRO Publishing, Melbourne.

Woinarski CZ, Burbidge AA, Harrisond PL (2015) Ongoing unravelling of a continental fauna: decline and extinction of Australian mammals since European settlement. *Proceedings of the National Academy of Sciences of the United States of America* **112**, 4531–4540. doi:10.1073/pnas.1417301112

Wong PP, Marone P, Lana P, Fortes M, Moro D, Agard J, Vicente L, Thonell P, Mulongoy KJ (2005) Island systems. In *Ecosystems and Human Well-being: Current State and Trends: Findings of the Condition and Trends Working Group, Volume 1.* (Eds R Hassan, R Scholes and N Ash) pp. 663–680. Island Press, Washington DC.

2

Values of islands across Australia's states and territories

Keith Morris, Dave Algar, David Armstrong, Derek Ball, Sally L. Bryant, Peter Canty, Peter B. Copley, Chris Dickman, Alaric Fisher, Graeme R. Gillespie, Michael Johnston and Dave Kelly

Summary

Australia's islands are an important component of the nation's conservation estate and support significant conservation assets. All states, the Northern Territory and the Australian government have responsibility for the management of these assets because of the values they hold to the nation, specifically values such as geological landforms, biodiversity and community use. However, these values are at risk from several threats and reliant on sustained management actions. Here we summarise key values, threats and management actions that are underway on islands across states and territories of Australia.

Introduction

Australia is often referred to as an island because it is surrounded by ocean. However, Australia is made up of more than 8000 islands, including the island state of Tasmania. These geologically and ecologically diverse islands have significant conservation values and are an integral part of the nation's conservation estate. Many of these values relate to the lack of development on, and isolation of, many islands which therefore offer safe havens to flora and fauna from a variety of threats that dominate on mainland areas.

All states and territories have some islands under their jurisdiction, and management is largely undertaken by Commonwealth and state government conservation agencies. In northern Australia, native title has been determined over many islands and these are now owned and managed by Traditional Owners, often in collaboration with government agencies. Elsewhere, only a small proportion of islands are freehold.

In this chapter, we summarise the inherent values, threats and management of islands across states and territories of Australia. We do so to assist island communities, managers and policy-makers to better understand the current status of Australia's island-based biodiversity, threats to biodiversity, and the management direction in place to conserve these insular landscapes.

Geography and geological landforms

The Cocos (Keeling) Islands are coral atolls located in the Indian Ocean at 120°12′S, 960°54′E, some 3000 km north-west of Perth, Western Australia (WA). The southern atoll is 14 km long and 10 km across and comprises 26 islands with a total area of 1300 ha, of which West Island is the largest island at 620 ha. All the islands are flat, their highest points being sand hills on the ocean side. The northern atoll comprises a single island, North Keeling Island (130 ha) which is a national park (Pulu Keeling National Park) established in 1995 (EA 1999; DNP 2004) and managed as a 'strict nature reserve' (IUCN category 1a).

Christmas Island is located in the Indian Ocean at 10°25′S and 105°40′E, some 2600 km north-north-west of Perth, and covers ~13 500 ha. It was formed from an undersea volcano that rose to the surface and has since subsided and risen again over geological time. The oceanic island is composed primarily of Tertiary limestone overlying volcanic andesite and basalt (EA 2002a). It rises steeply from the surrounding ocean and consists of a series of fringing limestone terraces, separated by rugged limestone cliffs and scree slopes, rising to an internal central plateau 200–360 m above sea level.

The Ashmore Reef National Nature Reserve is a coral cay located at 120°20′S, 123°00′E near the outer edge of the Sahul Shelf in the eastern Indian Ocean, 320 km off the north-west coast of WA. The Reserve includes three small, low-lying, uninhabited, vegetated coral islets within a reef – East, Middle and West Islands – covering a combined area of 112 ha. Cartier Island, a 0.4 ha coral cay, lies within the Cartier Island Marine Reserve 50 km to the south-east of Ashmore (EA 2002b; Russell *et al.* 2004). Cartier Island is nearly submerged during spring high tides and is swept during storm conditions. Consequently, few terrestrial taxa persist on the island.

The Coral Sea Islands Territory consists of several reefs and 24 small sand and/or shingle cays spread over an extensive geographic area in the Coral Sea, north-east of Queensland (~12–24°S, 144–156°E), extending east and south from the outer edge of the Great Barrier Reef (EA 2001). The cays are small, with a combined area of ~700 ha, and are split into two groups. The north-western group comprises ~17 named reefs and several islets and cays including Willis Islets, Coringa Islets and Lihou Cays. The south-easterly group comprises five reefs and associated cays.

The Heard Island and McDonald Island (HIMI) group are volcanic in origin and located in the Indian Ocean sector of the Southern Ocean at around 53°05′S, 73°30′E (DEH 2006). They are the only sub-Antarctic islands that are volcanically active. Off Heard Island, there are numerous outlying islets, rocks and reefs, the largest (Shag Island) lying some 10 km north of Heard Island. About 80% of Heard Island is glaciated, with ice up to 150 m deep and glaciers extending from 2745 m to sea level (Kiernan and McConnell 1999). Ice cliffs form a high percentage of the coastline. The McDonald Island group are ice-free, and composed of basaltic lava and tuffaceous material with little soil. McDonald Island consists of two sections – a northern sloping plateau and a southern steep-sided hill (Maxwell Hill) – both bounded by steep cliffs and joined by a narrow isthmus.

Approximately 100 islands lie off the coast of New South Wales (NSW); 47 support land plants, while the other 50 or so are little more than small bare rocks. Thirty-three coastal islands are less than 10 ha in area. Two groups of oceanic islands also lie within the jurisdiction of NSW. Lord Howe Island (LHI, 1455 ha) and 11 associated, much smaller islands are 780 km north-east of Sydney. Norfolk Island (NI, 3460 ha) plus two smaller islands and a few sea stacks are more remote, lying 1700 km north-east of Sydney.

There are 973 islands in the Northern Territory (NT), ranging in size from 0.08 ha to 578 617 ha. These include the second largest (Melville Island, 578 617 ha), fourth largest (Groote Eylandt, 232 600 ha) and fifth largest (Bathurst Island, 169 300 ha) islands in Australia. All NT islands lie on the continental shelf in the Timor and Arafura seas and the Gulf of Carpentaria, and were connected to the mainland before the Holocene. Most major islands in the NT are clustered in a series of archipelagos; the Tiwi Islands (including Bathurst and Melville Islands), Croker, Goulburn, Crocodile, Elcho, Wessel, English Co., Groote and Sir Edward Pellew groups.

Queensland (Qld) includes ~1955 islands (Geoscience Australia 2017) covering 673 073 ha within three regions:

1 the large sand islands off south-east Qld, notably K'gari-Fraser and Bribie Islands (Great Sandy Straits), and Moreton Island and the Stradbroke group (Moreton Bay);
2 continental islands and cays within both the Great Barrier Reef World Heritage Area (GBRWHA) and Torres Strait;
3 the Wellesley Islands in the Gulf of Carpentaria.

Fraser Island (184 000 ha) and North Stradbroke Island (27 500 ha) are the largest and second largest sand islands in the world, respectively. Within the GBRWHA and Torres Strait ~300 coral cays occur, about a third of which are vegetated to some degree. The Wellesley Island group consists of ~81 islands lying within the Gulf of Carpentaria, the largest semi-enclosed embayment of ocean in Australia. These range from large continental islands such as Mornington (97 020 ha), to small ephemeral islands of unconsolidated sediments on rock platforms.

South Australia (SA) has over 150 offshore islands, excluding minor near-coastal stacks and reefs submerged at high tide. Most are continental islands, high points on the former coastal plain progressively isolated by rising sea levels during the Holocene (Robinson *et al.* 1996). The total SA island area is ~467 200 ha. The largest, Kangaroo Island (441 600 ha), the third largest island in Australia after Tasmania and Melville Island (NT), separated from the mainland ~9500 years ago.

The island state of Tasmania has over 1000 islands referenced by Geoscience Australia (2017) but this number increases exponentially with islets, rock stacks and reefs. Flinders Island (130 000 ha) and King Island (110 000 ha) are Australia's sixth and seventh largest islands, 374 islands are >1 ha and 65 are >20 ha in size (TASMAP 2006; Parsons 2011). Rodondo Island and West Moncoeur in Bass Strait are less than 10 km from the Victorian border, while sub-Antarctic Macquarie Island lies 1500 km south of Hobart in the Southern Ocean. All Tasmanian islands except for Macquarie were formed by sea level rises during the Holocene; World Heritage-listed Macquarie is the only island in the world composed entirely of oceanic crust originating from deep below the Earth's surface (Dixon 1996).

There are some 138 islands >1 ha in Victorian waters, with most (48%) <10 ha in size (Geoscience Australia 2012). Total island area is 40 910 ha. The largest two islands are French (17 300 ha) and Phillip (10 300 ha) Islands in Western Port Bay. All Victorian islands are 'coastal' with the most isolated, Lady Julia Percy Island, located just 8 km offshore.

WA has an estimated 3747 islands, ~46% of the Australian total of 8222 islands (Geoscience Australia 2017). These are spread along most of the WA coast (absent from the Nullarbor coast); however, ~80% occur in the north of the state, off the Kimberley and Pilbara coasts. About 2000 of WA's islands are grouped into 11 archipelagos, with the Buccaneer Archipelago off the Kimberley coast containing around 1000 islands. Most WA islands are small with only 23% >1 ha, 7% >100 ha and 1% >1000 ha in area (Abbott and Burbidge 1995). The total area of WA islands is 305 790 ha, Dirk Hartog Island being the largest at 58 640 ha. Apart from Rowley Shoals and rocky islets on Scott and Seringapatam reefs, all WA islands are continental in origin and most are relics of coastal high ground separated from the mainland by rising sea levels during the Holocene (Abbott and Wills 2016).

Tenure

The Australian External Territories (AET) are a group of non-self-governing dependencies of Australia. On the inhabited islands of Cocos (Keeling) and Christmas Islands, several WA government departments and corporations assist local island shire councils in governance and provision of municipal services. National parks within the AET are Commonwealth Reserves managed by the Director of National Parks (Australian government).

General visitor access to Ashmore Reef was first restricted in 1988 to protect the biodiversity of the reserve. Public access is now restricted to West Island Lagoon and a small portion of West Island but commercial tourism activities are highly restricted and subject to permit. The area has been assigned to IUCN Category II (National Park), with the remainder of Ashmore Reef assigned to IUCN Category la (Strict Nature Reserve) (EA 2002b). Cartier Island Marine Reserve has been closed to the public for some decades as a Defence Practice Area.

Christmas Island has been an external territory of the Commonwealth of Australia since its transfer from British jurisdiction in 1958. A national park was established in 1980 and extended in 1986 and 1989 to include most of the rainforest and 63% of the island (EA 2002a).

The Coral Sea Islands became a territory of the Commonwealth in 1969. In 1982, the Lihou Reef Nature Reserve and Coringa-Herald Nature Reserve were declared by the Commonwealth government to protect the wildlife in that part of the territory. In 1987, the Elizabeth and Middleton Reefs Marine National Nature Reserve was declared. In 2012, this nature reserve was proclaimed as the Coral Sea Commonwealth Marine Reserve (Beeton *et al.* 2015).

The HIMI group were inscribed on the World Heritage list in December 1997 on the basis of their outstanding natural universal values, and comprise Australia's largest IUCN Category 1a (Strict Nature Reserve). Many of the animal species (mainly birds) known to inhabit the islands are listed as threatened and/or migratory under the *Environment Protection and Biodiversity Conservation Act 1999* (EPBC Act) and protected under international agreements. HIMI were added to the Commonwealth's National Heritage List in May 2007 and their wetlands are listed on the Directory of Important Wetlands in Australia. HIMI lie within Australia's Heard Island and McDonald Islands Marine Reserve, which covers ~7 100 000 ha of terrestrial and marine areas.

NSW islands fall primarily under the jurisdiction of local councils or the NSW government, with those of conservation significance managed by the state. Bowen Island, at Jervis Bay, is administered by the Australian government. LHI became part of NSW in 1855 when the colony was granted a measure of autonomous government from the British Crown. NI came under the jurisdiction of NSW on 1 July 2016, having previously been an external territory.

Most NT islands are Aboriginal freehold land and many are included within seven Indigenous Protected Areas (IPAs): Crocodile Islands IPA, Djelk IPA and proposed sea country extension, Dhimurru IPA, Laynhapuy IPA, Gumurr-Marthakal IPA, Anindilyakwa IPA and Yanyuwa IPA. Other major islands that are not within IPAs (such as the Tiwi Islands, Goulburn and Croker Islands) have local Indigenous ranger groups that undertake natural resource management. Twenty-two islands are within conservation reserves managed by the NT government (Garig Gunak Barlu, Limmen and Barranyi National Parks, Vernon Island, Channel Island and Indian Island Conservation Reserves) and Commonwealth government (part of Kakadu National Park).

In 2006, 44% of Qld islands (855 islands: 294 780 ha) were either conservation reserves or within protected areas (DEH 2006). Since that time there have been significant additions to island protected areas through the dedication of IPAs including all of the Wellesley Islands (Thuwathu/Bujimulla), some Torres Strait islands (Warul Kawa, Pulu Islet) and some offshore areas of the Wet Tropics region (Eastern Kuku Yalanji, Mandingalbay Yidinji and Girringun). These have added over 125 000 ha to the protected estate system, bringing the proportion of Qld islands with protection to more than 60%.

The majority of SA's islands are proclaimed as conservation reserves under the state's *National Parks and Wildlife Act 1972*. More recently, five areas on Kangaroo Island and major sections of the Nuyts Archipelago and Investigator Group were proclaimed as Wilderness Protection Areas under the *Wilderness Protection Act 1992*. Many SA islands are also now bounded by marine protected areas under the *Marine Parks Act 2007*, further protecting their terrestrial–marine ecological linkages. Only 10 islands are not fully protected. Four of the larger islands (Wedge, Thistle, Louth and Boston) are owned under freehold titles, three others (Spilsby, Taylor and Flinders) are perpetual leases. Island Point, Green and Wardang islands are Aboriginal lands. About 40% of Kangaroo Island is protected in 26 parks and the remaining 60% is privately owned, either as residential and holiday properties, or as farms and lifestyle blocks (many for nature conservation).

Most (89%) of Tasmania's 252 larger islands for which status is known, are statutory reserves, nine are private freehold, seven are Aboriginal lands and the remaining 12 have multiple land tenure. Reserved islands are managed under the *National Parks and Reserves Management Act 2002* by the Tasmanian Parks and Wildlife Service. Unreserved islands are managed by a range of mechanisms by state and local government agencies in accordance with Tasmania's Resource Management and Planning System and a variety of other legislations. Tasmania has 21 formally declared marine reserves covering 134 589 ha, equivalent to 2.6% of the state's coastal waters (Parsons 2011; DPIPWE 2016).

Most of Victoria's islands of conservation significance are managed by the Vic government. Five Vic islands (Sunday, Sandstone, Elizabeth, Little Dog and Big Dog Islands) are privately owned, three (Middle, Merri and Griffiths Islands) are managed by local government and one (Swan Island) by the Australian government. French and Phillip Islands in Western Port Bay both have extensive freehold components in addition to their conservation reserves (Johnston 2014).

About 980 WA islands (~26%) are managed by the state, and reserved for conservation purposes (CCWA 2009). Recent legislative changes allow Indigenous people to undertake customary activities on conservation reserves. Four of the island nature reserves along the Pilbara coast have leases excised for the purposes of oil and gas production. Of the ~2600 Kimberley islands, only 12 are vested in the state, with the majority of the others vested as unallocated crown land and covered by native title. The 122 islands in the Houtman Abrolhos archipelago, off the mid-west coast of WA, are currently vested in the Minister for Fisheries for conservation of flora and fauna, tourism, and purposes associated with fishing and aquaculture industries. Rottnest Island, 20 km off the coast from Perth, is vested in the Minister for Tourism for recreation and conservation. Garden Island is a naval base, HMAS Stirling, and managed by the Department of Defence. A few islands such as Dirk Hartog and Faure Islands were used for pastoral purposes until recently, but are now managed for conservation purposes. There are no privately owned islands in WA, although Dirk Hartog Island does have some blocks of freehold title and Faure Island is leased by the Australian Wildlife Conservancy.

Conservation values

Through their isolation, many Australian islands have avoided the threats that have impacted on the biodiversity of the mainland. They have conservation values as important sanctuaries for seabirds and migratory birds, roosting and breeding locations for marine turtles and pinnipeds, and refuges for endemic plant and animal species that have declined or become extinct on the mainland.

Important bird sanctuaries

One of the highest values of Australia's islands for biodiversity is as a refuge, roosting and nesting place for birdlife. Christmas Island is an internationally significant seabird rookery (Beeton *et al.* 2010) and declared a Key Biodiversity Area (KBA) (Birdlife International 2016d). Two of the Christmas Island wetlands, The Dales and Hosnies Spring, are listed as Ramsar Sites of International Importance (Misso and West 2014). Eight species of seabird breed on the island, six of which are commonly found on Pacific and Indian Ocean tropical islands: red-footed (*Sula sula*) and brown boobies (*S. leucogaster*); great (*Fregata minor*) and lesser frigatebirds (*F. ariel*), and red-tailed (*Phaethon rubricauda*) and white-tailed tropicbirds (*P. lepturus fulvus*). Two threatened species breed only on Christmas

Island: the Christmas frigatebird (*F. andrewsi*) and Abbott's booby (*Papasula abbotti*).

Nearby Pulu Keeling National Park (PKNP) supports 24 species of resident and migratory waterbirds and is listed as a Wetland of International Importance under the Ramsar Convention (Misso and MacRae 2014). PKNP regularly supports >20 000 waterbirds, >30 000 pairs of red-footed boobies, 15 000 pairs of common noddy (*Anous stolidus*) and 3000 pairs of great and lesser frigatebirds. Other species present include the masked booby (*S. dactylata*), brown booby, red-tailed tropicbird, white-tailed tropicbird and white tern (*Gygis alba*) (Carter 1994; North and Griffin 2006). PKNK supported the only remaining population of the Cocos buff-banded rail (*Gallirallus philipensis andrewsi*) (Misso and MacRae 2014) until its successful reintroduction to Horsburgh Island in the southern atoll (Woinarski *et al.* 2016).

The Ashmore islands support some of the most important seabird rookeries on the North West Shelf. On Ashmore Island, the nature reserve is a designated Ramsar Site due to its importance as a resting place for migratory shorebirds and for supporting large seabird breeding colonies. Large colonies of sooty terns (*Onychoprion fuscata*), crested terns (*Thalasseus bergii*), bridled terns (*Onychoprion anaethetus*) and common noddies (*Anous stolidus*) breed on East and Middle Islands. Smaller breeding colonies of little egrets (*Egretta garzetta*), eastern reef egrets (*E. sacra*) and black noddies (*A. minutus*) also occur. Many bird species found in the reserves are listed on international migratory bird agreements (EA 2002b). The islands are also an important staging point for many migratory wading birds: eastern curlew (*Numenius madagascariensis*), ruddy turnstone (*Arenaria interpres*), whimbrel (*Numenius phaeopus*), bar-tailed godwit (*Limosa limosa*), common sandpiper (*Actitis hypoleucos*), lesser sand plover (*Charadrius mongolus*), red-necked stint (*Calidris ruficollis*), grey-tailed tattler (*Tringa brevipes*) and wandering tattler (*T. incana*) (EA 2002b).

The Coral Sea Reserves (Coringa-Herald and Lihou Reef and Cays) are also recognised as a Ramsar Site. The reserves contain regionally significant seabird colonies, and many of the species recorded from the reserves are listed under international migratory bird agreements. Fourteen seabird species breed on the cays. While some species (red-footed booby, lesser and great frigatebirds, red-tailed tropicbird) have an extensive distribution outside of Australian waters, they are uncommon within mainland Australia, and these island reserves contain a significant proportion of the region's breeding populations (EA 2001).

The HIMI have also been identified as a KBA because they support large numbers of nesting seabirds (Woehler 2010). Thirty-four species have been recorded, of which 19 breed on Heard Island and 11 on the McDonald Islands. Penguins are the most abundant birds present, with populations of macaroni penguin (*Eudyptes chrysolophus*) the commonest species, reaching two million pairs within the islands, 16% of the world total. Gentoo penguin (*Pygoscelis papua*) is present, and its breeding population at Heard Island represents ~6% of the world population. Other breeding birds are black-browed albatross (*Diomedea melanophris*) (600–700 breeding pairs), light-mantled albatross (*Phoebetria palpebrata*) (200–500), southern giant petrel (*Macronectes giganteus*) (3000 pairs on Heard Island and an estimated 1400–1600 pairs on McDonald Island), cape petrel (*Daption capense*), Antarctic prion (*Pachyptila desolata*), fulmar prion (*P. crassirostris*), Wilson's storm petrel (*Oceanites oceanicus*), South Georgian diving petrel (*Pelecanoides georgicus*), common diving petrel (*P. urinatrix*), Heard Island cormorant (*Phalacrocorax atriceps nivalis*) (an endemic subspecies with a population of less than 100 pairs), Antarctic tern (*Sterna vittata*), subantarctic skua (*Catharacta lonnbergi*), kelp gull (*Larus dominicanus*) and Heard Island sheathbill (*Chionis minor nasicornis*) (an endemic subspecies). Wandering albatross (*Diomedea exulans*) was reported breeding for the first time in 1980. Most bird populations are reported as stable, although king penguin and black-browed albatross are increasing.

NSW islands, particularly the LHI and NI groups, are important breeding sites for at least 24 species of seabirds. Some land birds nest on coastal islands, but all breed primarily on the mainland. Coastal islands are also used as stopover points by ~40 species of migratory birds, and as roosting or foraging habitats by many more (Priddel and Wheeler 2014).

There are 147 seabird nesting colonies recorded in the NT and all, except for several little tern (*Sternula albifrons*) colonies on mainland beaches, are located on small offshore islands (Chatto 2001). Six island groups and one single island are recognised as KBAs (Dutson *et al.* 2009). Seagull Island, in the Tiwi group, supports the largest colony of crested tern in the world, and Haul Round Island near Maningrida supports one of the largest mixed-species seabird colonies in the NT, including globally significant numbers of roseate tern (*Sterna dougalli*) (Chatto 2001; BirdLife International 2016a). Several island groups in the Gulf are recognised nationally and internationally as important for nesting seabirds, including the Sir Edward Pellew Group, which supports more than 20 colonies with internationally significant numbers of crested and roseate tern, and nationally significant numbers of bridled, roseate and black-naped tern (*Sterna sumatrana*) (Chatto 2001; BirdLife International 2016b). Six tern species breed on Sandy Island and Low Rock, with internationally significant numbers of roseate terns and nationally significant numbers of several other species, including the only known lesser crested tern (*Thalasseus bengalensis*) breeding site in the NT (Chatto 2001; BirdLife International 2016c). Higginson, and neighbouring islands near Nhulunbuy, support internationally and nationally important numbers of bridled and roseate terns, smaller numbers of other terns and the only known NT nesting site for common noddy (Chatto 2001; Dutson *et al.* 2009).

Other islands, such as the Tiwi Islands and Milingimbi and Elcho Island, provide important seasonal feeding grounds for internationally significant numbers of migratory shorebirds including bar-tailed godwits, great knots (*Calidris tenuirostris*) and grey-tailed tattler (Chatto 2003). Large mixed-species flocks of shorebirds have been recorded on Crocodile Island in the Milingimbi Islands (Mahney *et al.* 2009). Tiwi and Milingimbi Islands are also recognised as KBAs for shorebirds (Dutson *et al.* 2009). North Peron Island is the only island-based nesting site for Australian pelicans (*Pelecanus conspicillatus*) in the NT (Chatto 2000) and is of national significance because it is one of the largest known regular breeding sites for this species in the world (Mahney *et al.* 2009).

Many of the Qld islands within the Great Sandy Strait, Moreton Bay and Shoalwater Bay/Corio regions lie within Ramsar Sites, and these provide extensive summer habitat of international significance for migratory shorebirds. More than 25% of Australia's tropical seabirds nest on islands in the GBRWHA, including more than 50% of Australia's roseate terns, lesser crested terns, black-naped terns and black noddies, and ~25% of the wedge-tailed shearwaters (*Ardenna pacifica*), brown booby, masked booby and red-tailed tropicbirds (Congdon *et al.* 2007). The Wellesley Islands also support extensive seabird breeding with nationally important numbers of crested tern, roseate tern, lesser frigatebird and brown booby.

Many islands off SA are breeding sites for tens of thousands of short-tailed shearwaters (*Ardenna tenuirostris*) and thousands of white-faced storm petrels (*Pelagodroma marina*). Most islands also provide critical feeding, roosting and/or breeding sites for many inshore seabird species such as little penguin (*Eudyptula minor*), pied cormorant (*Phalacrocorax varius*), gulls and terns and for resident and migratory shorebirds, such as the pied (*Haematopus longirostris*) and sooty oystercatchers (*H. fuliginosus*), dotterels, plovers, turnstone, stint, sandpipers and greenshank.

Tasmania's Bass Strait islands support the largest colonies (up to six million birds) of the migratory short-tailed shearwater and are the Australian stronghold for little penguin, pied oystercatcher, sooty oystercatcher and black-faced cormorant (*Phalacrocorax fuscescens*) (Brothers *et al.* 2001). King and Flinders Islands, and islands in Boullanger Bay, support the highest diversity of breeding and migratory shorebirds in Tasmania including >1% of global population for seven species. These values are reflected in Tasmania's 29 KBAs that are wholly or partly on islands, the highest number of island IBAs in any Australian state (Kirkwood and O'Connor 2010). Macquarie Island supports some of the world's largest sub-Antarctic bird breeding grounds, with an estimated 3.5 million breeding seabirds. Four species of penguin, four species of albatross, 11 species of petrel and two species of storm petrel breed on Macquarie Island, many of which are endemic and/or threatened species.

In Victoria, migratory shorebirds, resident waterbirds and seabirds use islands, rock stacks and reefs (Johnston 2014), and many of the islands or island groups are incorporated into five KBAs (Dutson *et al.* 2009). Surveys of most of the seabird islands have been undertaken over the last 30 years or so, for example Prescott (1980) and Harris *et al.* (1980).

Forty seabird species are known to nest on 591 islands off the WA coast (Burbidge and Fuller 1996, AH Burbidge *pers. comm.*). Of these, four are either Threatened or Priority (species of conservation concern listed by the state) species and 15 are migratory species listed by international agreements. Many islands are important nesting areas for threatened birds (Table 2.1). The Houtman Abrolhos are the most diverse group of islands for nesting seabirds in WA: 18 species are recorded nesting on each of Pelsaert and Wooded Islands, 15 species on West Wallabi Island and 13 species on Long Island.

Of the 73 listed KBAs identified in WA, 19 are islands or island groups (Dutson *et al.* 2009). Ten are along WA's northern coastline including islands in Exmouth Gulf, the Montebello, Lowendal and Barrow group of islands, Adele Island and Bedout Island. In southern WA, the Recherche Archipelago, Rottnest Island and the Houtman Abrolhos are KBAs. These are recognised internationally for their importance to bird conservation, particularly for seabird nesting and shorebird refuges.

The northern WA coastline supports some of Australia's largest migratory and resident shorebird populations, and ~47 species of migratory waders visit WA shores. Many larger islands along the north-west and Kimberley coasts provide safe feeding grounds for these, as well as important overwintering habitat for nonbreeding migrants (Bamford *et al.* 2011). More than 27 species have been reported as using the Exmouth Gulf, Ningaloo Coast and adjacent islands (Onton *et al.* 2013), and islands in the Exmouth Gulf have been listed as KBAs (Dutson *et al.* 2009).

Table 2.1. WA islands important as nesting sites for threatened birds

Island	Species
Adele Island – west Kimberley coast	Red-footed booby (*Sula sula*)
	Great frigatebird (*Fregata minor*)
	Lesser frigatebird (*Fregata ariel*)
Lacepede Islands – west Kimberley coast	Lesser frigatebird (*Fregata ariel*)
Bedout Island – west Kimberley coast	Masked booby (*Sula dactylatra bedoutii*)
	Lesser frigatebird (*Fregata ariel*)
Houtman Abrolhos (3 islands)	Australian lesser noddy (*Anous tenuirostris melanops*)
Double Island – Pilbara coast Troughton and Mulgudra Islands – Kimberley coast	Little tern (*Sterna albifrons*)

Marine turtles and pinnipeds

Of the world's seven species of marine turtle, five are known to breed on the coast and islands of the NT, Qld and WA. The northern islands, including coastal regions, have significant nesting sites for hawksbill (*Eretmochelys imbricata*) (Limpus 2008c; Hoenner 2012; FitzSimmons and Limpus 2014), Olive Ridley (*Lepidochelys olivacea*) (Limpus 2008d), green (*Chelonia mydas*) (FitzSimmons and Limpus 2014) and flatback (*Natator depressus*) turtles (Chatto and Baker 2008).

In the NT, significant nesting areas include the Groote Eylandt Archipelago, the Tiwi Islands, Cobourg Peninsula, NE Arnhem coast, Wessel, English Co. and Bromby Island groups and the Crocodile Islands (Chatto and Baker 2008; Limpus

2008a,b,c,d,e,f; Hoenner 2012). The NT flatback turtle population has long-term trend data from mark–recapture programs on Field and Bare Sand Islands and these indicate it is generally stable (Groom *et al.* 2016). Other NT marine turtle species are not systematically monitored and population trends are uncertain.

Qld islands provide nesting habitat for six of the world's species of marine turtles including three green turtle genetic breeding stocks (Limpus 2008b); the world's largest green turtle rookery at Raine Island, the Capricorn Bunker group of cays within the GBRWHA, and the Bountiful Islands within the Wellesley group. Key flatback turtle rookeries occur in the GBRWHA on Wild Duck, Peak and Avoid Islands, and on Bountiful and Pisonia Islands in the Wellesleys. One of the world's largest hawksbill turtle breeding areas is found at Milman Island (Limpus 2008c).

On the Pilbara coast of WA, 93% of green turtle nesting sites are islands, 76% of flatback turtle nesting sites are islands, and all hawksbill turtle nesting sites are on islands (Pendoley *et al.* 2016). Marine turtle species that nest in WA all represent distinct genetic stocks (DoEE 2016). They are significant for global marine turtle conservation because their breeding populations are estimated to be among the largest and most important in the Indo-Pacific region (Limpus 2008a,b,c,d,e,f).

Four species of pinniped have been recorded on islands around the Australian mainland and Tasmania: the New Zealand fur-seal (*Arctocephalus forsteri*), Australian fur-seal (*A. pusillus*), southern elephant seal (*Mirounga leonina*) and Australian sea-lion (*Neophoca cinerea*). SA islands are home to the majority (~86%) of all Australian sea-lions (DSEWPaC 2013), and provide numerous breeding sites for large and increasing numbers of New Zealand fur-seals (Shaughnessy and Goldsworthy 2015). All Australian fur-seal breeding colonies and haul-out sites are on islands in the south-east of Australia from Kangaroo Island in SA, the Bass Strait islands, to Montague Island in NSW (Kirkwood *et al.* 2010). The Australian sea-lion has been recorded breeding at 28 sites in WA (37% of Australian sites), most being on islands along the south coast of WA (Gales 1994) but also three islands off the mid-west coast and four in the Easter group of the Houtman Abrolhos group off Geraldton (Gales *et al.* 1992). Many more islands are used as haul-outs. The New Zealand fur-seal has been recorded breeding on 17 islands off the south coast of WA (Shaughnessy 1999); the largest WA colony is on Salisbury Island, supporting ~6% of the Australian population. Several more pinniped species occur on the sub-Antarctic HIMI group (Table 2.2), and sub-Antarctic Macquarie Island has four breeding and two non-breeding seal species.

Table 2.2. Seals of the HIMI group

Heard Island	Southern elephant seal (*Mirounga leonine*)
	Leopard seal (*Hydrurga leptonyx*)[a]
	Antarctic fur-seal (*Arctocephalus gazelle*)
	Weddell seal (*Leptonychotes weddelli*)
	Ross seal (*Ommatophoca rossi*)
	Crabeater seal (*Lobodon carcinophagus*)
	Sub-Antarctic fur-seal (*A. tropicalis*)
McDonald Islands	Antarctic fur-seal (*Arctocephalus gazelle*)
	Southern elephant seal (*Mirounga leonine*)
	Leopard seal (*Hydrurga leptonyx*)[a]

[a] Non-breeding.

Refuges for endemic and threatened species

Endemism

Through isolation, a wide range of taxa have evolved distinct genetic and physical forms on Australia's islands. Christmas Island is one of Australia's biodiversity icons in terms of endemism. Here, of 135 plant species, at least 18 are endemic (Table 2.3; Claussen 2005). There are 10 endemic bird taxa, four of which are listed as Endangered (Misso and West 2014; Table 2.3). Four native species of

Table 2.3. Endemic species of Christmas Island

Flora	Trees: Christmas Island palm (*Arenga listeria*), *Pandanus elatus, Dendrocnide peltata* Shrubs: *Abutilon listeria, Colubrina peduncalata, Grewia insularis, Pandanus christmatensis* Vines: *Hoya aldrichii, Zehneria alba* Herbs: *Asystasia alba, Diclipera maclearii, Peperomia rossi* Grass: Christmas Island duck beak (*Ischaemum nativitatis*) Fern: Christmas Island spleenwort (*Asplenium listeria*) Orchids: Ridley's orchid (*Brachypeza archytus*) *Phreatia listeria, Flickingeria nativitatis, Zeuxine exilis*
Mammals	Native bulldog rat (*Rattus nativitatus*)[a] Maclear's rat (*R. macleari*)[a] Christmas Island shrew (*Crocidura trichura*)[a] Christmas Island pipistrelle (*Pipistrellus murrayi*)[b] Christmas Island flying-fox (*Pteropus melanotus natalis*)
Birds	White-tailed tropicbird (*Phaethon lepturus fulvus*) Abbott's booby (*Papasula abbotti*) Christmas frigatebird (*Fregata andrewsi*) Christmas Island goshawk (*Accipiter hiogaster natalis*) Christmas Island hawk-owl (*Ninox natalis*) Christmas Island emerald dove (*Chalcophaps indica natalis*) Christmas imperial pigeon (*Ducula whartoni*) Christmas Island swiftlet (*Collocalia linchi natalis*) Christmas Island thrush (*Turdus poliocephalus erythropleurus*) Christmas Island white-eye (*Zosterops natalis*)
Terrestrial reptiles	Forest skink (*Emoia nativitatis*)[b] Blue-tailed skink (*Cryptoblepharus egeriae*), Lister's gecko (*Lepidodactylus listeri*) Giant gecko (*Cyrtodactylus sadleiri*)
Invertebrates	Land crabs: Christmas Island red crab (*Gecarcoidea natalis*), blue crab (*Discoplax hirtipes*), Jackson's crab (*Geograpsus stormi*) Butterflies: Christmas swallowtail (*Papilio memnon*)

[a] Extinct.
[b] Presumed extinct.

terrestrial reptiles (of which one is presumed extinct) and five mammal species (of which four are extinct) are endemic to Christmas Island (Table 2.3).

One of the world's greatest migration of invertebrates occurs on Christmas Island: the migration of land crabs. Christmas Island red crabs (*Gecarcoidea natalis*) are the most abundant of the 12 terrestrial land crab species, having an estimated population of 40–47 million (Misso and West 2014). Their annual migration occurs at the beginning of the wet season, when tens of millions make their way to the sea to breed in one of the world's most spectacular breeding migrations (Orchard 2012; Misso and West 2014).

One coastal seabird, Gould's petrel (*Pterodroma leucoptera leucoptera*), is endemic to NSW and breeds principally on Cabbage Tree Island, with smaller numbers on Boondelbah Island. The land fauna of the LHI (Box 2.1) and NI groups (Table 2.4) off NSW also boast large numbers of endemic species.

On Kangaroo Island (SA), there are 45 endemic plant species (Robinson and Armstrong 1999). No endemic fauna are recorded on Victorian islands, but there are three important species of endemic island flora: the French Island spider orchid (*Caladenia insularis*) and the Victorian distributions of island celery (*Apium insulare*) and coast boronia (*Boronia anemonifolia* subsp.).

Table 2.4. Endemic species of Norfolk Island

Taxonomic group	Species
Birds	Norfolk Island gerygone (*Gerygone modesta*)
	Green parrot (*Cyanoramphus cookii*)
	Slender-billed white-eye (*Zosterops tenuirostris*)
	White-chested white-eye (*Z. albogularis*) Norfolk Island scarlet robin (*Petroica multicolor multicolor*)
Reptiles	Lord Howe Island skink (*Oligosoma lichenigera*)[a]
	Gecko (*Christinus guentheri*)[a]
Plants	50 of 180 plant species, including Norfolk Island pine (*Araucaria heterophylla*)
Invertebrates	Over 100 insect species At least 20 species of gastropod

[a] Shared with Lord Howe Island.

Some of the larger NT islands, such as the Tiwi Islands and Groote Eylandt (Box 2.2), support endemic taxa.

Seven mammal subspecies, six bird subspecies, 11 reptile subspecies and 79 invertebrate species are found only on WA islands. Of the invertebrates, 73 of are camaenid land snails found only on islands off the Kimberley coast (Gibson and Kohler 2014). Barrow Island supports four endemic mammal subspecies, one endemic bird subspecies, one endemic reptile subspecies and several endemic stygofauna invertebrate species.

Many of Tasmania's islands are repositories for genetic diversity. For example, King Island supports six endemic subspecies of bird and 10 of Tasmania's 12 endemic resident breeding species, of which seven species are listed as threatened under state and/or Commonwealth legislation (Bryant 2014).

Refuges

Vegetation and flora

Some of the AET island vegetation communities reflect those of a northern hemisphere mesic

Box 2.1: Biodiversity values of Lord Howe Island

Lord Howe Island is the crown jewel of NSW islands. It is highly biodiverse, with a high degree of endemism across multiple taxa. The LHI group is World Heritage-listed. It supports extensive colonies of nesting seabirds, with at least 168 bird species recorded either living on or visiting the islands. The islands are one of two known breeding areas for the providence petrel (*Pterodroma solandri*) and support what is probably the world's largest breeding concentration of the red-tailed tropicbird (*Phaethon rubricauda*) and the most southerly breeding colony of the masked booby (*Sula dactylatra*). LHI is home to four endemic land birds: Lord Howe woodhen (*Gallirallus sylvestris*), Lord Howe currawong (*Strepera graculina crissalis*), Lord Howe golden whistler (*Pachycephala pectoralis contempt*) and Lord Howe silvereye (*Zosterops lateralis tephropleura*). The woodhen has been the subject of a successful recovery program (NPWS 2002) and the population is monitored annually. The endemic long-eared bat (*Nyctophilus howensis*) is now extinct.

LHI retains high cover of native vegetation (87%); there are 239 native vascular plant species, of which 47% are endemic to the LHI group (DECC 2007). The LHI flora has affinities with that of the Australian mainland, New Zealand, New Caledonia and NI.

More than 1600 terrestrial invertebrate species have been recorded, including 157 land and freshwater snails, 464 beetles, 27 ants, 183 spiders, 21 earthworms, 137 butterflies and moths and 71 springtails, with up to 60% of some groups comprising endemic species (DECC 2007).

Box 2.2: Groote Eylandt

Groote Eylandt is located off eastern Arnhem Land in the Gulf of Carpentaria and is the fourth largest island in Australia. Groote and its associated archipelago of 40 smaller islands have internationally significant conservation values. They support at least 900 plant and 330 vertebrate species (excluding marine fish), of which 12 are EPBC Act-listed threatened species. They also support the largest areas of marine turtle nesting in the NT, and are especially significant for green and hawksbill turtles. One islet supports more than 1% of the world population of roseate terns. Groote also supports most of the world population of the northern hopping mouse, and one of a few known extant populations of brush-tailed rabbit rat (*Conilurus penicillatus*). The island supports significant populations of a further seven animals and one plant listed as threatened under the *Territory Parks and Wildlife Conservation Act*.

Many of the threatening processes operating on terrestrial ecosystems across northern Australia are absent from, or occur at low levels, on Groote Eylandt. Groote remains free of cane toads, has no large introduced herbivores or pigs, no associated history of grazing pressure or land degradation, no extensive establishment of grassy weeds and retains a relatively benign fire regime. These conditions have enabled threatened mammal species and other biota to persist. With an absence of cane toads, Groote Eylandt is an important refuge for the northern quoll.

Feral cats are present on Groote Eylandt but to date appear to have had less impact on native mammal populations compared to the mainland, possibly due to retention of high-quality habitat. However, most Indigenous communities on Groote have adopted domestic cats and it is possible that this may result in population increases. The presence of the mining operation and human settlements, and associated sea and air traffic between the mainland and the island, provides a significant risk of introduction for non-indigenous species.

Many areas of Groote Eylandt have ecologically sound fire regimes (smaller, early dry-season burns); however, they are deteriorating in areas with easy human access. New roads associated with expanded mining activities may result in further alterations to fire regimes. Disturbances associated with roads and other developments on the island also increase the risk of establishment and spread of weeds.

Groote Eylandt falls entirely within the Anindilyakwa Land Council area. It is an IPA guided by a management plan developed in 2006. It has a well-established and active Indigenous ranger group that undertakes land and sea management including survey and removal of marine debris, protection of turtle nesting areas, collaborative biodiversity surveys with scientists, fire management, weed control, and cane toad surveillance and awareness. Rehabilitation of areas affected by mining operations is guided by the mine's rehabilitation program.

There is a large manganese mine and associated port and settlement on the western side of Groote Eylandt, which has been a significant driver of economic and infrastructure development. It is anticipated that biodiversity offsets resulting from planned mine expansion will support the implementation of a threatened species plan in collaboration with the Land Council and NT government.

community. For example, principal vegetation communities on Heard Island are tussock grassland, herbfield and feldmark, with smaller areas of meadow, pool complex and cushion carpet. Short tussock grass (*Poa cookii*) with cushions of the herb (*Colobanthus kerguelensis*) are present in coastal areas, with Kerguelen cabbage (*Pringlea antiscorbutica*) and *Azorella selago* cushions in established moraines and valleys up to 200 m. Above 200 m, mosses and lichens dominate ice-free regions and are important components of the flora at lower regions. On the McDonald Islands, tussock grass is common on eastern slopes and lower parts of the plateau, while cushions of *A. selago* cover higher areas, with Kerguelen cabbage and dwarf shrub.

Approximately 2500 plant taxa have been recorded from NT islands, including 200 introduced species. However, this is likely to be a conservative estimate as many islands have not been subject to comprehensive floristic surveys. Two islands (Melville and Groote Eylandt) have more than 1000 plant taxa, three (Bathurst, Croker, Elcho) have between 500 and 1000 taxa, and a further 40 have at least 100 taxa recorded. Information on threatened plants is less extensive than that on threatened fauna, due to a lower level of survey effort. Twenty-one plant taxa are listed as threatened either at a territory or national level. Threatened plant species have been recorded on nine NT islands with Melville Island supporting 17 and Bathurst Island supporting 13. This reflects the occurrence on these two islands of rare monsoon rainforest species.

Qld islands are representative of coastal mainland vegetation types and constituent flora, and are important places for conservation of its flora and vegetation communities. For example, the Torres Strait Islands contain 40% of all flora species known from the Cape York Bioregion and 13% of Qld vascular flora (Stanton *et al.* 2008), and islands in the GBRWHA contain ~23% of the Qld flora. Importantly, many islands are all but free of threatening processes – such as vegetation clearance – operating on the mainland.

Many Tasmanian islands have one or more threatened plant species or vegetation communities. King Island has 50 threatened plants and six threatened vegetation communities; many are endemic species or subspecies (TSS 2012). Bruny Island has 39 state or nationally threatened plant species (Cochran 2003), Rodondo Island supports climax communities of *Eucalyptus globulus* and *Melaleuca armillaris* which have evolved in the absence of fire, and the limestone flora on Prime Seal Island retains direct affinities with the Recherche Archipelago in WA (Harris *et al.* 2001). Island floras of national significance occur on Ile du Golfe, Maatsuyker Island and Flat Witch Island in Tasmania's South-west Wilderness World Heritage Area (Balmer *et al.* 2004). Tasmania's south-eastern region has the highest level of localised and endemic marine plants in Australia, with over 750 species of macro-algae recorded thus far. The nationally threatened kelp forests around Maria Island are world-renowned for their ecological and economic significance (Parsons 2011) but are now in critical decline. Macquarie Island has three endemic flora species, of which the cushion plant (*Azorella macquariensis*) is a keystone species of the feldmark community (PWS 2006).

While WA has several thousand islands, only one threatened plant taxon is known from one island (native fern *Asplenium obtusatum* ssp. *northlandicum* on Chatham Island). However, another 93 plant taxa are listed as Priority species in need of further survey and/or monitoring. Fifteen plant taxa are known only from WA islands.

Fauna

Approximately 330 Australian islands together provide refuge for at least two monotreme species, 70 marsupial species, 42 bat species and 30 rodent species (Abbott and Burbidge 1995), many of which have restricted distributions or have declined on the mainland.

Over 530 vertebrate species, including 16 introduced species, have been recorded on NT islands including 78 mammals, 295 birds, 133 reptiles and 26 frogs. Of these, 98 occur only on single islands and 207 occur on fewer than five islands. Thirteen vertebrates are restricted to islands within the NT, including Butler's dunnart (*Sminthopsis butleri*), golden bandicoot (*Isoodon auratus*), canefield rat (*Rattus sordidus*), Carpentarian skink (*Lerista carpentariae*), Arafura snake-eyed skink (*Cryptoblepharus gurrumul*), Tiwi Islands masked owl (*Tyto novaehollandiae melvillensis*) and Tiwi Island hooded robin (*Melanodryas cucullata melvillensis*). A further 28 have more than 50% of their NT records from islands, including Carpentarian antechinus (*Pseudantechinus mimulus*), northern hopping mouse (*Notomys aquilo*), mourning gecko (*Lepidodactylus lugubris*) and Arnhem shovel-nosed snake (*Brachyurophis morrisi*). Thirty-six fauna taxa recorded on 90 NT islands have been listed as threatened either under the *Territory Parks and Wildlife Conservation Act 2006* or EPBC Act; these include 13 mammals, 16 birds and seven reptiles.

Island mammal assemblages within the GBRWHA are depauperate despite habitats being similar to the adjacent mainland. The cause of this is unknown but possibly a result of extinctions due to a combination of small island area, relatively homogenous topography and fuel loading, and prevalence of historical island-wide fire events. Nevertheless, some islands provide refugia for several threatened species, for example the Proserpine rock-wallaby (*Petrogale persephone*) on Gloucester Island and Hayman Island (the latter being a translocated population), and coastal sheath-tail bats (*Taphozous australis*) on Brampton Island. Bentick Island in the Wellesley group supports what is possibly one of few island brush-tailed tree rat (*Conilurus penicillatus*) populations. K'gari-Fraser Island supports one of Australia's most important dingo (*Canis familiaris*) populations as it provides a refuge from further inbreeding with domestic dogs (Allen *et al.* 2015). Squirrel gliders (*Petaurus norfolcensis*) are known from five of seven large south-east Qld islands at densities like those of mainland areas (Bell *et al.* 2011).

Some SA islands support species that are now extinct on the mainland, such as the stick-nest rat (*Leporillus conditor*) on the Franklin Islands (Copley 1999). Others are home to species that have evolved independently of their mainland counterparts, such as the island southern brown bandicoot (Franklin and St Francis Islands), bush rat (Kangaroo and 13 other islands), black tiger snake (Kangaroo and 11 other islands), carpet python (St Francis Island) and the Pearson Island rock-wallaby. Many islands also provide refugia for species more threatened on the mainland, such as osprey, sea-eagle, little penguin, Cape Barren goose, southern bush stone-curlew, tammar wallaby and brushtail possum (Robinson *et al.* 1996). Some of the larger, modified islands have been used for the reintroduction and introduction of species that are threatened or extinct on the mainland (Copley 1994, Chapter 7). Most SA islands with biodiversity values are declared a Prohibited Area under the SA *National Parks and Wildlife Act 1972* and are accessible only under permit.

Coastal islands off NSW support populations of at least six species of terrestrial mammals including bats, three species of amphibians and nine species of reptiles. Of these vertebrates, the green and golden bell frog (*Litoria aurea*) is threatened; this species occurs on Bowen, Broughton and Kooragang Islands (White and Pyke 2008).

Twelve (~9%) of Victorian islands have had native terrestrial mammals recorded on them; of these, five islands have had the koala (*Phascolarctos cinereus*) introduced to them. The larger French (six terrestrial mammal species), Phillip (eight species) and Rotamah (eight species) Islands have the greatest diversity of native mammals (Abbott and Burbidge 1995). French Island also hosts populations of king quail (*Coturnix chinensis*) and swamp skink (*Egernia coventryi*) (Kirkwood and Johnston 2006).

WA islands have significant conservation value as refugia for threatened and endemic fauna species, particularly mammals. Sixty-five species of native mammals occur on 103 WA islands (3% of WA islands). Many are threatened or extinct on the mainland, or have declined significantly. Barrow Island with 13 mammal species, and Bernier and Dorre Islands with eight, are among the most important for mammal conservation. Without WA island populations, another five species of mammals (boodie *Bettongia lesueur*, western barred bandicoot *Perameles bougainville*, rufous hare-wallaby or mala *Lagorchestes hirsutus*, banded hare-wallaby *Lagostrophus fasciatus* and Shark Bay mouse *Pseudomys fieldi*) would have become extinct in the wild in Australia. Recovery plans have been prepared for several mammals found on WA islands including the Shark Bay marsupials, Shark Bay mouse, mala, quokka and dibbler. The Pilbara osprey (*Pandion haliaetus cristatus*) population is regarded as the only stable, potentially expanding population in Australia, and islands are important nesting sites for this species (Dennis and Clancy 2014).

Community values

Many of Australia's islands remain unpopulated or host few resident people. Some islands in Qld, NSW and Tasmania, however, support large residential populations. On Cocos (Keeling) Islands, only Home and West Islands in the southern atoll are inhabited, with a total population of some 590 residents of Malay and European origin. Europeans tend to reside on West Island and the Malay community on Home Island. The main commercial activity on the Cocos (Keeling) Islands is tourism. The Coral Sea

Islands have only one inhabited island (Willis), with three to four staff residing at the meteorological station. Economic activity is restricted to charter fishing and diving. The predominantly Chinese, Malay and European population of 2200 residents on Christmas Island live mainly on the northern tip of the island. Phosphate mining is the only significant economic activity on the island, although tourism through links with conservation is important and, in recent years, the island has become one of Australia's refugee detention centres.

Because most coastal islands in NSW are small there are no permanent settlements, although Broughton and Montague Islands continue to be used by Traditional Owners. These also receive small numbers of visitors as part of commercial tour operations, but most other islands are visited rarely. However, LHI and NI are permanently inhabited, their economies being based on tourism and attracting 20 000–30 000 visitors each year. The NI Regional Council provides local governance for the island's 1800 residents. About a quarter of the land area of NI is permanent pasture, and another 60% has been modified by settlements, agriculture and light industry, especially to encourage tourism. A 250 ha precinct, the Kingston and Arthur's Vale Historic Area, is World Heritage-listed.

There are permanent settlements on eight islands off the NT – Melville, Bathurst, Groote, Bickerton, Croker, Goulburn, Milingimbi and Elcho Islands – and intermittently occupied outstations on many others. Most NT islands continue to be used and managed exclusively or principally by Indigenous Traditional Owners. Intensive natural resource use has so far been limited to a large manganese mining operation and associated port and town on Groote Eylandt, and clearing of ~30 000 ha of eucalypt woodland for establishment of *Acacia mangium* plantations on Melville Island. All the large islands or island groups have active Indigenous ranger groups that undertake a variety of natural resource management activities including fire and weed management, biosecurity surveillance, patrols for ghost nets and marine turtle monitoring. Hunting by Indigenous owners also occurs on some islands. Some conservation values of NT islands are potentially vulnerable to human overutilisation, particularly more accessible seabird breeding colonies.

There has been some (relatively small-scale) mineral sand mining on Melville Island and a larger development is proposed for Bathurst Island. There have also been proposals for bauxite mining on the Wessel Islands. The manganese resource extends offshore from Groote Eylandt and seabed mining has been proposed, but this is opposed by Traditional Owners and there is currently a moratorium on seabed mining in NT waters.

There is little non-Indigenous use of NT islands, presumably due to remoteness, logistical constraints and access restrictions. There is some tourism to the Tiwi Islands and to Groote Eylandt, mostly for cultural tourism and fishing. There have been a few attempts to cultivate the high-end tourist market, such as the Dugong Resort on Groote and the Barra Base on Bathurst Island, with limited success.

The islands bordering Bynoe Harbour were used as a bombing range, and Bare Sand Island is an eco-tourism destination allowing observation of nesting and hatching marine turtles. Barranyi (North Island) National Park has boat-only access and basic camping facilities (no permit required). Waters around the (more accessible) islands are used by recreational fishers, with camping by permit allowed at designated sites on some islands.

Islands in south-east Qld including K'gari-Fraser, Bribie and the Moreton Bay Islands support large residential populations and attract high numbers of visitors. Mornington Island in the Wellesley group is the principal residential area in the group, although others are regularly visited as part of traditional use and custodianship. Torres Strait Islanders reside on many islands and also assume traditional custodianship. Tourism resorts remain dotted among islands in the GBRWHA although many have become defunct under the current poor tourism economic climate. Some islands off the central coast (e.g. Townsend) lie within the Australian Department of Defence.

Islands off SA that are large enough to support European occupation have been extensively modified by attempts to graze stock or grow crops. These enterprises led to significant ecological changes on many islands, primarily through the clearance and modification of vegetation and through heavy pressures from the hunting of wildlife for food and furs/skins. Whaling, sealing and farming also led to introduction and/or assisted arrival of weeds (e.g. African boxthorn and common ice-plant), pest animals (e.g. goats, pigs, rabbits, rats, mice, cats, pigeons, starlings) (Robinson *et al.* 1996), pathogens and diseases (e.g. toxoplasmosis and sarcosporidiosis). These have been equally destructive to both nature and farming.

The larger islands of King, Flinders, Cape Barren and Bruny Islands off mainland Tasmania have permanent populations, Maria Island has a staffed national parks centre, and Macquarie Island retains a permanently manned scientific station. The remaining islands are largely unpopulated although many have infrastructure including light stations, family homesteads or associated farming structures, and several have been developed for tourism or low-level visitation.

Tasmanian Aboriginal people's long occupation of islands remains integral to their cultural practices and is recorded in cave deposits, artefact scatters, middens, hut depressions, petroglyphs and quarries, of which only a small fraction have been documented or protected. In 1995, in an act of reconciliation, seven islands – Trefoil, Babel, Badger, Big Dog, Mt Chappell, Steep, lungtalanana (Clarke) – and most of Cape Barren Island were transferred by the Tasmanian government back to Aboriginal ownership. Murrayfield Station (4097 ha) on Bruny Island is owned by the Weetapoona Aboriginal Corporation.

French and Phillip Islands in Victoria have large areas of freehold ownership. Phillip Island has been extensively developed for agricultural, tourism and residential uses. The island has ~10 000 residents and receives 3.5 million visitors annually (Phillip Island Visitor Information Centre 2012). In contrast, the absence of a traffic bridge and tourist attractions has reduced development activity on French Island, where grazing is the main agricultural activity; two-thirds (12 100 ha) of the island is managed as national park and it receives fewer than 1000 visitors annually. Livestock grazing occurs on two crown-managed islands, Gabo and Snake, as well as on Phillip Island. Elsewhere, highly altered islands, such as Phillip Island, attempt to balance development alongside the management of biodiversity. For example, a housing estate on the Summerland Peninsula was purchased by the state to improve habitat for the little penguin population.

In WA, many of the islands from Exmouth Gulf north to the Kimberley are significant for Traditional Owners for hunting and cultural purposes (Vigilante *et al.* 2013; Abbott and Wills 2016). From Shark Bay south, islands had no value to Indigenous people. This is because none of the coastal tribes of Aborigines between Shark Bay and Eucla, on the SA border, had watercraft or could swim (Abbott and Wills 2016).

About 980 islands (26%) in WA are formally reserved for conservation purposes. Recent changes allow Indigenous people to undertake customary activities (camping, hunting, bringing dogs into areas, lighting campfires and undertaking ceremonies) on conservation reserves (DEC 2012a). Other WA islands have various tenures: leases excised for the purposes of oil and gas production (Barrow, Varanus and Airlie Islands) or mining (Koolan and Cockatoo Islands), native title (Kimberley islands), recreation and tourism (Montebellos and Dampier Archipelago, Houtman Abrolhos Archipelago, Rottnest Island) and defence activities (Montebello Islands atomic weapon testing 1952–56, Garden Island naval base). A few islands, such as Dirk Hartog and Faure Islands, were used for pastoral purposes until recently, but are now managed for conservation purposes.

Threats and management

The threats to Australia's external islands are essentially the same whether they are anthropogenic or natural events. Meteorological factors such as cyclones have a significant detrimental impact on island ecosystems (Goulding *et al.* 2016); their

Box 2.3: Island extinctions

Extinctions on Australian islands are well-documented. Tasmania, for example, has five island vertebrate taxa which are now extinct (Macquarie Island parakeet *Cyanoramphus erythrotis*, Macquarie Island rail *Gallirallus philippensis macquariensis*, King Island emu *Dromaius ater*, Tasmanian emu *Dromaius novaehollandiae diemenensis*, thylacine *Thylacinus cynocephalus*) and possibly a sixth, a Macquarie Island seal species, was extirpated before formal description.

Christmas Island (Table 2.3) also displays a poor record of extinctions of mammals and reptiles, primarily due to invasive species. Four of the five mammal species that were present on the island at settlement in 1888 have since become extinct. The diurnal native bulldog rat (*Rattus nativitatus*) and nocturnal Maclear's rat (*R. macleari*) were common at the time of settlement but disappeared soon after the introduction of black rats in 1899 (Wyatt *et al.* 2008); the Christmas Island shrew (*Crocidura trichura*) has not been recorded since 1985 and is believed extinct (Beeton *et al.* 2010). In 2009, the Christmas Island pipistrelle (*Pipistrellus murrayi*) became extinct (Martin *et al.* 2012). Only the threatened Christmas Island flying-fox (*Pteropus melanotus natalis*) remains extant. While several other factors are likely to have contributed to the demise of these native animals, including habitat destruction, natural catastrophes and the proliferation of the exotic yellow crazy ant (*Anoplolepsis gracilipes*), the introduction of exotic competitors and predators such as the feral cat and black rat are crucial factors (Beeton *et al.* 2010).

A changing climate and rising sea levels are becoming drivers of change on low-lying islands. Of note is Australia's most recent extinction – the Bramble Cays melomys (*Melomys rubicola*) – an extinction event presumed to have been associated with increased tidal surges (Gynther *et al.* 2016).

influence both in terms of frequency and severity is likely to increase as climate change progresses. Climate warming will also likely result in ecosystem change. For example, on the sub-Antarctic Heard Island climate warming will cause glacial contraction and resultant modification to vegetation structure and distribution (Commonwealth of Australia 2014a).

Introduced species remain a key threat to Australia's islands. Many were introduced inadvertently via shipwrecks, cargo and people, while others were deliberately released for various reasons such as food. One consequence has been local species extinctions, most of which are attributed to invasive species (Box 2.3). The ongoing potential for introducing invasive species to islands by both natural and human-induced means is high (Sakai *et al.* 2001; Reaser *et al.* 2007). The risks that invasive species present to island ecosystems can be exacerbated by transmission of diseases and the pathogens they carry.

On Cocos (Keeling) Islands, several rodent pests (brown rat *Rattus norvegicus*, black rat *R. rattus*, house mouse *Mus domesticus*) likely arrived before settlement from shipwrecks in the area (Wood Jones 1909; Bunce 1988), and over time spread to all islands in the southern atoll. Only house mice occur on Horsburgh Island. Domestic chickens (*Gallus gallus domesticus*) are now established on most but not all islands in the southern atoll (Carter 1994), and the green junglefowl (*Gallus varius*) of Java has been introduced to West Island. On nearby Keeling Island, four introduced terrestrial reptile species are present: mourning gecko (*Lepidodactylus lugubris*), four-clawed gecko (*Gehyra mutilata*), Asian house gecko (*Hemidactylus frenatus*) and blind snake (*Typhlops braminus*).

Land clearing is another threatening process for Cocos (Keeling) Island species. Most of the Cocos (Keeling) Islands' original vegetation has been cleared for coconut (*Cocos nucifera*) plantations (Misso and MacRae 2014), an industry which ceased in 1987. Additional damage, particularly the introduction of invasive species, occurred when the Cocos (Keeling) Islands group and Christmas Island became targets for military activity during the two World Wars.

Phosphate mining and other anthropogenic activities have impacted on the Ashmore Islands, with the deliberate or inadvertent introduction of several new plant species. House mice and black rats have been recorded and were still present in the early 1980s; however, a rodent baiting program conducted during the mid-1980s appears to have successfully eradicated the black rats (Russell *et al.* 2004). Fluctuating populations of house mice persist on East and Middle Islands. The introduced Asian house gecko (*Hemidactylus frenatus*) is widespread and abundant on West Island; however, it is not present on East or Middle Islands (Russell *et al.* 2004). A species of possible concern is the ginger ant (*Solenopsis geminata*), an aggressive species native to North America which has spread through much of the tropics and could potentially attack young and adult seabirds.

One of the most significant threats on NSW coastal islands is the establishment and spread of environmental weeds such as Kikuyu grass (*Pennisetum clandestinum*) and bitou bush (*Chrysanthemoides monilifera*), which outcompete native plants and decrease habitat quality for nesting seabirds and other animals. Kikuyu grass has been removed successfully from Montague Island, and invasive rodents and larger mammals such as the European rabbit (*Oryctolagus cuniculus*) and feral goat (*Capra hircus*) have been eradicated from most islands of conservation significance (Priddel *et al.* 2011). Additional potential threats are posed by the introduction, establishment and spread of new species such as the garden skink (*Lampropholis delicata*) and bleating tree frog (*Litoria dentata*) on LHI, as well as invertebrates such as slugs and earthworms (Cassis *et al.* 2003). Key threats on NI are the highly invasive Argentine ant (*Linepithema humile*), introduced rodents and feral cats. In addition, extensive clearing for agriculture and felling of large trees destroyed most suitable nest holes for the Norfolk Island owl (*Ninox novaeseelandiae undulata*; Garnett *et al.* 2011).

Eleven species of introduced vertebrates have been recorded on 30 NT islands. Most, such as the cat, cattle, donkey, goat, horse, pig, rusa deer, sambar deer and buffalo, were deliberate introductions. Cane toads and Asian house geckos arrived on some islands from unassisted colonisation or unintentional transport by visitors. Feral cats are now known from 12 NT islands, and cane toads from 15 islands. Larger islands and those close to the mainland tend to have more introduced species: Croker (six), Vanderlin (six), Melville (six) and Bathurst (five). In contrast, Groote Eylandt has only two introduced vertebrates (feral cat and Asian house gecko) despite its large size and relatively high human population.

Feral cats have been implicated as a major cause of mammal declines in Australia (Ziembicki *et al.* 2015). They are present on 12 NT islands, including all the largest islands. Mammal declines and extinctions have been reported on West and Southwest Islands in the Sir Edward Pellew group since the introduction of feral cats (Woinarski *et al.* 2011), and recent declines of the rabbit-eared rat (*Conilurus penicillatus*) on Melville Island appear to be related to feral cat activity (Davies *et al.*, unpub. data). Efforts to eradicate feral cats on West Island are underway (Paltridge *et al.* 2016).

Over 200 exotic plant species, including garden species, have been recorded across 85 NT islands. Numbers of weed species are strongly related to island size and extent of human habituation, with up to 133 exotic species on Melville Island. Twenty-five listed weed species have been recorded from 48 islands, for example 15 on Bathurst and 20 on Melville Island. Four of these are listed as Weeds of National Significance (WONS) and are major environmental weeds on NT islands: mimosa (*Mimosa pigra*), gamba grass (*Andropogon gayanus*), parkinsonia (*Parkinsonia aculeata*) and rubber vine (*Cryptostegia grandiflora*).

At least 19 of the 1854 Qld islands (1%) are known to have been colonised by exotic species such as ship rats and house mice (DEWHA 2009); however, a lack of survey effort suggests that this number may be much higher. Introduced plants, including invasive, habitat-altering species are prevalent on Qld islands. These include rubber vine which invades through wind-borne seed, and in recent decades has moved further south into northern areas of the Whitsunday Islands. Lantana (*Lantana camara*) is widespread and is continually spread and reintroduced by birds which consume its fruits and distribute seeds. Goats were historically introduced to many islands as a potential food source for stranded sailors. They have been removed from some smaller islands but feral populations remain on larger islands which contain better refuge, making removal much more difficult. These eradications have provided lessons in the need to make whole-of-system interventions in island pest control. For example, the most recent goat eradication on South Percy Island resulted in substantial increases in fuel loads and lantana cover, fuelling a wildfire several years later that caused significant damage to critically endangered littoral rainforest communities. Conversely, escaped herbaceous garden ornamentals from island resorts such as pink periwinkle (*Catharanthus roseus*), Singapore daisy (*Sphagneticola trilobata*) and others spread aggressively and suppress native grass growth, which in turn significantly reduces fuel loads and the capacity to introduce fire for ecological reasons (e.g. to reduce vegetation thickening or to prevent rainforest expansion into sclerophyll communities). These lessons have assisted efforts to develop more holistic island restoration regimes, for example on St Bees Island, where the regime combines goat removal, lantana and other weed control, fire application and programs to conserve its koala population.

The numbers of visitors and consequent impacts on K'gari-Fraser Island are considered contrary to maintenance of its World Heritage values (Wardell-Johnson *et al.* 2015). It is likely that pressure from residential use and visitation is also causing impact on other south-east Qld islands. Most islands in the GBRWHA and Wellesley group are protected areas, but increasing weed infestation and vegetation change from lack of appropriate fire management remains a concern.

Weeds such as African boxthorn and common ice-plant are significant problems on SA islands. Their seeds are transported readily by visiting birds and, without surveillance and subsequent control, they outcompete native species. Tree mallow (*Malva arborea*) on West Island reduced the breeding habitat for crested, fairy and Caspian terns until it was controlled. Annual grasses like rye-grass and barley-grass can overwhelm native plants and increase fire risks. New genetic research on an island pigface species (*Carpobrotus rossii*) and introduced mainland pigface species shows evidence of hybridisation (Biffin *et al.* 2016).

Several larger SA islands (e.g. Kangaroo, Althorpe, Wedge, Wardang, Thistle, Reevesby, Boston, Flinders and St Peter) have well-established populations of house mice. Only Kangaroo and Flinders Islands are currently known to be inhabited by black rats, although Thistle Island is being monitored following management of a recently discovered incursion. Whether either of these introduced rodents occurs on other islands is unknown due to lack of recent surveys.

Rabbits were introduced to numerous SA islands during early settlement as emergency food sources for stranded mariners. Most populations died out naturally, although rabbits continue to survive on a small number, including Wardang and Spilsby Islands. Sheep and cattle farms continue to exist on Kangaroo Island, but ungulates have been removed from most others except for sheep on Flinders and Boston Islands, and cattle and pigs on Flinders Island.

Feral cats are well-established on Kangaroo Island (where foxes and rabbits are absent), and trials are underway to assess the possibility of their eradication. Flinders, Wardang, Goose and Granite Islands also support feral cat populations (Robinson *et al.* 1996).

Tasmania has 44 species of exotic vertebrates, over 350 species of exotic invertebrates and at least 58 pest species in its marine environment, though this inventory is regularly revised (RPDC 2006; Mallick and Driessen 2009). The most numerous introductions have been to Maria Island, with over 90 exotic plant and 20 bird and mammal species deliberately released (Invasive Animals CRC 2012). Non-native mammals have been introduced to 69 Tasmanian offshore islands, the most common being rabbits and cats (*Felis catus*) (Terauds 2005; TPC 2009). There is no comprehensive list of exotic birds or invertebrates, but both these groups are widespread and expanding on many islands. European wasps (*Vespula germanica*) and bumblebees (*Bombus terrestris*) occur on Mewstone and De Witt Islands, and several species of introduced flatworms, earthworms, slugs and springtails occur on Macquarie Island.

Weeds, especially sea spurge (*Euphorbia paralias*), African boxthorn (*Lycium ferocissimum*), gorse (*Ulex europaeus*), canary broom (*Genista monspessulana*), Spanish heath (*Erica lusitanica*) and marram grass (*Ammophila arenaria*) now infest the coasts of most near-onshore islands. Weed removal programs are underway on over 20 Tasmanian islands, including 12 in the Furneaux group. Macquarie Island had at least five alien plant species; two (*Anthoxanthum odoratum* and *Rumex crispus*) have been removed (PWS 2006).

Climate change is recognised as a threat to many islands. Predictions show that more than 1440 km of Tasmania's coast is at risk of flooding, and more than 975 km of shoreline at risk of erosion, sand dune mobility, rock falls and slumping because of sea level rise and storm surges (Sharples 2006; DPIPWE 2010).

Due to their proximity to the mainland, most Victorian islands have incursions of exotic birds such as common blackbird (*Turdus merula*), common starling (*Sturnis vulgaris*) and European goldfinch (*Carduelis carduelis*) (DSE 2012). Common myna (*Acridotheres tristis*) and the Indian peafowl (*Pavo cristatus*) have recently established in the conservation estate on French Island. These act as vectors for weeds, such as African boxthorn (Noble and Adair 2014). Little attempt has been made to control these birds, as reinvasion is likely to occur. Paradoxically, introduced populations of hog deer (*Axis porcinus*) and fallow deer (*Dama dama*) are maintained on Sunday Island to support recreational hunting (Forsyth 2005).

Intentional releases of exotic species onto Victorian islands has occurred since the arrival of Europeans, and continues today. Brown hares (*Lepus apensis*) were released on Phillip Island in the 1860s (Lever 1992). Fallow deer were also released on Phillip Island in 1851, but had died out by ~1920 (Bentley 1998). Sambar (*Cervus unicolor*) were introduced to French Island in the 1850s and have been reported swimming from French or Phillip Islands to the mainland (Bentley 1998). Other deer species, such as hog and fallow, have been released onto some Victorian islands because deer species are afforded some protection under the *Wildlife Act 1975*, and deer hunting is permitted in season (DPI 2012). However, legal recreational deer hunting occurs only on Sunday Island.

WA islands have a sporadic history of disturbance and deliberate plant introductions. A total of 317 introduced plant species have been recorded from 206 (95%) of the islands along the west coast (Lohr and Keighery 2016). In general, introduced plants on WA islands are associated with human occupation and use, proximity to population centres and presence of breeding seabirds (Lohr and Keighery 2014, 2016; Lohr *et al.* 2016). Most are regarded as invasive and have potential to modify natural processes on islands. On some islands, plants have been deliberately introduced for shade and aesthetic reasons but are unlikely to become invasive.

Along the north coast, 139 introduced plant species have been recorded from 51 Kimberley islands (13% of named islands) and 61 from 100 Pilbara islands (~17% of named islands) (Table 2.5; Lohr *et al.* 2016). In the Kimberley, the low number of islands with weeds is partially an artefact of the low number of islands that have been surveyed (4% of total) and relatively low levels of seabird and human use. The most widespread weed on Kimberley islands is passion vine (*Passiflora foetida*), which is spread by birds. Buffel grass (*Cenchrus ciliaris*) and kapok (*Aerva javanica*) are probably the most widespread Pilbara island weeds, and are known from 49 and 28 islands in the region respectively (Lohr *et al.* 2016).

A disproportionately large number of weeds have been recorded on islands near Perth, estuarine islands, and islands with a history of intensive human activity (Table 2.5). The ice-plant (*Mesembryanthemum crystallinum*) and common sowthistle

Table 2.5. Islands off the WA coast with high numbers of recorded weeds

Island	Number of recorded weeds	History of island activity
Koolan	86	Mining
Cockatoo	36	Mining
Sunday	28	Mission for Aboriginal community
Varanus	27	Oil and gas
Barrow	22	Oil and gas
Thevenard	19	Oil and gas
Burrup Peninsula	17	Oil and gas
Hermite Island	6	British atomic tests
Garden	162	Early European use, holiday settlements and a naval base
Penguin	64	Tourism
Carnac	78	Tourism
Rottnest	101	Tourism
Dirk Hartog	64	Pastoral
Woody	57	Recreation
Middle	22	Early settlement

Source: Lohr *et al.* (2016).

(*Sonchus oleraceus*) are most commonly found on west coast islands (69% and 63% of islands, respectively).

A total of 116 introduced plant species have been recorded from 43 (17%) of islands along the south coast (Lohr and Keighery 2014), but not all islands have yet been surveyed. Those that pose greatest threat include bridal creeper (*Asparagus asparagoides*), pig's ear (*Cotyledon orbiculata*), African boxthorn and annual grasses in the *Avena*, *Bromus* and *Ehrharta* genera.

The Conservation Commission of Western Australia identified the establishment of weeds on island conservation reserves as a significant issue affecting island biodiversity (CCWA 2009). Management of these would be assisted by better knowledge of their status and distribution. Surveys of weeds on many islands, particularly in the Kimberley and along the south coast, have not yet been conducted. The recent development of a database (Lohr and Lohr 2016) of all existing records of weed occurrences on WA islands will assist in identifying survey gaps and planning for future work. An environmental weed strategy for WA (CALM 1999) provides a framework for ranking weeds on a statewide basis against three criteria – invasiveness, distribution and environmental impacts.

Twelve species of exotic mammals have been recorded on 121 islands off the WA coast (Abbott and Burbidge 1995). The black rat has been recorded on 32 islands (DEC 2012a), mostly along the Pilbara coast where they were accidentally introduced by the pearling industry in the 19th century (Morris 2002). This includes seven islands in the Montebello group; however, Burbidge *et al.* (2000) recorded black rats on almost all the 180 islands and vegetated islets in the archipelago. The black rat has been eradicated on 12 WA islands (Morris 2002) plus all the islands in the Montebello group (Burbidge 2003). The Pacific rat (*Rattus exulans*) is now known only from Adele Island, off the Kimberley coast, where it was likely introduced by Indonesian fishermen in the 19th century. Three attempts at eradicating rodents using ground and aerial baiting techniques have failed. The house mouse is known from 27 WA islands, having been introduced through use of the islands for settlement, tourist activities, the oil and gas industry or pastoral activity. Some introductions associated with the oil and gas industry (e.g. on Barrow, Thevenard and Varanus Islands) along the Pilbara coast have been recent events over the last 30 years. House mice have been eradicated from six islands and appear to have disappeared from three others following removal of the settlement (Bernier Island), competition from reintroduced native species (Faure Island) and a black rat eradication program (Hermite Island).

The cat has been recorded on 19 WA islands (Abbott and Burbidge 1995), its presence being largely associated with settlement of some islands as pastoral properties in the mid–late 19th century (e.g. Dirk Hartog and Faure Islands), or spread from nearby settlements or towns. Cats became feral on 13 islands, but have been eradicated from five of these. The largest island from which feral cats have been eradicated is Faure Island (5148 ha) and a program is underway to eradicate them from Dirk Hartog Island (58 640 ha) by 2018 (Algar *et al.* 2011). Some islands have had single cat records associated with early settlement (e.g. Bernier, Rosemary and Thevenard Islands), but the cat has disappeared or was removed when settlement ceased.

The European red fox (*Vulpes vulpes*) is known from 18 islands. All are close to the mainland or within estuaries, and are connected to the mainland by shallow water or infrastructure such as road and rail causeways. As such, foxes have probably been on the islands, or regularly accessing them, since their arrival on the adjacent mainland. In the south-west of the state this occurred in the 1920s, but in the Pilbara this was not until the mid-1930s (Long 1972). Regular fox control is undertaken on the northern end of the Burrup Peninsula and adjacent islands in the Dampier Archipelago where foxes have been recorded (Morris 1989; CALM 1990). This appears to be successful in reducing fox abundance on these islands.

The dingo/dog has been recorded on eight WA islands (Burbidge and McKenzie 1978; Abbott and Burbidge 1995). A recent survey of the Kimberley islands (Gibson and McKenzie 2012) found evidence only on Unwins Island. Rabbits were often released onto islands for food by sealers and whalers in the 19th century. Today, the rabbit is recorded from 24 WA islands, all in the southern part of the state.

Management

In the NT, active fire management is undertaken by Indigenous ranger groups on larger islands including Bathurst and Melville Islands. Island populations have been established for two threatened mammal species in the NT. Northern quolls (*Dasyurus hallucatus*) were established on Astel and Pambasso Islands in 2003 as insurance populations against a massive decline of mainland populations following invasion of cane toads (*Rhinella marina*) (Griffiths *et al.*, unpub. data). Golden bandicoots (*Isoodon auratus*) were successfully translocated from Marchinbar Island (the only NT location for this species) to Raragala and Guluwuru Islands between 2007 and 2009 (Brennan and Palmer 2009).

Other than legislative protection, most SA islands have little or no active management attention. There are three exceptions. Kangaroo Island has its own District Council and Natural Resources Management region and Board guided by an NRM plan and management investments, Granite Island is connected to the mainland by a short causeway and is a tourist attraction for the city of Victor Harbor, and there are four privately owned islands with varying levels of landowner management. Thistle Island, for example, has a management committee and an island caretaker; and Boston Island is managed for sheep grazing and as a tourist holiday retreat. Islands within Wilderness Protection Areas have regulations limiting access for recreational activities, in part to reduce the risk of accidental introductions of weeds. Except for Kangaroo Island, there are no biosecurity strategies or management programs for SA islands. This is a gap, especially given the recent arrival of black rats on Thistle Island.

On Macquarie Island, weka (*Gallirallus australis*) were extirpated by 1988, feral cats by 2002 and a multi-species program eradicated rabbits, black rats and house mice by 2014 (Commonwealth of Australia 2014b). The last feral cat was killed on Tasman Island in 2010 (PWS 2011). House mice were removed from Fischer Island in 2014 and a program to eradicate black rats from Big Green Island commenced in early 2016. A Bruny Island cat management program commenced in mid-2016 after receiving federal funding. A feasibility plan to remove feral pigs from Flinders Island has been developed but is yet to secure funding. Public access is restricted to a small number of islands unless authorised by permit (e.g. Macquarie Island, Judgement Rocks, North East Isle, South West Isle, Vissher Island, Ile de Phoques, Albatross, Rodondo) and partial restrictions apply to several islands within Freycinet National Park. A biosecurity protocol has been developed to reduce the spread of pests, weeds and diseases to Tasmania's islands, with additional biosecurity measures recommended for islands in Tasmania's South-west Wilderness World Heritage Area (Mallick and Driessen 2009). Biosecurity procedures are in place for people and equipment arriving on Macquarie Island (PWS 2006).

Marooning of native wildlife species for 'conservation' purposes has occurred historically on Victoria's islands, and continues (Johnston 2014). For example, koalas have been introduced to five Victorian islands since the late 19th century, and populations have increased rapidly in the absence of predation. These populations have required contentious management intervention, including chemical sterilisation (Menkhorst 2008). In 2012, the threatened eastern barred bandicoot (*Perameles gunnii*) was released on French Island as part of the recovery plan for the species (Hill *et al.* 2010), with further releases expected in the future (Coetsee 2016). The success of this translocation is uncertain at this stage – some bandicoots have been predated by feral cats and others have died from toxoplasmosis. This species was also released onto Churchill Island in 2015.

Currently, most Victorian islands are managed to protect their historic and biodiversity values. Regulations differ between the various tiers of declared conservation reserve but most generally prohibit transport of animals to the islands. An authorisation is required to land on wholly national park islands. Some islands and rock stacks (Lady Julia Percy, Twelve Apostles, Wilsons Promontory islands, Tullaberga Island) remain protected because of an absence of secure anchorage or beach access.

Management of introduced mammal predators on Victorian islands is undertaken as part of threatened species recovery programs. Feral cats have been removed from Gabo Island (Twyford *et al.* 2000). Similarly, a systematic program to remove foxes from Phillip Island is nearing completion (Kirkwood *et al.* 2014). Land managers have supported desexing of pets and removed >800 cats from French Island (Norvick 2015). Guardian dogs were used to protect little penguins and short-tailed shearwaters from predation by foxes and dogs on Middle Island (van Bommel 2010).

Fire is used to suppress growth of pine (*Pinus radiata*) in addition to achieving other ecological or fuel reduction (asset protection) objectives on French Island. Wildfires are infrequent on the smaller, unpopulated islands and are generally left unmanaged. Finally, biosecurity on Victorian islands is currently poor. Management plans are in place for the Victorian islands that are of conservation significance (Johnston 2014). These plans may include areas of mainland and neighbouring islands (e.g. Wilsons Promontory National Park), or include many islands (e.g. Nooramunga Marine and Coastal Park Draft Plan) or a component of an island (e.g. French Island National Park).

WA islands have been important sites for the reintroduction of threatened fauna to enable reconstruction of fauna assemblages where extinctions have occurred, and as sites to maroon threatened species to reduce their risk of extinction. A fauna reconstruction project in the Montebello group has returned the golden bandicoot, spectacled hare-wallaby, white-winged fairy-wren and spinifex bird to Hermite Island after the original populations were extirpated by feral cats and black rats. Similarly, the Barrow Island boodie was reintroduced to Boodie Island after black rats had been eradicated (Morris 2002). Other islands in the Montebello group have been used to maroon the Barrow Island boodie (Alpha Island), Shark Bay mouse (North West Island) and central Australian mala (Trimouille Island) to improve their conservation status. An ambitious plan to reintroduce 10 mammal species and one bird species, and introduce (for conservation reasons) an additional two mammal species, to Dirk Hartog Island National Park is underway (Morris *et al.* 2017).

There are no approved management plans for any of the ~2000 Kimberley islands; however, the majority are or soon will be surrounded by marine parks under joint management by Traditional Owner groups and state government (Parks and Wildlife 2013, 2016a). In addition, some Traditional Owner groups are preparing guidelines for development of management plans for conservation areas, including islands, in their areas (Hill 2008). In the Pilbara, there are approved management plans for the Dampier Archipelago nature reserves (CALM 1990), Barrow Island nature reserves (Parks and Wildlife 2015) and Muiron Islands nature/recreation reserves (CALM 1999), and management proposals for the Montebello Islands and surrounding waters (Morris 1991). Existing or proposed marine parks and management areas also encompass these island groups (CALM 2005; DEC 2007). A management plan for the Shark Bay terrestrial reserves covers 18 of the 40 island nature reserves in Shark Bay and Dirk Hartog Island National Park (DEC 2012b). One of the largest islands in Shark Bay, Faure Island, is a pastoral lease managed by Australian Wildlife Conservancy for conservation; however, no management plans for this island have yet been prepared. All islands in Shark Bay are included in the World Heritage Area and covered under the Shark Bay World Heritage Property Strategic Plan (DEC 2008). Many in the Henri Freycinet Harbour are included in the Freycinet Special Purpose Zone of the Shark Bay Marine Park, while all others (except Bernier and Dorre Islands) are in the General Use Zone of the marine park (CALM 1996).

Many island nature reserves along the mid-west coast have management plans, including the Houtman Abrolhos (DoF 2012), the Turquoise Coast Island nature reserves (CALM 2004), Carnac Island nature reserve (CALM 2003) and the Shoalwater Bay Islands nature reserves (CALM 1992). Waters around the Houtman Abrolhos are gazetted as a Fish Habitat Protection Area in recognition of their value to the commercial rock lobster and aquaculture industry. Many of the Turquoise Coast Island nature reserves are included in the Jurien Marine Park (DEC 2005). The Rottnest Island Authority is required to have a five-year management plan to guide management directions on Rottnest Island, including visitor services, facilities and conservation of the islands' environment and heritage values (RIA 2009). The South Coast Region Regional Management Plan proposes that area management plans be prepared for all island nature reserves in that region (CALM 1992). Management of the 105 islands in the Recherche Archipelago, plus Woody Island and some others nearby, is now guided by a management plan (Parks and Wildlife 2016b). The few small islands that lie between Cape Naturaliste and Cape Leeuwin, and off Augusta, are surrounded by the Ngari Capes Marine Park (DEC 2013); however, these do not have a terrestrial component.

Over the last 30 years, there has been good progress with eradicating introduced animals from WA islands, and recently the awareness of biosecurity programs needed to prevent further island introductions has increased. More recent island management plans reflect this, for example that for the Barrow Island nature reserves (Parks and Wildlife 2015). The Status Performance Assessment of biodiversity conservation on WA islands (CCWA 2009) recommended that preventing the establishment of exotic species on island conservation reserves should be a high priority, as should the eradication, where feasible, of exotics detected on island conservation reserves. A Quarantine Management System was established for the Gorgon Gas Development to reduce the risk of invasive species establishing on the Barrow Island nature reserves, and has been effective at preventing the establishment of any non-indigenous species because of the project (Scott *et al.* 2017). In addition, a decision support system for prioritising biosecurity actions on Pilbara islands is being developed. This has collated the biodiversity values of ~600 Pilbara islands and is developing modelling software that identifies the most cost-effective management actions for maintaining the islands' biodiversity values (Lohr *et al.* 2017, Wenger *et al.* 2017) and will guide managers in their operations across islands of high conservation value (Chapter 5). This decision support tool could be applied to other island groups.

Concluding comments

The management challenges confronting conservation of biodiversity on Australia's islands are enormous and a collaborative approach between government agencies, Traditional Owners and other island stakeholders is required for effective outcomes. Vulnerability to climate change, in addition to the impacts of invasive species, detrimental human activities such as further land clearing and biosecurity threats pose a significant and ongoing risk to native species persistence on Australia's islands. Unless the threats to native species on these islands can be reduced or ameliorated, ecosystem stability is increasingly threatened (Ives and Carpenter 2007). The risks can, in part, be reduced by urgently improving and enforcing biosecurity regulations, restricting or prohibiting damaging human actions, and removing threatening processes. Increasing public and political awareness of offshore island values are essential to ensure their ongoing protection.

Acknowledgements

We thank Dorian Moro for comments on earlier drafts of this chapter. Allan Burbidge, Kim Onton, Mark Blythman, Cheryl Lohr and Tony Tucker provided additional information on Western Australian aspects of this chapter.

References

Abbott I, Burbidge AA (1995) The occurrence of mammal species on the islands of Australia: a summary of existing knowledge. *CALMscience* **1**, 259–324.

Abbott I, Wills A (2016) Review and synthesis of knowledge of insular ecology; with emphasis on the islands of Western Australia. *Conservation Science Western Australia* **11**, 1–209.

Algar D, Johnston M, Hilmer SS (2011) A pilot study for the proposed eradication of feral cats on Dirk Hartog Island, Western Australia. In *Island Invasives: Eradication and Management.* Proceedings of the International Conference on Island Invasives. Occasional paper of the IUCN Species Survival Commission No. 42 (Eds CR Veith, MN Clout and DR Towns) pp. 10–16. IUCN, Gland, Switzerland and CBB, Auckland, New Zealand.

Allen BL, Higginbottom K, Bracks JH, Davies N, Baxter GS (2015) Balancing dingo conservation with human safety on Fraser Island: the numerical and demographic effects of humane destruction of dingoes. *Australasian Journal of Environmental Management* **22**, 197–215. doi:10.1080/14486563.2014.999134

Balmer J, Whinam J, Kelman J, Kirkpatrick JB, Lazarus E (2004) *A Review of the Floristic Values of the Tasmanian Wilderness World Heritage Area.* Nature Conservation Report 2004/3. Department of Primary Industries, Water and Environment, Tasmania.

Bamford M, Moro D, Craig M (2011) Barrow Island as an important bird area for migratory waders in the east Asian–Australasian flyway. *Stilt* **60**, 46–55.

Beeton B, Burbidge A, Grigg G, Harrison P, How R, McKenzie N, Woinarski J (2010) *Final Report of the Christmas Island Expert Working Group to Minister for the Environment, Water, Heritage and the Arts.* Department of the Environment, Water, Heritage and the Arts, Canberra.

Beeton RJS, Buxton CD, Cochrane P, Dittmann S, Pepperell JG (2015) *Commonwealth Marine Reserves Review: Report of the Expert Scientific Panel.* Department of the Environment, Canberra.

Bell SL, Brearley GK, Bradley AJ (2011) The distribution and density of the squirrel glider (*Petaurus norfolcensis*) on islands in southeast Queensland. *Proceedings of the Royal Society of Queensland* **117**, 297–308.

Bentley A (1998) *An Introduction to the Deer of Australia.* Australian Deer Research Foundation, Melbourne.

Biffin E, Toelken HR, van Dijk K, Kellermann J, Waycott M (2016) *An Assessment of Native and Introduced* Carpobrotus *spp. in the Adelaide and Mount Lofty Ranges Region. Final report for Natural Resources: Adelaide and Mt Lofty Ranges.* State Herbarium of South Australia, Department of Environment, Water and Natural Resources, Adelaide.

BirdLife International (2016a) *Important Bird and Biodiversity Area factsheet: Haul Round Island.* <http://www.birdlife.org>

BirdLife International (2016b) *Important Bird and Biodiversity Area factsheet: Pearce, Urquhart and Hervey Islands (Sir Edward Pellew Group).* <http://www.birdlife.org>

BirdLife International (2016c) *Important Bird and Biodiversity Area factsheet: Sandy Island and Low Rock (Gulf of Carpentaria).* <http://www.birdlife.org>

Birdlife International (2016d) *World Database of Key Biodiversity Areas.* <http://www.keybiodiversity areas.org>

Brennan K, Palmer C (2009) *All Good: Wankurra have Two New Island Homes.* Report to the Northern Territory Government, Darwin.

Brothers N, Pemberton D, Pryor H, Halley V (2001) *Tasmania's Offshore Islands: Seabirds and Other Natural Features.* Tasmanian Museum and Art Gallery, Hobart.

Bryant SL (2014) *Tasmania: A State of Islands.* Australia's State of the Islands report. Island Arks Symposium. <http://islandarks.com.au/island-reports/australias-state-of-the-islands-report/>

Bunce P (1988) *The Cocos (Keeling) Islands: Australian Atolls in the Indian Ocean.* Jacaranda Press, Brisbane.

Burbidge AA (2003) Montebello Renewal: Western Shield Review – February 2003. *Conservation Science Western Australia* **5**, 194–201.

Burbidge AA, Fuller PJ (1996) The Western Australian Department of Conservation and Land Management seabird breeding islands database. In *The Status of Australia's Seabirds. Proceedings of the National Seabird Workshop.* Canberra, 1–2 November 1993. (Eds GJB Ross, K Weaver, JC Greig) pp. 73–137. Biodiversity Group, Environment Australia, Canberra.

Burbidge AA, McKenzie NL (1978) *The islands of the north-west Kimberley: Western Australia.* Wildlife Research Bulletin Western Australia No. 7. Department of Fisheries and Wildlife, Perth.

Burbidge AA, Blyth JD, Fuller PJ, Kendrick PG, Stanley FJ, Smith LE (2000) The terrestrial vertebrate fauna of the Montebello Islands, Western Australia. *CALMscience* **3**, 95–107.

CALM (1990) *Dampier Archipelago Nature Reserves: Management Plan 1990–2000.* Management Plan No. 18. Department of Conservation and Land Management, Perth.

CALM (1992) *South Coast Region: Regional Management Plan 1992–2002.* Management Plan No. 24. Department of Conservation and Land Management, Perth.

CALM (1996) *Shark Bay Marine Reserves: Management Plan 1996–2006.* Management Plan No. 34. Department of Conservation and Land Management, Perth.

CALM (1999) *Jurabi and Bundegi Coastal Parks, and Muiron Islands: Management Plan 1999–2009.* Management Plan No. 43. Shire of Exmouth, Department of Conservation and Land Management and National Parks and Nature Conservation Authority, Perth.

CALM (2003) *Carnac Island Nature Reserve.* Management Plan No. 47. Department of Conservation and Land Management, Perth.

CALM (2004) *Turquoise Coast Island Nature Reserves: Management Plan.* Management Plan No. 50. Department of Conservation and Land Management, Conservation Commission of Western Australia, Perth.

CALM (2005) *Jurien Bay Marine Park: Management Plan 2005–2015.* Management Plan No. 49. Department of Conservation and Land Management, Perth.

Carter M (1994) Birds of the Cocos-Keeling Islands. *Wingspan* **September 1994**, 14–18.

Cassis G, Meades L, Harris R, Reid C, Carter G, Wilkie L, Jeffreys E (2003) *Lord Howe Island Terrestrial Invertebrate Biodiversity and Conservation.* Report to the NSW National Parks and Wildlife Service by the Australian Museum Centre for Biodiversity and Conservation Research, Sydney.

CCWA (2009) *Status Performance Assessment: Biodiversity Conservation on Western Australian Islands.* Conservation Commission of Western Australia, Perth.

Chatto R (2000) *Waterbird Breeding Colonies in the Top End of the Northern Territory.* Technical Report 69. Parks and Wildlife Commission of the Northern Territory, Darwin.

Chatto R (2001) *The Distribution and Status of Colonial Breeding Seabirds in the Northern Territory.* Technical Report 70. Parks and Wildlife Commission of the Northern Territory, Darwin.

Chatto R (2003) *The Distribution and Status of Shorebirds Around the Coast and Coastal Wetlands of the Northern Territory.* Technical Report 73. Parks and Wildlife Commission of the Northern Territory, Darwin.

Chatto R, Baker B (2008) *The Distribution and Status of Marine Turtle Nesting in the Northern Territory.* Technical Report 77. Parks and Wildlife Commission of the Northern Territory, Darwin.

Claussen J (2005) Native plants of Christmas Island. *Flora of Australia Supplementary Series* **22.** Australian Biological Resources Study, Canberra.

Cochran T (2003) *Managing Threatened Species and Communities on Bruny Island.* Threatened Species Unit, Department of Primary Industries Water and Environment, Tasmania.

Coetsee A (2016) Recovering the mainland eastern barred bandicoot. *Victorian Naturalist* **133**, 98–102.

Commonwealth of Australia (2014a) *Heard Island and McDonald Islands: Marine Reserve Management Plan 2014–2024.* Department of the Environment, Canberra.

Commonwealth of Australia (2014b) Macquarie Island is declared officially pest-free. Media release. The Hon. Greg Hunt MP, Minister for the Environment. Department of the Environment, Canberra <http://www.environment.gov.au/minister/hunt/2014/pubs/mr20140407.pdf>

Congdon BC, Erwin CA, Peck DR, Baker GB, Double MC, O'Neill P (2007) Vulnerability of seabirds on the Great Barrier Reef to climate change. In *Climate Change and the Great Barrier Reef: A Vulnerability Assessment.* Great Barrier Reef Marine Park Authority, Townsville.

Copley PB (1994) Translocations of native vertebrates in South Australia: a review. In *Reintroduction Biology of Australian and New Zealand Fauna.* (Ed. M. Serena) pp. 35–42. Surrey Beatty and Sons, Sydney.

Copley P (1999) Natural histories of Australia's stick-nest rats, genus *Leporillus* (Rodentia: Muridae). *Wildlife Research* **26**, 513–539. doi:10.1071/WR97056

DEC (2005) *Jurien Bay Marine Park Management Plan 2005–2015.* Management Plan No. 49. Department of Conservation and Environment, Perth.

DEC (2007) *Management Plan for the Montebello/Barrow Islands Marine Conservation Reserves 2007–2017.* Management Plan No. 55. Department of Environment and Conservation, Perth.

DEC (2008) *Shark Bay World Heritage Property Strategic Plan 2008–2020.* Department of Environment and Conservation (Perth), Department of Environment, Water, Heritage and the Arts (Canberra).

DEC (2012a) *Guide to Aboriginal Customary Activities on DEC-managed Lands and Waters.* Department of Environment and Conservation, Perth.

DEC (2012b) *Shark Bay Terrestrial Reserves and Proposed Reserve Additions.* Management Plan No. 75. Department of Environment and Conservation, Conservation Commission Western Australia, Perth.

DEC (2013) *Ngari Capes Marine Park Management Plan 2013–2023.* Management Plan No. 74. Department of Environment and Conservation, Marine Parks and Reserves Authority, Perth.

DECC (2007) *Lord Howe Island Biodiversity Management Plan.* Department of Environment and Climate Change, Sydney.

DEH (2006) *Australia's World Heritage: Australia's Places of Outstanding Universal Value.* Department of Environment and Heritage, Canberra.

Dennis TE, Clancy GP (2014) The status of the osprey (*Pandion haliaetus cristatus*) in Australia. *Journal of Raptor Research* **48**, 408–414. doi:10.3356/JRR-OSPR-14-01.1

DEWHA (2009) *Background Document for the Threat Abatement Plan to Reduce the Impacts of Exotic Rodents on Biodiversity on Australian Offshore Islands of less than 100 000 Hectares.* Department of Environment, Water, Heritage and the Arts, Canberra.

Dixon G (1996) *A Reconnaissance Inventory of Sites of Geo-Conservation Significance on Tasmanian Islands.* Report to Parks and Wildlife Service, Tasmania.

DNP (2004) *Pulu Keeling National Park Management Plan*. Director of National Parks, Parks Australia North, Darwin.

DoEE (2016) *Draft Recovery Plan for Marine Turtles in Australia*. Department of Environment and Energy, Canberra.

DoF (2012) *The Houtman Abrolhos Islands Management Plan*. Fisheries Management Paper No. 260. Department of Fisheries, Perth.

DPI (2012) *Victorian Hunting Guide 2012*. <http://www.dpi.vic.gov.au/game-hunting/about-game- hunting/publications/hunting-guide>

DPIPWE (2010) Vulnerability of Tasmania's natural environment to climate change: an overview. Unpublished Report. Resource Management and Conservation Division, Department of Primary Industries, Parks, Water and Environment, Tasmania.

DPIPWE (2016) Parks and Wildlife Service Tasmania. <www.parks.tas.gov.au/index.aspx?base=5806>

DSE (2012) *The Victorian Biodiversity Atlas*. Department of Sustainability and Environment, Melbourne. <https://vba.dse.vic.gov.au/vba/.

DSEWPaC (2013) *Issues Paper for the Australian Sea Lion (Neophoca cinerea)*. Department of Sustainability, Environment, Water, Population and Communities, Canberra.

Dutson G, Garnett S, Gole C (2009) Australia's Important Bird Areas: key sites for conservation. *Birds Australia Conservation Statement*, No. 15. Birds Australia, Melbourne.

EA (1999) *Pulu Keeling National Park Plan of Management*. Environment Australia, Darwin.

EA (2001) *Coringa-Herald National Nature Reserve and Lihou Reef National Nature Reserve Management Plan*. Environment Australia, Canberra.

EA (2002a) *Third Christmas Island National Park Management Plan*. Environment Australia, Canberra.

EA (2002b) *Ashmore Reef National Nature Reserve and Cartier Island Marine Reserve (Commonwealth Waters) Management Plans*. Environment Australia, Canberra.

FitzSimmons NN, Limpus CJ (2014) Marine turtle genetic stocks of the Indo-Pacific: identifying boundaries and knowledge gaps. *Indian Ocean Turtle Newsletter* **20**, 2–18.

Forsyth DM (2005) How can research contribute to the management of wild deer in Australia? In *Proceedings of the National Feral Deer Management Workshop*, Canberra, November 2005. (Ed. S McLeod). Invasive Animals Cooperative Research Centre, Canberra.

Gales NJ, Cheal AL, Pobar GJ, Williamson P (1992) Breeding biology and movements of Australian sea-lion Neophoca cinerea, off the west coast of Western Australia. *Wildlife Research* **19**, 405–416.

Gales NJ, Shaughnessy PD, Dennis TF (1994) Distribution, abundance and breeding cycle of the Australian sea-lion Neophoca cinerea (Mammalia: Pinnipedia). *Journal of Zoology, London*, **234**, 353–370.

Garnett S, Szabo J, Dutson G (2011) *The Action Plan for Australian Birds 2010*. CSIRO Publishing, Melbourne.

Geoscience Australia (2012) *Coastline Lengths*. <http://www.ga.gov.au/education/geoscience-basics/dimensions/coastline-lengths.html>

Geoscience Australia (2017) <www.gov.ga.gov.au/education/geoscience>

Gibson LA, Kohler F (2014) Determinants of species richness and similarity of species composition of land snail communities on Kimberley islands. In *Biodiversity Values on Selected Kimberley Islands, Australia*. (Eds LA Gibson, S Yates and P Doughty). *Records of the Western Australian Museum Supplement* **81**, 41–66.

Gibson LA, McKenzie NL (2012) Occurrence of non-volant mammals on islands along the Kimberley coast of Western Australia. In *Biodiversity Values on Selected Kimberley Islands, Australia*. (Eds LA Gibson, S Yates and P Doughty). *Records of the Western Australian Museum Supplement* **81**, 15–40.

Goulding W, Moss PT, McAlpine CP (2016) Cascading effects of cyclones on the biodiversity of south-west Pacific Islands. *Biological Conservation* **193**, 143–152. doi:10.1016/j.biocon.2015.11.022

Groom RA, Dunshea GJ, Griffiths AD, Griffiths KM (2016) *The Distribution and Abundance of*

Dugong and Other Marine Megafauna in Northern Territory, August 2016. Department of Land Resources Management, Flora and Fauna Division, Darwin.

Gynther I, Waller N, Leung LK (2016) Confirmation of the extinction of the Bramble Cay melomys, *Melomys rubicola,* on Bramble Cay, Torres Strait: results and conclusions from a comprehensive survey in August–September. Unpublished Report. Department of Environment and Heritage Protection, Brisbane.

Harris MP, Brown RS, Wainer J (1980) Great Glennie Island, Glennie Group, Victoria – Seabird Islands No. 94. *Corella* **4**, 93–95.

Harris S, Buchanan A, Connolly A (2001) *One Hundred Islands: The Flora of the Outer Furneaux.* Department of Primary Industry, Parks, Water and Environment, Hobart.

Hill R, Winnard A, Watson M (2010) *National Recovery Plan for the Eastern Barred Bandicoot (mainland) Perameles gunnii unnamed subspecies.* Department of Sustainability and Environment, Melbourne.

Hill R, Miriuwung and Gajerrong peoples, Hill DG, Goodson S (2008) *Miriuwung-Gajerrong Cultural Planning Framework. MG Guidelines for Developing Management Plans for Conservation Parks and Nature Reserves under the Ord Final Agreement.* WA Department of Environment and Conservation, Yawoorroong Miriuwung Gajerrong Yirrgeb Noong Dawang Aboriginal Corporation and CSIRO.

Hoenner X (2012) Spatial and behavioural ecology of hawksbill turtles nesting on Groote Eylandt, northern Australia. PhD thesis. Faculty of Engineering, Health, Science and the Environment, Charles Darwin University.

Invasive Animals CRC (2012) Moving Tassie devils to Maria Island. *Island Net Newsletter* **8**, 8.

Ives AR, Carpenter SR (2007) Stability and diversity of ecosystems. *Science* **317**, 58–62. doi:10.1126/science.1133258

Johnston M (2014) Victoria: the state of Victoria's islands. In Island Arks Symposium II, 7–9 February. University House, Canberra. <http://islandarks.com.au/victoria-state-of-the-islands>

Kiernan K, McConnell A (1999) Geomorphology of the sub-Antarctic Australian territory of Heard Island–McDonald Island. *Australian Geographer* **30**, 159–195. doi:10.1080/00049189993693

Kirkwood R, Johnston M (2006) Terrestrial mammals of Phillip and French Islands, Western Port, Victoria. *Victorian Naturalist* **123**, 146–156.

Kirkwood J, O'Connor J (2010) The state of Australia's birds 2010: islands and birds. *Supplement to Wingspan* **20**, 1–52.

Kirkwood R, Pemberton D, Gales R, Hoskins A, Mitchell T, Shaughnessy PD, Arnould JPY (2010) Continued population recovery by Australian fur seals. *Marine and Freshwater Research* **61**, 695–701. doi:10.1071/MF09213

Kirkwood R, Sutherland DR, Murphy S, Dann P (2014) Lessons from long-term predator control: a case study with the red fox. *Wildlife Research* **41**, 222–232. doi:10.1071/WR13196

Lever C (1992) *They Dined on Eland: The Story of the Acclimatisation Societies.* Quiller Press, London.

Limpus CJ (2008a) *A Biological Review of Australian Marine Turtle Species. 1. Loggerhead Turtle, Caretta caretta (Linnaeus).* Queensland Environmental Protection Agency, Brisbane.

Limpus CJ (2008b) *A Biological Review of Australian Marine Turtle Species. 2. Green Turtle, Chelonia mydas (Linnaeus).* Queensland Environmental Protection Agency, Brisbane.

Limpus CJ (2008c) *A Biological Review of Australian Marine Turtle Species. 3. Hawksbill Turtle, Eretmochelys imbricata (Linnaeus).* Queensland Environment Protection Agency, Brisbane.

Limpus CJ (2008d) *A Biological Review of Australian Marine Turtle Species. 4. Olive Ridley Turtle, Lepidochelys olivacea (Eschscholtz).* Queensland Environmental Protection Agency, Brisbane.

Limpus CJ (2008e) *A Biological Review of Australian Marine Turtle Species. 5. Flatback Turtle, Natator depressus (Garman).* Queensland Environmental Protection Agency, Brisbane.

Limpus CJ (2008f) *A Biological Review of Australian Marine Turtle Species. 6. Leatherback Turtle, Dermo-*

chelys coriacea (Vandelli). Queensland Environmental Protection Agency, Brisbane.

Lohr M, Keighery G (2014) The status and distribution of alien plants on the islands of the south coast of Western Australia. *Conservation Science Western Australia* **9**, 181–200.

Lohr M, Keighery G (2016) The status and distribution of naturalised alien plants on the islands of the west coast of Western Australia. *Conservation Science Western Australia* **10**, 1–43.

Lohr M, Lohr C (2016) *Weeds on Western Australia's Islands: A Historical Database with Records Spanning from 1913 to 2014*. Version 1. AEKOS Data Portal. doi:10.4227/05/56D79D59C41D3. <http://www.aekos.org.au/dataset/229076>

Lohr MT, Lohr CA, Keighery G, Long V (2016) The status and distribution of non-native plants on the gazetted and territorial islands off the north coast of Western Australia. *Conservation Science Western Australia* **10**, 1–73.

Lohr CA, Hone J, Bode M, Dickman CR, Wenger A, Pressey RL (2017) Modelling dynamics of native and invasive species to guide prioritization of management actions. *Ecosphere* **8**, e01822. doi:10.1002/ecs2.1822

Long JL (1972) *Introduced Birds and Mammals in Western Australia*. Western Australian Agricultural Protection Board Technical Series No. 1.

Mahney T, McKay L, Ziembicki M, Westaway J, Brennan K (2009) Northern Territory Seabird Breeding Colony Monitoring. Unpublished Report to Department of Natural Resources, Environment, the Arts and Sport, Darwin.

Mallick SA, Driessen MM (2009) *Review, Risk Assessment and Management of Introduced Animals in the Tasmanian Wilderness World Heritage Area*. Nature Conservation Report 10/01. Department of Primary Industries, Parks, Water and Environment, Tasmania.

Martin TG, Nally S, Burbidge AA, Arnall S, Garnett ST, Hayward MW, Lumsden LF, Menkhorst P, McDonald-Madden E, Possingham HP (2012) Acting fast helps avoid extinction. *Conservation Letters* **5**, 274–280. doi:10.1111/j.1755-263X.2012.00239.x

Menkhorst P (2008) Hunted, marooned, re-introduced, and contracepted: a history of koala management in Victoria. In *Too Close for Comfort: Contentious Issues in Human–wildlife Encounters* (Eds D Lunney, A Munn and W Meikle) pp. 73–92. Royal Zoological Society of NSW, Sydney.

Misso M, MacRae I (2014) The conservation management of Pulu Keeling National Park: challenges and perspectives. *Raffles Bulletin of Zoology* **30**(Supplement), 24–28.

Misso M, West J (2014) Conservation management of the terrestrial biodiversity of Christmas Island: challenges and perspectives. *Raffles Bulletin of Zoology* **30**(Supplement), 17–23.

Morris KD (1989) Feral animal control on Western Australian islands. In *Australian and New Zealand Islands: Nature Conservation Values and Management*. (Ed. AA Burbidge) pp. 105–111. Proceedings of a Technical Workshop, Barrow Island, Western Australia, 1985. CALM Occasional Paper 2/89. Department of Conservation and Land Management, Perth.

Morris KD (1991) *Management Proposals for the Montebello Islands and Surrounding Waters*. CALM Occasional Paper 3/91. Department of Conservation and Land Management, Perth.

Morris KD (2002) The eradication of the black rat on Barrow and adjacent islands off the north-west coast of Western Australia. In *Turning the Tide: The Eradication of Invasive Species*. (Eds CR Veitch and MN Clout) pp. 219–225. IUCN SSC Invasive Species Specialist Group. IUCN, Gland, Switzerland.

Morris K, Page M, Thomas N, Ottewell K (2017) *A Strategic Framework for the Reconstruction and Conservation of the Vertebrate Fauna of Dirk Hartog Island 2017–2030*. Department of Park and Wildlife, Perth.

Noble M, Adair R (2014) African boxthorn (*Lycium ferocissimum*) and its vertebrate relationships in Australia. *Plant Protection Quarterly* **29**, 80–83.

North P, Griffin M (2006) *Pulu Keeling National Park Plan*. Director of National Parks and Fusion Films.

Norvick J (2015) Catching cats on French Island. *Park Watch*, June, No. 261.

NPWS (2002) *Approved Recovery Plan for the Lord Howe Woodhen*. National Parks and Wildlife Service, Sydney.

Onton K, Maurer G, Weller D (2013) *North West Cape Welcomes Shorebirds: Identifying Habitat and Building Conservation Capacity*. BirdLife Australia, Melbourne.

Orchard M (2012). *Crabs of Christmas Island*. Christmas Island Natural History Association.

Paltridge R, Johnson A, Fitzpatrick S, Goodman C (2016) Reversing the Decline of Mammals in Northern Australia: Response of Native Mammals to Cat Management on the Pellew Islands 2011–2015. Unpublished Report to Territory NRM. Desert Wildlife Services, Alice Springs.

Parks and Wildlife (2013) *Lalang-garram/Camden Sound Marine Park Management Plan 73 2013–2023*. Department of Parks and Wildlife, Perth.

Parks and Wildlife (2015) *Barrow Group Nature Reserves Management Plan 82 2015*. Department of Parks and Wildlife, Perth.

Parks and Wildlife (2016a) *Lalang-garram/Horizontal Falls and North Lalang-garram Marine Park Joint Management Plan*. Management Plan No. 88. Department of Parks and Wildlife, Perth.

Parks and Wildlife (2016b) *Esperance and Recherche Parks and Reserves*. Management Plan No. 94. Department of Parks and Wildlife, Perth.

Parsons KE (2011) *Nowhere Else on Earth: Tasmania's Marine Natural Values*. Report for Environment Tasmania. Aquenal, Tasmania.

Pendoley KL, Whittock PA, Vitenbergs A, Bell C (2016) Twenty years of turtle tracks: marine turtle nesting activity at remote locations in the Pilbara, Western Australia. *Australian Journal of Zoology* doi:10.1071/ZO16021.

Prescott TW (1980) Lawrence Rocks, Victoria: Seabird Islands No. 100. *Corella* **4**, 107–109.

Phillip Island Visitor Information Centre (2012) *Phillip Island: Official Visitor Guide*.

Priddel D, Wheeler R (2014) New South Wales: state of the islands. In *Australia's State of the Islands Report*, pp. 24–40.<http://islandarks.com.au/files/2014/01/State-of-New-South-Wales-Islands.pdf>

Priddel D, Carlile N, Wilkinson I, Wheeler R (2011) Eradication of exotic mammals from offshore islands in New South Wales, Australia. In *Island Invasives: Eradication and Management*. Proceedings of the International Conference on Island Invasives. Occasional Paper of the IUCN Species Survival Commission No. 42 (Eds CR Veith, MN Clout and DR Towns) pp. 337–344. IUCN Gland, Switzerland, and CBB, Auckland, New Zealand.

PWS (2006) *Macquarie Island Nature Reserve and World Heritage Area Management Plan*. Parks and Wildlife Service, Department of Tourism, Arts and the Environment, Tasmania.

PWS (2011) Tasman Island feral cats eradicated. *Buttongrass Newsletter*, July, 2.

Reaser J, Meyerson LA, Cronk Q, De Poorter M, Eldrege LG, Green E, Kairo M, Latasi P, Mack RN, Mauremootoo J, O'Dowd D, Orapa W, Sastroutomo S, Saunders A, Shine C, Thrainsson S, Vaiutu L (2007) Ecological and socioeconomic impacts of invasive alien species in island ecosystems. *Environmental Conservation* **34**, 98–111. doi:10.1017/S0376892907003815

RIA (2009) *Rottnest Island Management Plan 2009–2013*. Rottnest Island Authority, Fremantle.

Robinson AC, Armstrong DM (1999) *A Biological Survey of Kangaroo Island, South Australia, 1989 and 1990*. Heritage and Biodiversity Section, Department for Environment, Heritage and Aboriginal Affairs, Adelaide.

Robinson T, Canty P, Mooney T, Rudduck P (1996) *South Australia's Offshore Islands: Biological Survey of South Australia*. Resource Management Branch, Department of Environment and Natural Resources, Adelaide.

RPDC (2006) *State of the Environment Tasmania*. Resource Planning and Development Commission. <http//www.rpdc.tas.gov.au/RPDC>

Russell BC, Neil K, Hilliard R (2004) *Ashmore Reef National Nature Reserve and Cartier Island Marine Reserve. Marine and Terrestrial Introduced Species Intervention and Management Strategy*. Report for

Department of Environment and Heritage, Canberra.

Sakai AK, Allendorf FW, Holt JS, Lodge DM, Molofsky J, With KA, Baughman S, Cabin RJ, Cohen JE, Ellstrand NC, McCauley DE, O'Neil P, Parker IM, Thompson JN, Weller SG (2001) The population biology of invasive species. *Annual Review of Ecology and Systematics* **32**, 305–332. doi:10.1146/annurev.ecolsys.32.081501.114037

Scott JK, McKirdy SJ, van der Merwe J, Green R, Burbidge AA, Pickles G, Hardie D, Morris K, Kendrick P, Thomas M, Horton KL, O'Connor SM, Downs J, Stoklosa R, Lagdon R, Marks B, Nairn M, Mengersen K (2017) Zero-tolerance biosecurity protects high-conservation-value island nature reserve. *Scientific Reports* doi:10.1038/s41598-017-00450-y

Sharples C (2006) *Indicative Mapping of Tasmanian Coastal Vulnerability to Climate Change and Sea-level Rise: Explanatory Report.* 2nd edn. Consultant Report for Department of Primary Industries, Water and Environment, Tasmania.

Shaughnessy PD (1999) *The Action Plan for Australian Seals.* Environment Australia, Canberra.

Shaughnessy PD, Goldsworthy SD (2015) Increasing abundance of pups of the long-nosed fur seal (*Arctocephalus forsteri*) on Kangaroo Island, South Australia, over 26 breeding seasons to 2013–14. *Wildlife Research* **42**, 619–632. doi:10.1071/WR14209

Stanton DJ, Fell DG, Gooding DO (2008) *Vegetation Communities and Regional Ecosystems of the Torres Strait Islands, Queensland, Australia: An Accompaniment to Land Zone, Vegetation Community and Regional Ecosystem Maps.* Final Report to Torres Strait Regional Authority Land and Sea Management Unit. <http://www.tsra.gov.au/__data/assets/pdf_file/0010/2008/3dtorresstraitreport_finalversion_1nov2008.pdf>

TASMAP (2006) *Islands of Tasmania Map, Scale 1: 600 000.* © State of Tasmania.

Terauds A (2005) *Introduced Animals on Tasmanian Islands.* Biodiversity Conservation Branch, Department of Primary Industries, Water and Environment, Tasmania.

TPC (2009) *Natural Values Chapter and Introduced Species on Offshore Islands Indicator. State of the Environment Report: Tasmania 2009.* Tasmanian Planning Commission, Tasmania.

TSS (2012) *King Island Biodiversity Management Plan.* Threatened Species Section, Department of Primary Industries, Parks, Water and Environment, Tasmania.

Twyford BL, Humphrey PG, Nunn RP, Willoughby L (2000) Eradication of feral cats (*Felis catus*) from Gabo Island, south-east Victoria. *Ecological Management and Restoration* **1**, 42–49. doi:10.1046/j.1442-8903.2000.00007.x

van Bommel L (2010) *Guardian Dogs: Best Practice Manual for the Use of Livestock Guardian Dogs.* Invasive Animals Cooperative Research Centre, Canberra.

Vigilante T, Toohey J, Gorring A, Blundell V, Saunders T, Mangolamara S, George K, Oobagooma J, Waina M, Morgan K, Doohan K (2013) Island Country: Aboriginal connections, values and knowledge of the Western Australian Kimberley islands in the context of an island biological survey. *Records of the Western Australian Museum* **81**(Supplement), 145–183. doi:10.18195/issn.0313-122x.81.2013.145-182

Wardell-Johnson G, Schoeman D, Schlacher T, Wardell-Johnson A, Weston MA, Shimizu Y, Conroy G (2015) Re-framing values for a World Heritage future: what type of icon will K'gari-Fraser Island become? *Australasian Journal of Environmental Management* **22**, 124–148. doi:10.1080/14486563.2014.985267

Wenger AS, Adams VM, Iacona GD, Lohr C, Pressey RL, Morris K, Craigie ID (2017) Comprehensive, fine-resolution estimation of management costs, and the liabilities of simplistic assumptions. Submitted to *Biological Conservation.*

White AW, Pyke GH (2008) Green and golden bell frogs in New South Wales: current status and future prospects. *Australian Zoologist* **34**, 319–333. doi:10.7882/AZ.2008.010

Woehler EJ (2010) Status and conservation of the seabirds of Heard Island. In *The State of Australia's Birds 2010 – Islands and Birds.* (Eds J Kirkwood and J O'Connor). Supplement to *Wingspan 20*, Birds Australia.

Woinarski J, Ward S, Mahney T, Bradley J, Brennan K, Ziembicki M, Fisher A (2011) The mammal fauna of the Sir Edward Pellew Islands, Northern Territory, Australia: refuge and death-trap. *Wildlife Research* **38**, 307–322. doi:10.1071/WR10184

Woinarski JCZ, MacRae I, Flores T, Detto T, Reid J, Pink C, Flakus S, Misso M, Hamilton N, Palmer R, Morris K, Znidersic L, Hill B (2016) Conservation status and reintroduction of the Cocos Buff-banded Rail, *Gallirallus phillipensis andrewsi*. *Emu* **116**, 32–40. doi:10.1071/MU15052

Wood Jones F (1909) Fauna of the Cocos-Keeling Atoll. *Proceedings of the Zoological Society of London* **1909**, 132–159.

Wyatt KB, Campos PF, Gilbert MTP, Kolkotronis S-O, Hynes WH, DeSalle R, Daszak P, MacPhee RDE, Greenwood AD (2008) Historical mammal extinction on Christmas Island (Indian Ocean) correlates with introduced infectious disease. *PLoS One* **3**, doi:10.1371/journal.pone.0003602.

Ziembicki M, Woinarski JCZ, Webb J, Vanderduys E, Tuft K, Smith J, Ritchie E, Reardon T, Radford I, Preece N, Perry J, Murphy B, McGregor H, Legge S, Leahy L, Lawes M, Kanowski J, Johnson C, James A, Griffiths T, Gillespie GR, Frank A, Burbidge A, Fisher A (2015) Stemming the tide: progress towards resolving the causes of decline and implementing management responses for the disappearing mammal fauna of northern Australia. *Therya* **6**, 169–225. doi:10.12933/therya-15-236

3

Enhancing island conservation outcomes: the policy and legal context, need, and options

John C.Z. Woinarski, Andrew A. Burbidge and April E. Reside

Summary

Collectively, Australian islands have many outstanding biodiversity values, and play a significant role in the conservation of Australia's biodiversity. However, many island species are highly susceptible to novel threats. Hence, although islands comprise <0.5% of the Australian landmass, about 30% of Australian extinctions since European settlement have been of island-endemic species and the three most recent Australian extinctions (all within the last decade) have been island-endemic species. This disproportionate rate of island biodiversity loss is partly due to characteristics of island species, but also reflects the inadequate consideration to date in national legislation and strategy of the special conservation needs for island biodiversity. For example, islands are not recognised as having particular conservation significance or as meriting conservation priority in the nation's primary conservation policy, *Australia's Biodiversity Conservation Strategy 2010–2030*. Here, we propose nine recommendations for amendments to legislation, strategy and management to better reflect and protect the significant biodiversity found on Australia's islands:

1. Implement more effective policy to constrain global climate change.
2. Include islands with outstanding conservation values as a Matter of National Environmental Significance.
3. Include 'significant wildlife breeding aggregations' as a Matter of National Environmental Significance.
4. Include a Data Deficient category within the conservation status categorisations of the *Environment Protection and Biodiversity Conservation Act 1999*.
5. Develop a systematic policy approach to biodiversity conservation on Australian islands, within a revised national biodiversity strategy.
6. Implement an ongoing program to eradicate priority threats from islands with significant biodiversity values.
7. Develop and implement an ongoing program for island biodiversity survey and monitoring.
8. Develop a national risk assessment for biodiversity values on all island.
9. Establish an enduring national management fund for the conservation of island biodiversity.

Introduction

The conservation need

Australia's many islands support much that is most distinctive in Australian biodiversity. These values include many endemic species, refuges for species that have declined elsewhere in their range, largely unmodified environments, a high potential to act as translocation sites, and important colonial breeding sites for seabirds, marine turtles, seals and some non-marine species such as pied imperial pigeon (*Ducula bicolor*).

However, although islands (here considered to be those smaller than Tasmania) comprise <0.5% of the Australian land area (a total island area of 32 921 km^2 within a total Australian land area of 7 692 617 km^2: Geoscience Australia 2004), island species have experienced a disproportionate share of Australia's biodiversity loss. Of 54 animal taxa considered Extinct under Australia's *Environment Protection and Biodiversity Conservation Act 1999* (EPBC Act), 21 (39%) occurred only on islands (Table 3.1). The proportion is especially high for birds, with 18 of the 22 (82%) Australian bird taxa listed as Extinct formerly occurring only on islands. For plants, island species have been less affected, with only one of the 36 (2.8%) Australian plant species listed as Extinct being formerly restricted to islands. In total, of all 90 Australian plant and animal species listed as Extinct, 22 (24%) occurred only on islands. However, these figures are an under-estimate, as the official EPBC Act list does not yet include several other recognised extinctions (Table 3.1). A further seven island-endemic taxa are extinct; therefore, 30% of all known Australian plant and animal extinctions since European settlement were of island-endemic species. Furthermore, at least two other island-endemic species have recently become extinct in the wild (the blue-tailed skink *Cryptoblepharus egeriae* and Lister's gecko *Lepidodactylus listeri*), two others have probably become extinct recently (Christmas Island shrew *Crocidura trichura* and Tiwi Island hooded robin *Melanodryas cucullata melvillensis*), and at least two more are perilously close to extinction (the King Island scrubtit *Acanthornis magnus greenianus* and King Island brown thornbill *Acanthiza pusilla archibaldi*) (Garnett *et al.* 2011; Eldridge *et al.* 2014; Woinarski *et al.* 2014b; Andrew *et al.* in press).

Except for a few notable pulses associated mostly with periods of introductions to islands of black rats (*Rattus rattus*) or cats (*Felis catus*) (and in some cases, establishment of human settlement), these island extinctions have been relatively continuous since the 1780s (Fig. 3.1), and the rate of extinctions of Australia's island-endemic species shows no sign of diminishing. Indeed, the three most recent Australian extinctions have all been island-endemic species – the Christmas Island pipistrelle (*Pipistrellus murrayi*) in 2009, Bramble Cay melomys (*Melomys rubicola*) between 2009 and 2014, and the Christmas Island forest skink (*Emoia nativitatis*) in 2014 (Woinarski *et al.* 2017).

It is not a statistical artefact that Australia's island-endemic species have suffered such a disproportionate rate of loss. Many island species have a range of traits that render them particularly susceptible to extinction: they occupy a small area, often have small population sizes, have limited genetic diversity and low reproductive rates, lack immunity to introduced diseases, are naive to introduced predators and/or have few effective defences against introduced herbivores. These characteristics have provided little resilience for island-endemic species to the introduction of invasive species or disease, and to the land use changes that have affected many islands since human settlement. Consequently, at the global level, island species are disproportionately overrepresented among the world's recent extinctions (Alcover *et al.* 1998; Blackburn *et al.* 2004; Steadman 2006; Duncan and Blackburn 2007), with nearly two-thirds of recent global extinctions being island-endemic species (Jones *et al.* 2016).

While many island species may be more extinction-prone than mainland species, conservation management may also be simpler and more likely to succeed on islands. This is because islands often have fewer threats and, once controlled, threats may be less likely to reinvade from surrounding areas. Islands also often have fewer competing land uses than do mainland areas. Hence, biodiversity conservation on islands presents not only a challenge, but also a significant conservation opportunity.

Given the well-recognised biodiversity values of islands generally and of many Australian islands

Table 3.1. Australian species (and subspecies) known to have become extinct since 1788 that were formerly restricted to islands (smaller than Tasmania)

Scientific name	Common name	Island(s)	Extinction date
Persoonia prostrata		Fraser	unknown
Solanum bauerianum[a]	Bridal flower (Lord Howe Island)	Lord Howe, Philip	?1950s (last collected 1949)
Streblorrhiza speciosa[a]	Phillip Island glory pea	Phillip (Norfolk group)	?1830s (last collected 1830)
Emoia nativitatis	Christmas Island forest skink	Christmas	2010s
Dromaius ater	King Island emu	King	1800s
Dromaius baudinianus	Kangaroo Island emu	Kangaroo	1820s
Columba vitiensis godmanae	Lord Howe white-throated pigeon	Lord Howe	1850s
Gallicolumba norfolciensis[a]	Norfolk Island ground-dove	Norfolk	1800s
Hemiphaga novaeseelandiae spadicea	Norfolk Island New Zealand pigeon	Norfolk	1900s
Porphyrio albus	White gallinule	Lord Howe	?1790s (last record 1788)
Hypotaenidia philippensis macquariensis	Macquarie Island buff-banded rail	Macquarie	1890s
Nestor productus	Norfolk Island kaka	Norfolk, Phillip	1850s
Cyanoramphus novaezelandiae subflavescens	Lord Howe Tasman parakeet	Lord Howe	1860s
Cyanoramphus novaezelandiae erythrotis	Macquarie Island red-fronted parakeet	Macquarie	1890s
Ninox novaeseelandiae albaria	Lord Howe southern boobook	Lord Howe	1950s
Gerygone insularis	Lord Howe gerygone	Lord Howe	1920s
Lalage leucopyga leucopyga	Norfolk Island long-tailed triller	Norfolk	1940s
Rhipidura fuliginosa cervina	Lord Howe grey fantail	Lord Howe	1920s
Zosterops strenuus	Robust white-eye	Lord Howe	1920s
Zosterops albogularis	White-chested white-eye	Norfolk	?1980s
Turdus poliocephalus poliocephalus	Norfolk Island thrush	Norfolk	1970s
Turdus poliocephalus vinitinctus	Lord Howe thrush	Lord Howe	1920s
Aplonis fusca fusca[b]	Norfolk Island Tasman starling	Norfolk	1950s
Aplonis fusca hulliana[b]	Lord Howe Tasman starling	Lord Howe	1920s
Nyctophilus howensis	Lord Howe long-eared bat	Lord Howe	1920s
Pipistrellus murrayi[a]	Christmas Island pipistrelle	Christmas	2000s
Melomys rubicola[a]	Bramble Cay melomys	Bramble Cay	2000s
Rattus macleari	Maclear's rat	Christmas	1900s
Rattus nativitatis	Bulldog rat	Christmas	1900s

[a] Extinct taxa not yet formally listed as Extinct under the EPBC Act.
[b] Listed as Extinct at species level.

specifically, the disproportionate and ongoing rate of loss of Australian island-endemic biodiversity and the broader (global) recognition of the susceptibility of island species generally, there is a good case to consider that islands and island biodiversity merit particular conservation priority, and that such priority should be reflected and embedded in effective legislation, policy and management.

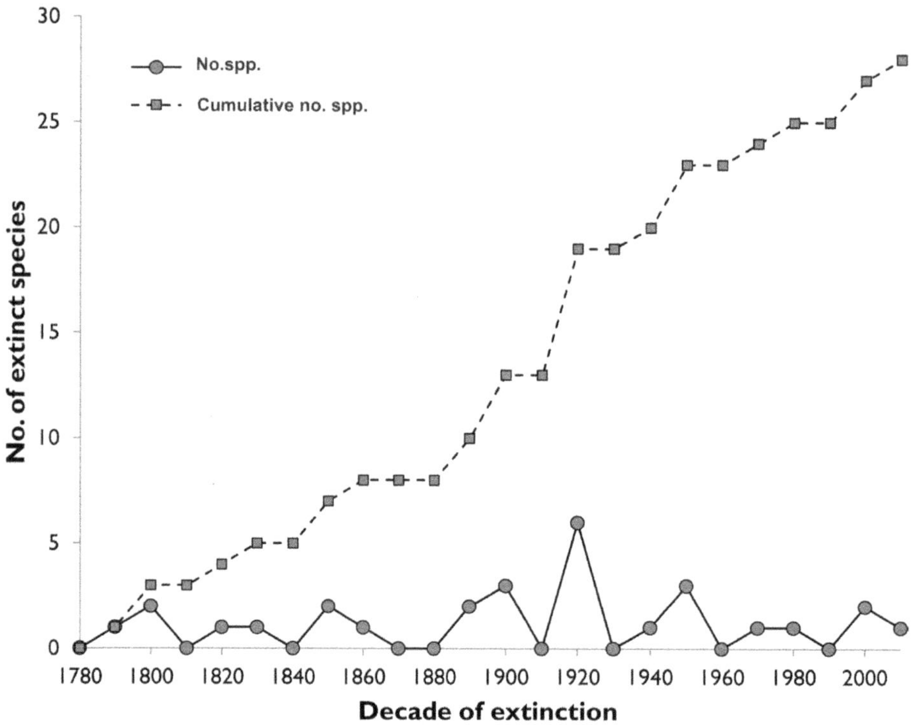

Fig. 3.1: Estimated dates of extinctions for Australian island-endemic species. Note that most of these dates relate to the last known record, and in many cases the species may have persisted long after this date. Of species listed in Table 3.1, this graph excludes *Persoonia prostrata*, for which the decade of extinction cannot be reliably estimated. Squares and dashed line mark cumulative number of extinctions of island-endemic species; filled circles and solid line represent the number of extinctions of island-endemic species within a given decade (e.g. 1920 refers to the decade from 1920 to 1929).

Current legal, policy and management context for islands

Australia's environmental policies and legislation are complex and suboptimal. The Australian constitution vests primary responsibility for most environmental matters with the states and, subsequently, the territories. The notable exception is for 'Matters of National Environmental Significance', which are determined in part by international treaties to which the national government is a signatory: these are currently restricted to a tightly circumscribed set of nine features. Many of these features are managed cooperatively by the Australian and relevant state/territory governments. The state/territory and national governments jointly develop and implement some environmental policies, with the most substantial relevant example being *Australia's Biodiversity Conservation Strategy 2010–2030* (Natural Resource Management Ministerial Council 2010).

Consequently, the conservation management of biodiversity in Australia involves a somewhat haphazard set of idiosyncratic state/territory policies and legislation, overlaid in some instances by national policies and legislation. Furthermore, some environmental responsibilities are held in some cases by local governments (although this tier of administration is absent on many islands). Collectively, these multiple responsibilities result in little coherence in, but rather markedly disparate, conservation approaches, concern and outcomes for Australian islands (as they do for Australia's mainland areas). Further complicating the policy and legal framework for island conservation management is the substantial variation among islands in terms of ownership and tenure, development history, biosecurity settings, extent of exploitation, threatening processes and human population size.

This policy potpourri partly mirrors the remarkable environmental range spanned by Australian islands – in origin, size, climate, isolation, geomorphology and topography, among other factors. Accordingly, there is comparable variation in their conservation values and threats. This chapter considers the conservation consequences of this disjointed policy and legislative context across the set of Australia's islands, and suggests options for a more coherent and effective approach to the conservation of biodiversity on Australian islands. Much of the content relates to national policy and legislation, because a comprehensive review of all state and territory policy and legislation is impractical.

National and collaborative approach, policy and legislation

Australia's islands are mostly within state and territory jurisdictions. However, several island groups with significant biodiversity values are the primary responsibility of the Australian Commonwealth government, for historical, strategic and other reasons. Collectively, the tally of islands managed by the Commonwealth comprises about 85 islands of area >1 ha (2% of Australia's islands) with a total area of 562 km^2 (1.7% of the total area of Australia's islands) (Geoscience Australia 2004). There is administrative variation even among these Commonwealth-administered groups. Norfolk Island (and its satellite islands) is currently managed as an Australian external territory. Christmas Island has been managed as an Australian external territory since being acquired by the Australian government in 1958, although with varying levels of local (shire) administrative responsibility. The Cocos (Keeling) Islands group is a relatively recent acquisition by the Australian government, and is managed as an Australian external territory, with limited local responsibilities. However, for Christmas Island and the Cocos (Keeling) Islands group, the Western Australian government provides some environmental services and responsibilities under specific bilateral agreements with the Commonwealth government. The isolated, small and uninhabited Ashmore and Cartier Islands are managed as Australian external territories, as is a small set of uninhabited islands in the Coral Sea. The sub-Antarctic Heard and McDonald Islands are also managed as Australian external territories. While sub-Antarctic islands have some management responsibility vested in the Australian Antarctic Division, most other Australian external territories are vested in the Department of Infrastructure and Regional Development. Some small uninhabited islands around the continental shelf of Australia's claimed Antarctic territory are also managed by the Australian government, likewise with some administrative responsibilities vested in the Australian Antarctic Division. The Commonwealth also has some direct responsibilities for Bowen Island in Booderee National Park (jointly managed by Indigenous landholders and the Australian government's Parks Australia) in Jervis Bay, and Barron and Field Islands in Kakadu National Park (also jointly managed by Indigenous landholders and Parks Australia).

National environmental policy and legislation extends to these islands, and to other islands where a Matter of National Environmental Significance is involved. The variable distribution of these matters across Australian islands is summarised in the following three paragraphs.

World Heritage listing (under the *Convention Concerning the Protection of the World Cultural and Natural Heritage*) provides some profile and protection to biodiversity on Norfolk Island, islands of the Great Barrier Reef, Fraser Island, islands in the Heard-McDonald group, Macquarie Island and Lord Howe Island. Islands are also included as part of other broader World Heritage listings, notably the Shark Bay area (which includes Bernier, Dorre, Faure and Dirk Hartog Islands, all with significant biodiversity values, particularly for the conservation of threatened mammal species); the Tasmanian World Heritage area (which includes Maatsuyker and other islands); Kakadu National Park; and the Ningaloo Coast World Heritage area (which includes Muiron Islands) (Chapter 10). This set encompasses only a portion of those Australian islands with highest biodiversity value.

Species or ecological communities listed as threatened under the EPBC Act are afforded some

Table 3.2. Nationally listed threatened species for some taxonomic groups on Australian islands

Taxonomic group	No. of threatened species (excluding extinct species)	No. (%) of these species occurring only on islands	No. (%) of these species occurring on both mainland and islands	No. of islands with threatened spp.
Plant	1263	73 (5.8)	48 (3.8)	44
Terrestrial invertebrate	53	12 (22.6)	2 (3.8)	12
Frogs	29	0 (0)	3 (10.3)	12
Terrestrial reptiles	52	10 (19.2)	5 (9.6)	26
Terrestrial mammals	96	12 (12.5)	37 (38.5)	110[a]

Note that for these tallies, Tasmania is not considered an island; the set of terrestrial mammals here excludes primarily marine species; tallies relate to extant EPBC Act listed taxa only; island occurrences relate to ongoing presence and may include some recent introductions of threatened species; bird tallies are not included.

[a] Does not include records for the vulnerable grey-headed flying-fox (*Pteropus poliocephalus*), which is a dispersive visitor to many additional islands.

protection by the Australian government, mostly against acute threats, even on islands primarily managed by state and territory governments. There is no current account of nationally listed threatened species by islands, but our assessment for some taxonomic groups (Table 3.2) indicates that many Australian islands support nationally listed threatened species, which are offered at least notional protection due to that listing. However, given that the three most recent extinctions of Australian species were all listed as threatened nationally, this protective mechanism is self-evidently inadequate (Woinarski *et al.* 2017). In contrast to the occurrence of many threatened species on islands, only one of Australia's 75 listed terrestrial threatened ecosystems occurs primarily on islands: Kangaroo Island narrow-leaved mallee (*Eucalyptus cneorifolia*) woodland.

Of Australia's 65 wetland sites listed with international significance (and hence afforded some protective status) under the *Convention on Wetlands of International Importance* ('Ramsar convention'), 22 include all or parts of some islands, with notable examples including Ashmore Reef, islands in the Coral Sea, Moreton Island and other nearby islands within the Moreton Bay site, Pulu Keeling (North Keeling Island) and small sites on Christmas, King and Flinders Islands.

Of the 21 Key Threatening Processes listed under the EPBC Act, only one relates directly to islands: 'Predation by exotic rats on Australian offshore islands of less than 1000 km^2 (100,000 ha)'. The explicit recognition of this threat prompted a planning response (DEWHA 2009) and some funding to eradicate introduced rats on islands.

As with legislation, relevant national strategies and policies have no specific island consideration. The nation's primary conservation policy, *Australia's Biodiversity Conservation Strategy 2010–2030* (Natural Resource Management Ministerial Council 2010), has no mention of islands (other than in two figure captions), let alone recognition of the biodiversity importance of islands or of the conservation susceptibility of island biota, or any consideration of a strategic response to the conservation of biodiversity on Australian islands. Australia's primary strategy for the development of its conservation reserve system, *Directions for the National Reserve System: A Partnership Approach* (National Reserve System Task Group 2010), likewise a collaboration of states/territories and Commonwealth governments, has a brief mention of the need for conservation on islands and uses Kangaroo Island as one case study, but otherwise provides no strategic or coherent approach to island conservation. Indeed, many islands were not included in initial versions of the Interim Biogeographical Regionalisation of Australia (Thackway and Cresswell 1995), the basis for planning the national reserve system (Hobbs 2014). This omission is notwithstanding that their often high level of endemism renders some islands irreplaceable in any conservation reserve design that seeks comprehensive inclusion of biodiversity elements (Pressey *et al.* 1994).

Many islands have high biodiversity value coupled with extreme pressures such that a substantial component of biodiversity is highly susceptible to loss: this is essentially the definition of a biodiversity hotspot (Myers *et al.* 2000). However, when the Australian government sought in 2003 to identify hotspots as a prioritisation filter for conservation resourcing (https://www.environment.gov.au/biodiversity/conservation/hotspots/national-biodiversity-hotspots#hotspot15), it included (in part) only one island (Mt Lofty/Kangaroo Island) among the 15 defined hotspots. More recently (in 2015), the Australian government's *Threatened Species Strategy* (Commonwealth of Australia 2015) recognised an explicit priority for island conservation, and included specific targets to eradicate feral cats from five islands of high biodiversity value.

Some bilateral agreements (between Commonwealth and state governments) influence conservation management on some islands. A good example is the shared responsibility for the eradication of invasive species on Macquarie Island (managed by the Tasmanian government as a conservation reserve). There are also shared arrangements between the Commonwealth and Queensland governments for the management of the Great Barrier Reef, including many islands. A perhaps less successful example is the Regional Forest Agreement process, embedded within the EPBC Act, that allows bilateral provisions ostensibly to balance timber production and biodiversity conservation in some designated areas (including parts of Bruny Island, Tasmania), but which has had some notable failures in conservation outcomes (Lindenmayer *et al.* 2015).

State/territory approaches, policy and legislation

As with national shortcomings, for state and territory jurisdictions there is little or no legislation or policy that recognises the particular biodiversity values of islands and targets island conservation management (Table 3.3). Western Australia has been exceptional (Box 3.1), having long managed its many islands as important conservation assets, with systematic programs of survey, pest eradication, translocation and some biosecurity practice (Algar *et al.* 2002; Burbidge and Morris 2002; Gibson *et al.* 2012). However, this formidable effort (with significant conservation outcomes) has not been firmly rooted in law or policy. Some elements of these conservation management actions have also occurred, albeit typically less substantially, in most other states and the Northern Territory.

States and territories maintain their own lists of threatened species (including many species that occur entirely or partly on islands). The associated state/territory legislation and policy settings provide some protection for these species and some prioritisation for resourcing for their conservation management – although, as with national legislation, listing *per se* offers little conservation safeguard or assurance of management investment. In a welcome step, there is a current collaborative Commonwealth/states/territories process to harmonise the individual jurisdictional listings of threatened species.

Conservation management on individual Australian islands is highly variable. Conservation reserves occur in, or encompass, many Australian islands (Chapter 2). For many of these reserved islands, conservation management plans (with some statutory authority) provide a description of island conservation values and management priorities designed to maintain or recover those values. Notable state examples include management plans for Hinchinbrook Island National Park (Queensland Parks and Wildlife Service 1999), Shark Bay islands (DEC 2012) and Barrow Island (DPW 2015). For some islands included within Indigenous Protected Areas, Healthy Country plans have been developed by landholder groups, in part as a basis for ongoing Commonwealth investment (Wunambal Gaambera Aboriginal Corporation 2010; Moorcroft *et al.* 2012). In other cases, only part of an island is reserved, but island-wide recovery plans for sets of co-occurring threatened species have been developed and implemented: examples include those for Lord Howe Island (DECC (NSW) 2007), King Island (Threatened Species Section

Box 3.1: Legal and policy settings and management for the conservation of biodiversity on Western Australian islands

Western Australia has more islands than any other state or territory – 3747 – of which 2285 are >1 ha and 300 >100 ha. Most (2633) are in the Kimberley but, apart from the Great Australian Bight, islands exist along most of the coastline. South of the Kimberley, major groups include the Dampier Archipelago (42 islands), the Montebello Islands (180 islands), Houtman Abrolhos (150 islands) and the Archipelago of the Recherche (232 islands).

Despite no explicit mention in legislation, islands have long been a major plank in the state's conservation policy and practice as many are critical for mammal, seabird and marine turtle conservation, and as examples of ecosystems unaltered by the many changes wrought on mainland Australia since European settlement. Of particular importance for mammals are Bernier and Dorre Islands in Shark Bay (inhabited by four species that became extinct on the mainland) and Barrow Island (including one species that became extinct on the mainland). Seabirds have been recorded breeding on >500 Western Australian islands and marine turtles nest on most sandy island beaches north of Shark Bay (Burbidge *et al.* 1996).

Most islands in the Kimberley are exclusive native title. Many of these have been or will be declared as Indigenous Protected Areas and are increasingly covered by Healthy Country plans (Moorcroft *et al.* 2012) managed by Aboriginal rangers.

Most islands south of the Kimberley are nature conservation reserves, declared under the *Land Administration Act 1997* and managed pursuant to the *Conservation and Land Management Act 1984*, while the *Biodiversity Conservation Bill 2015* covers the protection and conservation of flora and fauna state-wide. Islands with other tenures include Dirk Hartog (a national park), the Houtman Abrolhos (a multi-purpose reserve controlled by the Department of Fisheries and a proposed national park), Rottnest (a recreation area managed by a Board) and Garden Islands near Perth (a naval base, HMAS *Stirling*).

There is a long history of biological surveys on Western Australian islands. The first major one was of the Archipelago of the Recherche (Willis 1953) and the most recent was of the larger Kimberley islands undertaken in association with Traditional Owners (Gibson and McKenzie 2012). There has also been a long history of island management, especially the eradication of invasive mammals, including goat (*Capra hircus*), rabbit (*Oryctolagus cuniculus*), black rat, house mouse (*Mus domesticus*), fox (*Vulpes vulpes*) and feral cat (Algar *et al.* 2002; Burbidge and Morris 2002; Morris 2002). Feral cats have now been eliminated from almost all Western Australian islands. Biosecurity planning is increasingly important (Chapter 6). The current and potential value of Western Australian islands as refuges for biodiversity threatened by factors widespread on mainland areas has long been recognised (Abbott 2000) and the state's islands have been widely but strategically used for translocations, mostly of mammals, with such cases carefully regulated by specific translocation guidelines.

Resourcing remains a major issue. Islands are often remote from towns and require vessels or helicopters for access. Without regular surveillance, problems such as the arrival of invasive species, species loss and island degradation are likely. Because of the importance of islands to biodiversity conservation, there is a need for better legislative and policy protection for Western Australian islands and, particularly, regular surveillance for invasive species.

2010), Bruny Island (Cochran 2003) and Kangaroo Island (Taylor 2008).

To some extent, the conservation effectiveness of island-wide recovery plans is dependent upon the extent to which the island's land use is directed towards biodiversity conservation: conservation management plans on populated islands with many other land uses and interests almost inevitably involve many more compromises than for unpopulated islands that are entirely conservation reserves. Although there are many bad examples, there are also some examples of 'best practice' planned cohabitation of resource extraction and biodiversity conservation on island. In these, development is constrained to less sensitive areas, and biodiversity impacts are offset by substantial investments in targeted conservation management (Chapter 9).

The main risk to most island species is the introduction of invasive species. For many and perhaps most Australian islands, there are no

Table 3.3. Main relevant policy settings relating to biodiversity conservation on islands in Australian states and territories

Jurisdiction	No. of islands >1 ha	Main biodiversity conservation policy or legislation	The extent to which islands are specifically considered in that policy
New South Wales	87	*Biodiversity Conservation Bill 2016* *Local Land Services Amendment Bill 2016* *Coastal Protection Act 1979* *Coastal Management Act 2016 No. 20* *Environmental Trust Act 1998* *Heritage Act 1977* *Biosecurity Act 2015 No. 24* *Lord Howe Island Act 1953 No. 39* *National Park Estate (Reservations) Act 2002 No. 137* *Wilderness Act 1987 No. 196* *Saving Our Species 2016–21*	Generally no specific recognition of island biodiversity, although some special provisions for Lord Howe Island; no specific island conservation objectives
Northern Territory	612	No biodiversity conservation strategy *Territory Parks and Wildlife Conservation Act 2000*	No specific consideration of island conservation; no specific island conservation objectives
Queensland	1713	*Nature Conservation Act 1992* *Queensland Marine Parks Act 2004* *Great Barrier Reef Marine Park Act 1975*	Some provisions for Torres Strait Islander land; no specific island conservation objectives
South Australia	238	*No Species Loss: A Nature Conservation Strategy for South Australia 2007–17*	Some recognition of biodiversity values of islands (especially Kangaroo Island, and nesting seabird colonies); no specific island conservation objectives
Tasmania	289	*Threatened Species Protection Act 1995* *Nature Conservation Act 2002* *National Parks and Reserves Management Act 2002* *Natural Heritage Strategy for Tasmania 2013-2030: Securing Our Natural Advantage*	Values of 'islands' permeate the natural heritage strategy, and some specific recognition of biodiversity values of particular islands (Macquarie, Bruny, Maria); no specific island conservation objectives
Victoria	138	*Flora and Fauna Guarantee Act 1988* *Victoria's Biodiversity: Our Living Wealth* *Protecting Victoria's Environment: Biodiversity 2037*	Some recognition of biodiversity value of Phillip Island; no specific island conservation objectives
Western Australia	2285	*Biodiversity Conservation Act 2016* *Conservation and Land Management Act 1984*	No specific island policies, but long history of island biological surveys and eradication of invasive species

existing legal constraints on the importation of plants or animals (e.g. cats) from nearby mainland areas. This may be particularly the case for populated islands and where all or most land is privately owned. There are some notable exceptions: in Western Australia, regulations prohibit the introduction of invasive species to islands that are nature reserves, as are most islands south of the Kimberley. However, enforcement of such regulations has generally been limited.

Options for improving the policy and legal settings

Australia's current legal, policy and management approach to the conservation of island biodiversity is piecemeal and inadequate (Woinarski *et al.* 2014a). The distinctive and important, but highly susceptible, conservation values of the large set of Australian islands have not been appropriately recognised. The history of many island extinctions shows that island species can rapidly be rendered extinct following introductions of invasive species, such that policy (and management) needs to be proactive and precautionary (Martin *et al.* 2012). Here, we present a set of nine recommendations for better recognising and safeguarding the nationally and internationally significant biodiversity values occurring across Australian islands.

Changes needed to legislation and policy

1 *Implement more effective policy to constrain global climate change.*
Global climate change threatens the future of much of the world's biodiversity. Island biota may be particularly susceptible, as many island species have little capability for adaptation because of their limited genetic variation and islands may offer few altitudinal or other options for distributional shifts to track changing climate conditions. The distinctive (and endemic-rich) but limited extent of cloud forest on the peaks of Mounts Gower and Lidgbird on Lord Howe Island is one example of an island environment at severe risk from climate change (DECC (NSW) 2007). Many islands are low-lying and may be inundated in extreme weather events or more gradually diminished by rising sea levels. The recent extinction of the Bramble Cay melomys due to episodic inundations of the only island on which it occurred may be the forerunner of the loss of many other island species as a consequence of global climate change (Watson 2016). Currently, measures taken internationally and nationally to reduce greenhouse gas emissions will be insufficient to stave off much climate change and a consequential increased rate of extinctions, especially in island species. Much loss of Australia's island biota will occur unless Australian policy on climate change and greenhouse gas emissions is significantly improved.

2 *Include islands with outstanding conservation values as a Matter of National Environmental Significance.*
To better reflect and more effectively address the significant potential for islands to contribute to the conservation of Australia's (and indeed the world's) biodiversity, and in light of the ongoing loss of Australian island biodiversity, the EPBC Act needs to be amended to include islands with outstanding biodiversity values as a Matter of National Environmental Significance. Wherever possible, amendments to the EPBC Act should be complemented by inclusion of appropriate provisions in state and Northern Territory legislation. With criteria relating to numbers of endemic species, species richness and/or breeding aggregations, it is relatively straightforward to identify such priority islands (Ecosure 2009).

3 *Include 'significant wildlife breeding aggregations' as a Matter of National Environmental Significance.*
Some national protection is afforded, albeit indirectly, for sites supporting significant aggregations of some groups of species (notably waterfowl and other aquatic species) through established thresholds and protocols used in defining wetland sites of international significance (Ramsar Convention Bureau 2002), and for shorebird species (Gallo-Cajiao 2014). However, there is no comparable recognition and protection offered to sites that support significant breeding aggregations of other wildlife groups (Fig. 3.2). In Australia, most such sites (e.g. for seabirds, seals and marine turtles) are now on islands. Such sites may be critical for the conservation of many species because, at some time of their lifecycle, much of the species' total population is concentrated in a small area. Threshold criteria for defining the significance of such sites can be readily transferred from existing shorebird and waterfowl site criteria (e.g. >1% of the world population breeding at the site). Some of these aggregations are cur-

Fig. 3.2: The southern end of Pelsaert Island in the Houtman Abrolhos, Western Australia, has significant breeding aggregations of wedge-tailed shearwater (*Ardenna pacifica*) that nest in burrows, sooty tern (*Onychoprion fuscatus*) that nest on the ground and common noddy (*Anous stolidus*) that nest on low shrubs. Most birds in this photograph are common (or brown) noddies. Over one million seabirds occur here during the breeding season. Photo: Andrew Burbidge.

rently afforded protection because the species concerned are listed as threatened; however, many of the aggregations comprise species not listed as threatened.

4 *Include a Data Deficient category within the conservation status categorisations of the EPBC Act, with appropriate management implications and commitment to redress the lack of information.*

Such a change would be consistent with the protocols of the IUCN (IUCN Standards and Petitions Subcommittee 2016), which largely form the basis for conservation status assessment nationally and in most states and territories. This category may be particularly relevant for island species, because remoteness renders much island biota poorly known and documented. The remoteness of many islands from major population centres also means that island species are less likely to be regularly monitored and hence their population trends are unlikely to be well resolved.

5 *Develop a systematic policy approach to biodiversity conservation on Australian islands, within a revised national biodiversity strategy.*

This should include:

- development of specific conservation management planning for all of the most significant islands in every state;
- development and implementation of substantially enhanced biosecurity

protocols for all islands, especially those with high conservation values (Chapter 6);
- appropriate levels of constraints on new developments on any currently undeveloped islands;
- evaluating the potential for, and risks of, islands as translocation sites for threatened species unlikely to persist on the mainland (Chapter 7).

6 *Implement an ongoing program to eradicate priority threats from islands with significant biodiversity values.*
Australia's *Threatened Species Strategy* has identified the need to control (or preferably eradicate) priority threats to threatened species on some islands. To date (January 2017), five islands have been proposed as sites for the eradication of feral cats. However, feral cats occur on nearly 100 Australian islands (Legge *et al.* 2017), and many other islands have uncontrolled populations of other invasive species. The national control and eradication program presaged in the *Threatened Species Strategy* should be sequentially and systematically expanded to propose that invasive species that are major threats to island biodiversity are eradicated from substantially more Australian islands.

7 *Develop and implement an ongoing program for island biodiversity survey and monitoring.*
The biota of many islands is unknown or poorly known. For example, there are more than 2600 islands in the Kimberley but fewer than 30 have been surveyed reasonably comprehensively (Gibson *et al.* 2012). Biodiversity inventory of Australian islands has been neither systematic nor comprehensive, and it is likely that there is much unrecognised endemism. This may be particularly likely for some of the less conspicuous invertebrate groups, which typically are underrepresented in current listings of Australian threatened species (Walsh *et al.* 2012). Targeted surveys of some invertebrate groups on some islands have demonstrated remarkable levels of radiation and endemism, including many taxa of great antiquity (Iredale 1944; Ponder 1982; Criscione and Köhler 2013). Yet there have been few such surveys for relatively uncharismatic groups of species, and it is likely that further surveys on more islands will reveal many more island-endemic species, many of which may have requirements for urgent conservation actions. Furthermore, it is not simply less charismatic taxonomic groups that have been undersampled or undiscovered on islands – genetic analyses show much previously unrecognised endemism in island representatives of what were formerly considered single widespread species (Rosauer *et al.* 2016).

8 *Develop a national risk assessment for biodiversity values on all islands.*
To better understand, safeguard and manage the biodiversity values of Australian islands, a national program for risk assessment of Australian islands should be developed. Islands with high risks of biodiversity loss, and susceptible island species, should be priorities for preventative and remedial management.

9 *Establish an enduring national management fund for the conservation of island biodiversity.*
Although there have been recent examples of substantial investments in the management of threats on Australian islands (notably eradication programs for invasive species on Macquarie and Dirk Hartog Islands), there is generally limited or no resourcing for conservation management on most Australian islands, and many remain entirely unmanaged. A strategic and long-term program initiated by governments to conserve wildlife on Australian islands is warranted. Many islands, particularly in northern Australia, are now managed by Indigenous ranger groups as Indigenous Protected Areas (Chapter 8). Currently, these programs have only short-term investment commitments, and require far more substantial longer-term security.

Concluding comments

The Australian and state and territory governments have statutory obligations to protect biodiversity and prevent extinction. Clearly this obligation is not being met on Australian islands. Islands present a special conservation opportunity. In many cases (because of their relatively small size and unconnected nature) conservation may be far more tractable and cost-effective on islands than on mainland areas. Furthermore, islands do not yet have many of the threats facing mainland environments, and relatively simple biosecurity precautions may long prevent the arrival or establishment of such threats. Indeed, it is clear that eradication of invasive species from islands and the subsequent restoration of island biotas is extremely cost-effective compared with many other conservation actions (Genovesi 2011; Jones *et al.* 2016). Australia has a long history of treating its island estate with disdain or inadequate care, of not recognising islands for their biodiversity significance and susceptibility. The extraordinary biodiversity of Australia's islands has suffered severely from such a history of ill-use and neglect. It is time to redress the damage.

References

Abbott I (2000) Improving the conservation of threatened and rare mammal species through translocation to islands: case study Western Australia. *Biological Conservation* **93**, 195–201. doi:10.1016/S0006-3207(99)00144-5

Alcover JA, Sans A, Palmer M (1998) The extent of extinctions of mammals on islands. *Journal of Biogeography* **25**, 913–918. doi:10.1046/j.1365-2699.1998.00246.x

Algar DA, Burbidge AA, Angus GJ (2002) Cat eradication on Hermite Island, Montebello Islands, Western Australia. In *Turning the Tide: The Eradication of Invasive Species. Proceedings of the International Conference on Eradication of Island Invasives.* (Eds CR Veitch and MN Clout) pp. 14–18. IUCN, Gland, Switzerland.

Andrew P, Cogger H, Driscoll D, Flakus S, Harlow P, Maple D, Misso M, Pink C, Retallick K, Rose K, Tiernan B, West J, Woinarski JCZ Somewhat saved: a captive breeding program for two endemic Christmas Island lizard species, now extinct in the wild. *Oryx* (In press).

Blackburn TM, Cassey P, Duncan RP, Evans KL, Gaston KJ (2004) Avian extinction and mammalian introductions on oceanic islands. *Science* **305**, 1955–1958. doi:10.1126/science.1101617

Burbidge AA, Morris KD (2002) Introduced mammal eradications for nature conservation on Western Australian islands: a review. In *Turning the Tide: The Eradication of Invasive Species. Proceedings of the International Conference on Eradication of Island Invasives.* (Eds CR Veitch and MN Clout) pp. 64–70. IUCN, Gland, Switzerland.

Burbidge AA, Johnstone RE, Fuller PJ (1996) The status of seabirds in Western Australia. In *The Status of Australia's Seabirds: Proceedings of the National Seabird Workshop.* Canberra, 1–2 November. (Eds G Ross, K Weaver and J Greig) pp. 57–71. Biodiversity Group Environment Australia, Canberra.

Cochran T (2003) *Managing Threatened Species and Communities on Bruny Island.* Threatened Species Unit, Department of Primary Industries, Water and Environment, Hobart.

Commonwealth of Australia (2015) *Threatened Species Strategy.* Canberra.

Criscione F, Köhler F (2013) More on snails and islands: molecular systematics and taxonomic revision of *Setobaudinia* Iredale (Gastropoda: Camaenidae) from the Kimberley, Western Australia, with description of new taxa. *Invertebrate Systematics* **27**, 634–654. doi:10.1071/IS13027

DEC (2012) *Shark Bay Terrestrial Reserves and Proposed Reserve Additions.* Management Plan No. 75. WA Department of Environment and Conservation, Perth.

DECC (NSW) (2007) *Lord Howe Island Biodiversity Management Plan.* NSW Department of Environment and Climate Change, Sydney.

DEWHA (2009) *Threat Abatement Plan to Reduce the Impacts of Exotic Rodents on Biodiversity on Australian Offshore Islands of less than 100 000 Hectares.* Department for Environment, Water, Heritage and the Arts, Canberra.

DPW (2015) *Barrow Group Nature Reserves.* Management Plan No. 82 2015. WA Department of Parks and Wildlife, Perth.

Duncan RP, Blackburn TM (2007) Causes of extinction in island birds. *Animal Conservation* **10**, 149–150. doi:10.1111/j.1469-1795.2007.00110.x

Ecosure (2009) *Prioritisation of High Conservation Status Offshore Islands.* Report to the Department of the Environment, Water, Heritage and the Arts. Ecosure, Cairns.

Eldridge MDB, Meek PD, Johnson RN (2014) Taxonomic uncertainty and the loss of biodiversity on Christmas Island, Indian Ocean. *Conservation Biology* **28**, 572–579. doi:10.1111/cobi.12177

Gallo-Cajiao E (2014) *Review of the International Policy Framework for Conserving Migratory Shorebirds in the East Asian–Australasian Flyway.* East Asian–Australasian Flyway Partnership, Brisbane.

Garnett ST, Szabo JK, Dutson G (2011) *The Action Plan for Australian Birds 2010.* CSIRO Publishing, Melbourne.

Genovesi P (2011) Are we turning the tide? Eradications in times of crisis: how the global community is responding to biological invasions. In *Turning the Tide: The Eradication of Invasive Species. Proceedings of the International Conference on Eradication of Island Invasives.* (Eds CR Veitch and MN Clout) pp. 5–8. IUCN, Gland, Switzerland.

Geoscience Australia (2004) *Geodata Coast 100K 2004.* Geoscience Australia, Canberra.

Gibson LA, McKenzie NL (2012) Identification of biodiversity assets on selected Kimberley islands: background and implementation. *Records of the Western Australian Museum* (Supplement 81), 1–14. doi:10.18195/issn.0313-122x.81.2012.001-014

Gibson LA, Yates S, Doughty P (2012) *Biodiversity Values on Selected Kimberley Islands, Australia.* Western Australian Museum, Perth.

Hobbs J-PA (2014) A glaring omission in Australia's marine conservation planning. *Marine Policy* **44**, 149–151. doi:10.1016/j.marpol.2013.08.020

Iredale T (1944) The land mollusca of Lord Howe Island. *Australian Zoologist* **10**, 299–330.

IUCN Standards and Petitions Subcommittee (2016) *Guidelines for Using the IUCN Red List Categories and Criteria. Version 12.* IUCN, Gland, Switzerland.

Jones HP, Holmes ND, Butchart SHM, Tershy BR, Kappes PJ, Corkery I, Aguirre-Muñoz A, Armstrong DP, Bonnaud E, Burbidge AA, Campbell K, Courchamp F, Cowan PE, Cuthbert RJ, Ebbert S, Genovesi P, Howald GR, Keitt BS, Kress SW, Miskelly CM, Oppel S, Poncet S, Rauzon MJ, Rocamora G, Russell JC, Samaniego-Herrera A, Seddon PJ, Spatz DR, Towns DR, Croll DA (2016) Invasive mammal eradication on islands results in substantial conservation gains. *Proceedings of the National Academy of Sciences* **113**, 4033–4038.

Legge S, Murphy BP, McGregor H, Woinarski JCZ, Augusteyn J, Ballard G, Baseler M, Buckmaster T, Dickman CR, Doherty T, Edwards G, Eyre T, Fancourt B, Ferguson D, Forsyth DM, Geary WL, Gentle M, Gillespie G, Greenwood L, Hohnen R, Hume S, Johnson CN, Maxwell N, McDonald P, Morris K, Moseby K, Newsome T, Nimmo D, Paltridge R, Ramsey D, Read J, Rendall A, Rich M, Ritchie E, Rowland J, Short J, Stokeld D, Sutherland DR, Wayne AF, Woodford L, Zewe F (2017) Enumerating a continental-scale threat: how many feral cats are in Australia? *Biological Conservation* **206**, 293–303.

Lindenmayer DB, Blair D, McBurney L, Banks SC (2015) The need for a comprehensive reassessment of the Regional Forest Agreements in Australia. *Pacific Conservation Biology* **21**, 266–270. doi:10.1071/PC15042

Martin TG, Nally S, Burbidge AA, Arnall S, Garnett ST, Hayward MW, Lumsden LF, Menkhorst P, McDonald-Madden E, Possingham HP (2012) Acting fast helps avoid extinction. *Conservation Letters* **5**, 274–280. doi:10.1111/j.1755-263X.2012.00239.x

Moorcroft H, Ignjic E, Cowell S, Goonack J, Mangolomara S, Oobagooma J, Karadada R, Williams D, Waina N (2012) Conservation planning in a cross-cultural context: the Wunambal Gaambera Healthy Country Project in the Kimberley, Western Australia. *Ecological Management and Restoration* **13**, 16–25. doi:10.1111/j.1442-8903.2011.00629.x

Morris KD (2002) The eradication of the black rat (*Rattus rattus*) on Barrow and adjacent islands off the north-west coast of Western Australia. In *Turning the Tide: The Eradication of Invasive Species. Proceedings of the International Conference on Eradication of Island Invasives.* (Eds CR Veitch and MN Clout) pp. 219–225. IUCN, Gland, Switzerland.

Myers N, Mittermeier R, Mittermeier C, da Fonseca G, Kent J (2000) Biodiversity hotspots for conservation priorities. *Nature* **403**, 853–858. doi:10.1038/35002501

National Reserve System Task Group (2010) *Australia's Strategy for the National Reserve System 2009–2030.* Canberra.

Natural Resource Management Ministerial Council (2010) *Australia's Biodiversity Conservation Strategy 2010–2030.* Department of Sustainability, Environment, Water, Population and Communities, Canberra.

Ponder WF (1982) Hydrobiidae of Lord Howe Island (Mollusca: Gastropoda: Prosobranchia). *Marine and Freshwater Research* **33**, 89–159. doi:10.1071/MF9820089

Pressey RL, Johnson IR, Wilson PD (1994) Shades of irreplaceability: towards a measure of the contribution of sites to a reservation goal. *Biodiversity and Conservation* **3**, 242–262. doi:10.1007/BF00055941

Queensland Parks and Wildlife Service (1999) *Hinchinbrook National Park Plan of Management.* Queensland Parks and Wildlife Service, Brisbane.

Ramsar Convention Bureau (2002) *What is the Ramsar Convention on Wetlands?* Ramsar Convention Bureau, Gland, Switzerland.

Rosauer DF, Blom MPK, Bourke G, Catalano S, Donnellan S, Gillespie G, Mulder E, Oliver PM, Potter S, Pratt R (2016) Phylogeography, hotspots and conservation priorities: an example from the Top End of Australia. *Biological Conservation* **204**, 83–93. doi:10.1016/j.biocon.2016.05.002

Steadman DW (2006) *Extinction and Biogeography of Tropical Pacific Birds.* University of Chicago Press, Chicago.

Taylor DA (2008) *Draft Recovery Plan for 15 Nationally Threatened Plant Species on Kangaroo Island, South Australia 2003–2013.* 2nd edn. SA Department for Environment and Heritage, Adelaide.

Thackway R, Cresswell ID (1995) *An Interim Biogeographic Regionalisation for Australia: A Framework for Establishing a National System of Reserves.* Version 4. Australian Nature Conservation Agency, Canberra.

Threatened Species Section (2010) *King Island Biodiversity Management Plan.* Department of Primary Industries, Parks, Water and Environment and NRM South, Hobart.

Walsh JC, Watson JEM, Bottrill MC, Joseph LN, Possingham HP (2012) Trends and biases in the listing and recovery planning for threatened species: an Australian case study. *Oryx* **47**, 131–143.

Watson J (2016) Bring climate change back from the future. *Nature* **534**, 437. doi:10.1038/534437a

Willis JH (Ed.) (1953) *The Archipelago of the Recherche.* Australian Geographical Society, Melbourne.

Woinarski J, Ball D, Burbidge AA (2014a) Islands. In *Ten Commitments Revisited: Securing Australia's Future Environment.* (Eds D Lindenmayer, S Dovers and S Morton) pp. 117–128. CSIRO Publishing, Melbourne.

Woinarski JCZ, Burbidge AA, Harrison PL (2014b) *The Action Plan for Australian Mammals 2012.* CSIRO Publishing, Melbourne.

Woinarski JCZ, Garnett ST, Legge SM, Lindenmayer DB (2017) The contribution of policy, law, management, research, and advocacy failings to the recent extinctions of three Australian vertebrate species. *Conservation Biology* **31**, 13–23.

Wunambal Gaambera Aboriginal Corporation (2010) *Wunambal Gaambera Healthy Country Plan: Looking after Wunambal Gaambera Country 2010—2020.* Wunambal Gaambera Aboriginal Corporation, Kalumburu.

4

The 4Cs approach to island management

Derek Ball, Jodie Williams and Margaret Christian

Summary

We contend that improved outcomes for biodiversity conservation will require different and more holistic conservation models than those used in the past and present. The fundamental reason is that community support for conservation is simply too low for it to be effective in the long term, and a much greater level of community support is required. We propose that the 4Cs approach to conservation is a contender worthy of serious consideration. This approach integrates conservation interventions with the needs of communities, their cultures and their commercial (economic) prerogatives. However, the 4Cs approach will need to include a new range of specialist professions for it to be effective – and that raises additional challenges. We suggest the collective impact approach might be of benefit in coordinating the new conservation management teams. Finally, we provide a case study of Norfolk Island where the 4Cs approach has been in place for several years, with what we believe to be encouraging results.

Introduction

For millennia, Indigenous peoples held landscapes and their component natural resources in custodianship. Those resources and their management practices sustained their communities, moulded their cultures and provided the basis for commerce in goods and services. This custodianship, in most parts of the world, was severely disrupted by progressive global expansion of the influence of European colonialists across land and seascapes. This era of disruption is most easily and simply described as a shift to exploitation of and trade in natural resources as the basis of political and economic growth and influence. This era continued largely unabated for centuries, although in relatively recent times, natural resources have begun to be seen in the light of finiteness.

Re-adoption of some principles of custodianship began almost 150 years ago. In 1872, the US gazetted the world's first national park, Yellowstone. That was quickly followed in 1879 when Australia declared the 'National Park' (now Royal National Park) near Sydney to provide respite from overcrowded and insanitary conditions in large cities. The Royal Society for the Protection of Birds (originally the Plumage League) aimed to offer birdlife better protection through its 1889 decree 'That Members shall discourage the wanton destruction of Birds, and interest themselves generally in their protection'. Dedicated conservation organisations were formed from the mid twentieth century: the Nature Conservancy in 1951, the World Wide Fund for Nature in 1961, the Australian Conservation Foundation in 1963 and Greenpeace in 1971.

From an Australian perspective, the humble beginning of Australia's protected area network from a single national park in 1879 has grown to cover 137 501 544 ha (17.88%) of the Australian landmass (CAPAD 2014). More broadly, renewed interest in custodianship might be reflected in the timeline of a more modern era of environmentalism. The 1970s and 1980s in Australia witnessed a new age of interest in the environment, manifested in successful public campaigns such as those that resulted in protection of the Queensland Wet Tropics, the Great Barrier Reef and the Franklin River in Tasmania. Governments created ministerial environment portfolios and dedicated management institutions such as National Parks and Wildlife Services in Tasmania (1971) and Queensland (1975). From those campaigns arose the world's first environment-focused political party, the Australian Greens. In 1987, then Prime Minister Bob Hawke gave his now famous statement on the environment, 'Our Country Our Future' and declared the 1990s the decade of Landcare. In 1995, initial investments were made to facilitate and support renewed custodianship arrangements by Indigenous Australians through establishment of the Indigenous Protected Areas program. Conservative government established the Natural Heritage Trust in 1997.

Current status of biodiversity conservation in Australia

It would seem that, in Australia, custodianship of landscapes and their component natural resources is well under repair. However, the imperative question must be, is it effective? A review of financial statements in 2014–15 annual reports published by state, territory and federal government departments (with charters for management of protected areas and wildlife, and nature conservation matters off protected areas) demonstrated a spend of A$1.9 billion in that year. That figure is broadly similar to investments made over the last five years or more. It does not include the A$451 million administered by the federal Department of Environment, which largely comprised money devolved and granted to, for example, the 56 regional organisations funded by the Natural Heritage Trust. In addition, considerable but unquantified investment was made by the plethora of non-government organisations, industry and philanthropic foundations operating in Australia, or in support of Australia's newly realised custodianship. In reviewing the likely impact of what might be considered impressive investment, perhaps a useful example might be found in Australia's 2006 State of the Environment Report commentary on the status and projection for Australia's mammals:

> *Twenty-two species of mammals are extinct in Australia; a further eight species are found only on islands. A further two, possibly three, have become extinct on Christmas Island. Arid and semi-arid parts of Australia have generally lost a high proportion of their mammals. What these examples show is a continuing serious decline in biodiversity across Australia. This declining trend may be at different rates for various taxa. As bird species are often quite mobile, the full effect of habitat contraction and change in condition has yet to become evident. In contrast, many mammal species have high attrition rates with changes in habitat. The contraction of mammal species and mammal extinction is unparalleled in any other component of Australia's biodiversity. Significant decline is continuing and further extinctions are likely.*

Sadly, the last prediction has been realised in the decade since the report was published, with another two mammal extinctions – the Christmas Island pipistrelle (Lunney *et al.* 2011) and the Bramble Cays melomys (Gynther *et al.* 2016) – both of which had their sole area of occupancy within an island protected area. A broader assessment of the status of threatened species in Australia (Taylor *et al.* 2011) showed that 'declines overwhelmingly dominated species trends, with 606 (72%) of the 841 species declining, or rapidly declining, in all of their subregional populations, while 12% had some but not all populations with increasing or stable trends, and finally, 16% had all populations showing increasing or stable trends'. Since that study, dozens more species have been listed as threatened under the *Environment Protection and Biodiversity Conservation*

Act 1999 (EPBC Act). Clearly, in terms of security of biodiversity at species level, there is little to support a notion of effective custodianship.

A similar outlook exists for Threatened Ecological Communities (TECs) listed under the EPBC Act. At the end of 2013, 67 were listed, another five were added in 2014 and a further six in 2015. None have been recovered nor removed from the listing. At the landscape scale, State of the Environment reporting continues to paint a bleak picture for environmental sustainability in Australia.

Based on this, it seems abundantly clear that a thorough and urgent review is required of how Australia effects custodianship of its biological diversity. Such a review should include such topics as the cost-effectiveness and efficiency of government institutions, appropriate prioritisation of investments in conservation interventions, including an overall consideration of the principle of triage, and a greater focus on protection of remaining high-value areas over repair of damaged low-value systems. However, there is a far more fundamental question. How could such an unsuccessful program of publicly funded activities, that are demonstrably failing their purpose, continue for so long without more widespread public examination?

The future pathway

We contend that the pathway to more effective conservation of Australia's biological diversity will not be found in conservation ecology alone, nor in its applied principles and practices. Endeavours in these fields can and do provide successes, as shown in the many examples in other chapters. However, clearly this technical capacity has not proven sufficient. The pathway to more effective conservation lies in the answers to some simple questions: why don't more people care? How and why, in what has been a tiny slice of the timescale of human evolution, have we moved from a realisation that we are totally dependent on the natural world, to the current attitude of general indifference?

Of course, this predicament is not restricted to Australia. It is global, and often compounded by real and/or perceived conflicts between conservation and human need. For example, Newmark and Hough (2000) reviewed the approach of the integrated conservation and development programs (ICDPs) in Africa that arose in the 1980s and 1990s. Such programs attempted to link conservation goals to socio-economic development and the consequent benefits to local communities. More recently, Rands *et al.* (2010) proposed that radical changes are required to promote conservation of biological diversity as something of global worth. They recommended integrating conservation into policy instruments dealing with broader socio-economic needs and aspirations, and the creation of enabling conditions of appropriate institutions, governance and behaviours. However, such proposals to broaden the practice of conservation to include or embrace socio-economic programs ('new conservation') have not gone uncontended. For example, Miller *et al.* (2011) defined the dispute as the 'new conservation debate' between the strong protected areas approach ('nature protectionists') and the development-oriented conservationists ('social conservationists'). We propose that consideration of this dispute is critical in designing new programs for biodiversity conservation that might achieve desired goals. Table 4.1 shows the characteristics of each approach (Miller *et al.* 2011).

The two approaches appear quite different on examination, but as described are probably more likely to reflect the most extreme expression. We believe it would be enlightening to consider something of a middle-ground approach. This is described by Kareiva (2014) in response to criticism by Soule (2013), noting that a key element of new conservation is not necessarily placing people above conservation goals, but in recognising the value of conservation to people and ensuring people benefit from it. We would add, ensuring that both the intrinsic and the realised value of conservation to people are communicated well. This is critical because, as Kareiva (2014) points out, most of the world's people are city-dwellers who are largely and increasingly isolated from

Table 4.1. Characteristics of conservation approaches

Nature protectionists	Social conservationists
Policy goal • Protection of biodiversity Policy instruments • PAs, especially 'people-free parks' Empirical claims • PAs are the most effective means to protect biodiversity • PAs can bring considerable benefits to local people • The human presence is a threat to biodiversity conservation • ICDPs fail to protect biodiversity • Conservation must focus on biodiversity protection to succeed • The goals of conservation and development are often, if not inherently, incompatible Normative claims • Conservation should focus primarily on biodiversity protection • Biodiversity should be protected for its intrinsic value • In some cases, biodiversity conservation is more important than poverty alleviation Ethical foundations • Non-anthropocentrism, preservationism Primary disciplines • Conservation biology, environmental philosophy	Policy goal • Sustainable development/conservation with poverty alleviation Policy instruments • ICDPs, extractive reserves Empirical claims • PAs often fail to protect biodiversity • Social costs of PAs are real and substantial • Humans and biodiversity can co-exist productively • ICDPs will work if properly reformed • Conservation will fail unless poverty is addressed • Human livelihoods can improve with effective conservation efforts (e.g. ecosystem services) Normative claims • Conservation should be focused on human welfare • Biodiversity should be protected for its instrumental value to humans • Poverty alleviation trumps biodiversity protection as a conservation goal Ethical foundations • Enlightened anthropocentrism, social justice Primary disciplines • Anthropology, political ecology, development economics

PAs = protected areas.

nature. Thus, communicating the benefits they indirectly receive from biodiversity will be crucial in gaining their support. Another key aspect of the debate is the track record of each side to date. As outlined earlier using Australia as an example, it should be clear that the nature protectionist approach simply has not worked – even after at least 150 years of development and application. While some would contend that neither has the social conservationist approach delivered as promised, this approach is relatively in its infancy and continues to evolve. We propose that the middle-ground pragmatism outlined by Kareiva (2014) is a sensible approach. Given the urgency in conservation of biodiversity in the face of continued rapid loss, it is now critical to move from academic debate and develop or find a model with broad applicability. We are of the opinion that the urgency is clear for island conservation in Australia as there is much to lose and, due to the inherent characteristics of islands (Chapter 1), there is extremely high potential for them to act as protective arks for much of Australia's biodiversity.

Applying the 4Cs approach to island conservation initiatives requires a deep understanding of the social organisation of that island's community and its priorities, sense of belonging, cultural practices, lifestyle choices and economic challenges and opportunities, both real and perceived. Box 4.1 provides a brief example of some key characteristics of the Norfolk Island community that have been recognised and guide the design of conservation initiatives on that island. We suggest that undertaking similar studies on other islands, to guide conservation initiatives, would be of great benefit in designing 4C approaches.

Box 4.1: Norfolk Island case study

Our community has an infamous and known legacy weaving from the UK to Tahiti to Pitcairn Island, sensationalised in numerous Hollywood feature films of *Mutiny on the Bounty*. We are the descendants of a distinct community who migrated 6000 km westwards from Pitcairn Island to resettle on Norfolk Island. Today, as Norfolk Islanders of Pitcairn descent, we have a unique ethnicity of common ancestry, shared historical memories, community solidarity, a common culture and homeland. Upon arrival on the island in 1856 the original Pitcairner families were granted 50 acre blocks of land. They began to settle and build their own homes. We are a community with a resolute backbone of our own laws, customs and traditions. Family land is bequeathed to subsequent generations. A monetary value is not placed upon the land, as we see ourselves simply as custodians for the next generation.

Our ecological footprint is indebted to our foremothers, the Tahitian women, who brought to Pitcairn Island detailed knowledge of Pacific plants, medicines, fish and the sea. Such wisdom was a foundation of our community's survival and growth on Pitcairn. This knowledge was transferred throughout the generations and henceforth to our Norfolk Islander community as they resettled on the island. Traditional farming and whaling industries were established. Skills of self-sufficiency and the availability of rich natural resources on the island established an early economy based upon sharing and gifting. Our strong cohesion of extended families sustained a dense, multiplex social network within our island's compact geographical size of 40 km^2. Our community developed in relative isolation until the arrival of advanced communication technology and increasing numbers of settlers from the mainland and further afield. The greatest shift in our traditional lifestyle was the construction of the airstrip in World War II. The island's accessibility by air brought a growing tourism industry. The community's traditional fair-trade economy of sharing produce and skills moved into a primarily cash subsistence economy. Tourism is our main revenue stream today.

Self-sufficiency and resilience is our way of life and provides a safety net as we are fully aware of our vulnerability due to our geographic isolation. We understand the finiteness of our available natural resources – on the land and sea – with customs that respect the fragility of the cycles of life. For example, fishing is limited to family needs, and islanders adhere to voluntary fishing bans during spawning seasons. Local tourism promotion shares protocols and necessities of daily life on the island, such as our reliance upon tank water for household use. Accommodation guidebooks discourage long showers, for example, and visitors are made aware of water limits, especially during summer months.

Norfolk Island has firm ecological parallels with New Zealand. It is one of the few islands founded on a seamount in a warm temperate zone. Our island supports an intricate range of endemic plants, animals and ecosystems, for which global biodiversity conservation efforts are now being recognised and recommended.

Mindful all-encompassing community consultation is essential as conservation management plans and programs are established for Norfolk Island. A unified commitment to the protection of our environment, unique ecosystem, flora and fauna species and geodiversity features as connected to our island heritage and traditions is imperative. The island is a valued breeding site for several hundred thousand seabirds including the masked booby (*Sula dactylatra*), grey ternlet (*Procelsterna albivitta*), white tern (*Gigis alba*), sooty tern (*Onychoprion fuscatus*) and wedge-tailed shearwater (*Ardenna pacifica*). Philip Island supports one of the largest breeding populations of red-tailed tropicbirds (*Phaethon rubricauda*) in Australia.

Respect for and engagement of our traditional knowledge and experience contributes vital hidden layers of understanding as initiatives of biodiversity conservation are applied. The growing worldwide exploration of new ways to support and engage people in conservation projects is showcased by the global promotion of volunteer trips to isolated places. A richly rewarding experience is guaranteed for the conscious traveller by their contribution of physical work and by first-hand experience of local community networks and way of life. The benefit is two-way, and outcomes are productive and sustainable.

On Norfolk Island, we are fortunate beneficiaries of Wildmob volunteer groups who visit and work towards the restoration of our critical habitat, with activities to protect our endangered species. Norfolk Island is home to the endemic and endangered Norfolk Island green parakeet (*Cyanoramphus cookii*) and other endemic birds such as the Norfolk Island golden whistler (*Pachycephala pectoralis xanthoprocta*) and Norfolk Island scarlet robin (*Petroica multicolor multicolor*).

Hands-on work plans include bushland restoration and pest eradication within the Norfolk Island National Park, other reserves and on private conservation land. Projects aim to improve the natural habitat for over 55 threatened or endangered species, and up to 18 species of seabirds that breed within the Norfolk Island group. Brainstorming and collaborating with our local ecologists and land managers to improve wildlife and habitat management has been a thoroughly successful approach to engaging and receiving ongoing community support for the Wildmob initiatives. Local endorsement at all levels of planning for Wildmob volunteer group visits include on-island travel arrangements, local shopping and employment.

Wildmob's student volunteer groups visit Norfolk Island to undertake a conservation adventure and to explore our remote and spectacular island. These student trips contribute voluntary hours towards their further qualifications such as Certificate III – Active Volunteering. Work plans include not only restoring valuable tracts of native habitat, but also clearing marine debris and taking audits on the island. This has been a precursor to our local community and Norfolk Island students of worldwide participation in recording processes and statistical gathering of environmental impacts. Our relative geographic isolation is not unaffected by increasing global impacts, which are transported upon the waves and currents of the surrounding Pacific Ocean. Wildmob's approach to conservation initiatives on Norfolk Island is focused upon collaboration and building our community's sense of self-empowerment. Supporting locals to more effectively identify conservation tasks that need action and to gain the necessary skills to complete the jobs, means cultivating sustainability. This is a way of life that we 'know' traditionally but which cannot remain static. Building community awareness of international environmental practices and issues while sharing opportunities for locals to contribute and strengthen their traditional knowledge is invaluable.

The 4Cs approach

We suggest that a new model for biodiversity conservation, regarding Australia's islands (and also relevant elsewhere), should have the following characteristics: it should reject approaches that have demonstrably not worked in the past, contain the key element of new conservation (recognition of the potential benefit of conservation to people), include programs to ensure that potential is realised, and include sound communication of that realisation to people. We have defined the fundamental problem in improving biodiversity conservation outcomes in terms of the principal question: 'why don't more people care'? In our experience, there are few to no human communities, on islands or elsewhere, where nobody is interested or involved in conservation. Almost invariably there are people with an interest in natural history, perhaps in a non-government conservation organisation or maybe in a bird-watching club or other social grouping. However, this is not our target audience; these people are already supporters of conservation. New allies are needed and, in forging these new partnerships, we need to ask 'what would motivate more people to support conservation?' There are many approaches to this question, with Maslow's Hierarchy of Needs (Maslow 1943) being perhaps the simplest and most often used. This hierarchy encapsulates the five basic needs of humans: in order, physiological needs, safety needs, love and belongingness needs, self-esteem needs and fulfilment needs. To grow supporters for conservation, we need to consider these five basic needs and to frame a new model for biodiversity conservation that will be widely viewed as contributing to people's needs and aspirations.

Insofar as we are aware, the Zeitz Foundation first developed the 4Cs approach and has continued to promote its use globally. Its essence is that conservation interventions need to be undertaken holistically, in a way that they transparently offer benefits for human communities, their cultures and their economies. In the interests of clarity, we repeat the components of the 4Cs as set out by the Zeitz Foundation.

1 Biodiversity is life. Conservation is safeguarding this biodiversity and the integrity of the

ecosystem services it provides which support global needs.

2 People matter. It is the right of every person to have their basic needs met, and enhancing the well-being of communities is a fundamental obligation of all.

3 Our world is culturally diverse. Celebrating commonality is what nurtures understanding. Respecting difference is crucial to our future. Our ability to innovate and evolve is what makes us uniquely human. Drawing on our uniqueness, we enrich each other and contribute to a greater common good.

4 Trading and the accumulation of wealth have been central to the development of civilisations over thousands of years and are likely to remain so. Uncontrolled, this commerce has had negative impacts, but if conducted in a more holistic and sustainable way it can be a positive contributor to the long run.

We restate these 4Cs in more general terms.

1 Conservation of biodiversity is essential and is the primary goal of our middle path new conservation model.

2 Conservation needs to be done in a way that supports and nurtures human communities, rather than creating divisiveness.

3 In doing conservation, cultures need to be respected and seen as a source of new knowledge, ideas and innovation.

4 Conservation needs to recognise commercial prerogatives, and become a source of economic prosperity rather than a burden.

A biodiversity conservation model that works across these 4Cs provides a clear framework to deliver outcomes which match the five motivators, as defined by Maslow (1943) (Table 4.2).

Table 4.2. Motivator links to the 4Cs

Motivator	4Cs
Physiological (air, food, water, sleep, warmth etc.)	Conservation
Safety (security of environment, resources, health etc.)	Conservation and Commerce
Love and belongingness (friendship, family, love etc.)	Community
Self-esteem (achievement, respect etc.)	Community and Culture
Fulfilment (morality, creativity etc.)	Culture

A 'new conservation' model for Australia's islands

Expanding from the above summary, we consider how the 4Cs approach could be used specifically as the basis of a new model for island conservation programs, that is able to succeed and be sustained in the long term, through gaining broad support from communities. Conservation issues often result in significant dispute and consequent divisiveness. Well-known Australian examples include forest conservation *v.* logging in Tasmania and on Fraser Island, rat eradication *v.* human health and tourism on Lord Howe Island, conservation and eco-tourism interests *v.* sand mining on Fraser Island and Stradbroke Island. These disputes are complicated and, we suggest, based on multiple issues of perceived and/or real conflict of interests. For islands, there are key socio-economic characteristics that need to be clearly understood to unravel these conflicts and when proposing and designing conservation programs. This understanding is essential for taking the 4Cs approach.

Communities who inhabit islands are typically small, are often relatively isolated and have a strong sense of place and ownership. Island communities often have distinct cultures, Indigenous or otherwise. Of critical importance, uninhabited islands may lie in claim to a broader community of interest that accesses and uses them temporarily or intermittently and they may have similar characteristics to that of an inhabitant island community. Islands often have undiversified economies, relying on only one or a few revenue streams (e.g. tourism, mining or fishing) and thus they are extremely susceptible to economic downturn. Resources such as water and issues such as food security are not abstract concepts – for islanders they are often day-to-day considerations. Consequently, islanders are strongly self-sufficient. A level of communal governance might exist and a hierarchy for decision-making may be accepted for convenience for some but not all issues that affect islanders. And that might change as circumstances change. In summary, it is not surprising that islanders can be reticent towards 'outsiders' and can react strongly against conservation interventions that are seen as imposed from the 'outside'. Simply, islanders, either permanent inhabitants or more loosely a community of interest around one or more islands, are often in a position of vulnerability. Protection of their place and their way of life is a strong motivating factor for them.

Armed with these insights, we make these specific recommendations for developing island conservation 4Cs programs.

Conservation

We find little fundamental fault in contemporary conservation ecology or its applied principles and practices. There is a high level of technical capacity to develop and implement actions that conserve biodiversity – and that capacity is rapidly increasing. This capacity is necessary but not sufficient. The spatial and temporal application of conservation programs clearly needs prioritisation, as demonstrated by Australia's multi-billion dollar investments that have not achieved measurable gains. Australia needs a national prioritisation framework that explicitly determines where the best return on investment lies among its islands, and among the opportunities that islands provide as arks of biodiversity. There is a large and growing body of academic work that could guide such prioritisation, including that designed to make cross-realm considerations (i.e. across terrestrial, marine and freshwater systems and boundaries) and to incorporate socio-economic issues, for example Álvarez-Romero *et al.* (2015). But such academic work needs to be translated into simpler frameworks that can be readily operationalised. We contend that conservation programs cannot be undertaken in isolation from other socio-economic factors, and that each conservation program needs to deliver clear outcomes within the other 3Cs – community, culture and commerce.

Community

We strongly recommend that the next step in developing a new conservation model for Australia's islands not be one that is embodied in conservation ecology nor its applied practice; indeed, it is a step that such practitioners are often least qualified to take. Understanding the interplay between conservation goals, and the needs and aspirations of island communities, is essential to long-term success. We have outlined some important socio-economic characteristics of island communities; however, it needs to be understood that these are generalisations. A thorough understanding needs to be developed of island communities and/or communities of broader interest that exist around islands, so that programs can be tailored to maximise benefits to those communities and ensure effective communication of potential and realised values of biodiversity. This does not necessarily mean that the broad conservation management priorities are overridden. Rather, from the perspective of the desire to support and nurture communities, this understanding will assist the design of adjunct and supporting activities. The conservation manager's toolkit can include identification of local leaders who might champion conservation, use of existing support and consultative networks, youth ambassadorships, learning and teaching opportunities, local employment and exchange programs. Finally, there needs to be a frank assessment of likely conflict between community values and aspirations and the desired conservation practice or outcome, whether it is likely to involve the whole community or a sector, and what the root causes might be.

Culture

Within the science and practice of conservation management, Indigenous cultures are increasingly being considered and traditional ecological knowledge is being encouraged and utilised to enhance outcomes (Chapter 8). In a broader sense, non-Indigenous islanders also hold significant and valuable knowledge about their environment, its change over time and the drivers that have caused that change. This traditional knowledge, including oral histories, is an invaluable addition to the conservation manager's portfolio of learning. A consideration of culture is required, to achieve the long-term success of conservation programs. There is a distinct culture among many island communities with broad themes such as self-reliance and the desired capacity to scale up and scale down governance structures. In practical terms, arguably the most important cultural considerations needed are respect for the knowledge and practices of a culture, and the willingness to work within its context. The conservation manager's toolkit might include consultation within existing cultural protocols such as Councils of Elders, demonstrated willingness to include traditional ecological knowledge and oral histories in the design of conservation programs, recognition that stakeholder consultation will need to be different for different issues (i.e. different scaling of governance), and allowing culturally appropriate time for the community to engage, discuss, debate and decide upon proposed conservation actions. Perhaps most important is the need to understand that reticence towards outside influence is based on cultural influences. It should not necessarily be interpreted as opposition, but as the desire of islanders to proceed in a way that offers protection to their place and their way of life.

Commerce

Business is often viewed as the antithesis to biodiversity conservation, and even as a root cause of its demise through poorly controlled or uncontrolled exploitation of natural resources. It is not surprising that conservation ecologists and practitioners are often distrustful of business and rarely see the practice of commerce as a useful tool. It is perhaps worthwhile to reiterate the Zeitz Foundation's view, as part of its 4Cs approach, that 'Uncontrolled, this commerce has had negative impacts, but conducted in a more holistic and sustainable way, it can be a positive contributor to the long run'. In this context, eco-tourism, while sometimes flawed in that it can have negative impacts on biodiversity if not appropriately controlled, has demonstrated the potential for development of revenue streams from nature. Some eco-tourism models actively contribute some of that revenue to conservation management (Chapter 11). From a 4Cs approach, we take that principle further and have two goals: generation of revenue or other resources to support biodiversity conservation, and creation of commercial benefits to communities to garner their support for conservation, i.e. to demonstrate the financial value of biodiversity. In those respects, commercial products that could be developed as adjuncts to core conservation management might include volunteer tourism to provide both human resources and monetary income, and profit-for-purpose eco-tourism, particularly in emerging high-value low-impact markets. These products also ensure that local communities reap financial benefits through providing visitor services such as travel arrangements, accommodation, meals, tours and the like. Both 'conservation tourism' and conservation managers can provide commercial benefits to communities through employment and by sourcing materials and the like from local providers.

The fifth C and a caution

Readers will appreciate that the 4Cs approach requires skills, knowledge and experience that may lie outside the capacity of many conservation ecologists and practitioners. To be effective in its design and execution, the 4Cs approach will require people who have high-level capacities in community engagement, communication, anthropology and business development. Creation and coordination of a partnership of people with these capacities will have its own challenges. Here we would like to introduce the concept of 'collective impact', defined by Kania and Kramer (2011) as the 'commitment of a group of important actors from different sectors to a common agenda for solving a specific social problem'. While collaboration and partnership are nothing new, Kania and Kramer (2011) provide a compelling case that there is a high risk of failing to achieve goals unless five key elements are present:

1 a truly common agenda as a shared vision for change;
2 shared measurement systems, i.e. a shared view of what success looks like;
3 mutually reinforcing activities, each participant taking ownership and charge of what they do best;
4 continuous communication so that each recognises and appreciates their common motivation;
5 a supporting organisation that facilitates and coordinates, so that each participant is free to focus on what they do best.

Concluding comments

It appears that conservation of biological diversity remains a low priority for the Australian community – in the case of Australia's island arks, this means the loss of many opportunities to improve that conservation. We contend that drawing more explicit and transparent links between conservation outcomes, and the consequent benefits for communities, their cultures and their economies (the 4Cs approach), is a way of raising support for conservation. This approach will be challenging, because conservation ecologists and practitioners will need to look outside their profession and embrace others. In doing so, better models for collaboration (the collective impact approach) will need to be developed to build and implement more holistic and socially progressive agendas.

References

Álvarez-Romero JG, Adams VA, Pressey RL, Douglas M, Dale AP, Augé AA, Ball D, Childs J, Digby M, Dobbs R, Gobius N, Hinchley D, Lancaster I, Maughan M, Perdrisat I (2015) Integrated cross-realm planning: a decision-makers' perspective. *Biological Conservation* **191**, 799–808. doi:10.1016/j.biocon.2015.07.003

CAPAD (2014) *Collaborative Australian Protected Area Database – CAPAD 2014*. <http://www environment.gov.au/land/nrs/science/capad/2014>

Gynther I, Waller N, Leung LK (2016) *Confirmation of the Extinction of the Bramble Cay Melomys melomys rubicola on Bramble Cay, Torres Strait: Results and Conclusions from a Comprehensive Survey in August–September 2014*. Department of Environment and Heritage Protection, Brisbane.

Kania J, Kramer M (2011) Collective impact. *Stanford Social Innovation Review* (Winter), 36–41.

Kareiva P (2014) New conservation: setting the record straight and finding common ground. *Conservation Biology* **28**, 634–636. doi:10.1111/cobi.12295

Lunney D, Law B, Schulz M, Pennay M (2011) Turning the spotlight onto the conservation of Australian bats and the extinction of the Christmas Island pipistrelle. In *The Biology and Conservation of Australasian Bats* (Eds B Law, P Eby, D Lunney and L Lumsden) pp. 485–498. Royal Zoological Society of NSW, Sydney.

Maslow AH (1943) A theory of human motivation. *Psychological Review* **50**, 370–396. doi:10.1037/h0054346

Miller TR, Minteer BA, Malan Leon C (2011) The new conservation debate: the view from practical ethics. *Biological Conservation* **144**, 948–957. doi:10.1016/j.biocon.2010.04.001

Newmark WD, Hough JL (2000) Conserving wildlife in Africa: integrated conservation and development projects and beyond. *Bioscience* **50**, 585–592. doi:10.1641/0006-3568(2000)050[0585:CWIAIC]2.0.CO;2

Rands MRW, Adams WM, Bennun L, Butchart SHM, Clements A, Coomes D, Entwistle A, Hodge I, Valerie Kapos V, Scharlemann JPW, Sutherland WJ, Bhaskar Vira B (2010) Biodiversity conservation: challenges beyond 2010. *Science* **329**, 1298–1303. doi:10.1126/science.1189138

Soule M (2013) The 'new conservation'. *Conservation Biology* **27**, 895–897.

Taylor FJT, Sattler PS, Evans M, Fuller RA, James EM, Possingham HP (2011) What works for threatened species recovery? An empirical evaluation for Australia. *Biodiversity and Conservation* **20**, 767–777. doi:10.1007/s10531-010-9977-8

5

What's next on the list? Prioritising management actions on Australia's islands

Cheryl A. Lohr, Christopher M. Baker, Michael Bode and Robert L. Pressey

Summary

Invasive non-indigenous species are among the most powerful drivers of biotic extinction, particularly on islands. While controlling invasive species can provide significant conservation benefits, planning control actions even across relatively few islands and threatening species offers hundreds, if not thousands, of alternative management action combinations and places to apply them. This makes it difficult to plan actions efficiently – to have the most benefit for a given investment – across groups of islands. This is exactly the kind of problem that systematic conservation planning is designed to solve. The Islands Decision-Support System (DSS) is a new decision-support software, designed to prioritise the myriad of possible actions we could take to protect island species from invasive non-indigenous species and other threats. The Islands DSS uses data drawn from user-friendly spreadsheets, spatially and temporally dynamic models of ecosystems, and models of the efficacy and cost of management actions. The Islands DSS incorporates the newest advances in systematic conservation planning but was designed for practical applications, in collaboration with managers of islands off the Pilbara coast and in the southern Great Barrier Reef. The software is interactive and participatory, and is intended for day-to-day use to refine and support decisions related to island management that can be implemented over both short and long timeframes.

Introduction: why set priorities?

Decision-making was once considered more art than science, because there were no hard and fast rules for arriving at a good decision. We were asked to simply trust in the unspoken, unjustified and intuitive thinking behind a leader's decisions on complex matters (Saaty 2008). However, in today's world, we demand efficiency, transparency and defensibility in decisions that affect our lives. Decision-making has become a science.

That statement is especially relevant for conservation. Setting aside areas for the conservation of biodiversity, ecosystem services and recreational or scenic values is an ancient and widespread human practice (Margules and Pressey 2000). Globally, protected areas now cover some 14.7% of land and 4.1% of the oceans (UNEP-WCMC and IUCN 2016). Decision-making about conservation does not end with the establishment of protected areas. Maintenance of biodiversity and other natural values is necessary in places outside formal protection and, within protected areas, decisions are constantly needed on how to allocate limited resources to management through activities such as control of invasive species, regulation of fire regimes and maintenance of visitor facilities.

Decisions about investment in management activities within protected areas typically lack an explicit basis, often reflecting political expediency or limited perceptions of values and threats, and are thereby limited in cost-effectiveness in promoting the persistence of natural values. Unstructured approaches to decisions about conserving and managing natural resources led to attempts in the 1970s to produce explicit ranking systems for priorities (Smith and Theberge 1986), which were superseded from the early 1980s (Pressey 2002) by the burgeoning field of systematic conservation planning (Kukkala and Moilanen 2013; Alvarez-Romero *et al.*, *pers. comm.*).

Islands, and the need for prioritisation of management actions

Invasive non-indigenous species are among the most powerful drivers of biotic extinction, particularly on islands. On Australian islands, the eradication or control of some of the worst animal and plant pests (black rat *Rattus rattus*, red fox *Vulpes vulpes*, feral cat *Felis catus*, goat *Capra hircus*, Mossman river grass *Cenchrus echinatus*) is possible (Veitch and Clout 2002). But we should not confuse the technical and practical ability to eliminate or control pests with strategic decision-making about where, when and how control actions should be applied. The future of native biodiversity on Australian islands depends on strategic interventions that maximise outcomes for native species and ecosystems, with the limited resources available. Australia does not, however, have a comprehensive national plan of action for removing established non-indigenous species and preventing invasion or reinvasion (Nias *et al.* 2010).

To make robust decisions concerning complex systems, recent research (2012–17) has advanced systematic conservation planning software, from software used to design conservation networks to include software that can prioritise threat mitigation activities. This new decision-support software (Box 5.1) helps planners and managers make decisions in interactive participatory ways, facilitates the exploration of alternative scenarios and the effect of different budgets and timeframes, and investigates the potential effects of invasive species.

Systematic conservation planning

Systematic conservation planning has several main characteristics that ensure the most efficient and effective areas are prioritised for conservation. First, decision-makers must make explicit choices about the valued resources or 'features' to be considered in the planning process. Second, decision-makers must set explicit and quantitative conservation objectives for those features. Objectives could relate to recovery (e.g. what increase in abundance of a native species is needed?) or protection (e.g. how much reduction in abundance is acceptable without the species being at risk?). Third, systematic conservation planning uses mathematical formulae to account for the cost and relative effectiveness of actions taken to mitigate threats and to prioritise actions, and places to achieve the quantitative objectives given certain constraints (Fig. 5.1). As a framework for decision support, systematic conservation planning draws on data, theory and practical experience from diverse fields, including spatial modelling, geographic information systems, population dynamics, economics, social science and the accumulated experience of on-ground managers. Key benefits of a systematic approach are defensibility in the face of competing uses or competing demands for funding, and replicability and accountability in the face of critical review (Margules and Pressey 2000).

Several systematic conservation planning tools have been developed in the last 20 years. MARXAN is currently the most widely used conservation planning software in the world. Developed in 2000, MARXAN was designed to prioritise areas for inclusion in a conservation reserve system, extending the previous, similar system known as SPEXAN (Ball *et al.* 2009). MARXAN selects areas that meet objectives for biodiversity conservation while minimising the total cost of reserve acquisition, allowing the user to adjust the amount of connectivity between selected areas (Ball *et al.* 2009). The system is flexible, and the inclusion of dynamic rather than static benefit and cost functions can allow for the scheduling of reserve acquisition or threat removal (Visconti *et al.* 2010; Januchowski-Hartley *et al.* 2011). Similarly, C-Plan (Pressey *et al.* 2009) and Zonation (Moilanen 2007) are primarily designed for the selection of new reserve networks. C-Plan uses a measure of irreplaceability to select areas, while Zonation works in reverse – it starts with all areas in a region, then removes areas in reverse order of conservation priority, producing a hierarchical series of important places. RobOff is a non-spatial tool (C-BIG 2016) designed to allocate resources to land-use alternatives or conservation actions, some of which can have uncertain effects on biodiversity over time (Pouzols and Moilanen 2013).

Even with all these tools, there remain some limitations of planning software for real-world applications. For example, existing systems do not use continuous models of population dynamics that allow us to differentiate between species that will recover slowly after threat removal and those that may recover quickly. The Pilbara islands project presented the opportunity to develop a new decision-support system that overcame these limitations. One new development was the ability to allocate multiple actions, if needed, to the same places (islands, in our case). A second development was temporal explicitness, recognising that responses of invasive species to management actions are not immediate but follow trajectories over time, with initial reductions in abundance followed by ongoing decline to eradication, stability at a reduced level, or another increase if funding for actions is withdrawn. Understanding species population dynamics helps to allocate budgets across years (who needs help now *v.* who can wait a little longer), decide on levels of control (*v.* eradication) and predict the consequences of discontinuing management. To add these two new aspects of functionality, we designed a decision-support tool that could do what existing systems can do while extending capability to better deal with the realities of conservation decision-making. Our interim name for this decision-support software tool is the Islands Decision-Support System (DSS) (Box 5.1). This tool also has potential for application to mainland and marine ecosystems.

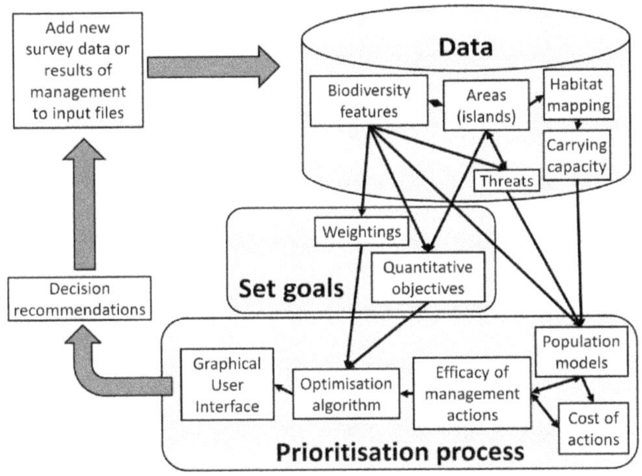

Fig. 5.1: Islands DSS, showing the modules that manage inputs in the form of raw data and models, and the logical flow of the prioritisation approach. The models and optimisation algorithm work simultaneously in the software, so the prioritisation process has just five steps: collect the input data, decide what management outcomes you want, run the software, interpret and implement the decision recommendations, feed any new data back into the input files.

Essential components of systematic conservation planning

If we were to strip back conservation planning to its basics – just a few of the modules in Fig. 5.1 – six components would be needed:

1 delineation of the areas to be assessed and compared for possible conservation investment;
2 data on biodiversity features of concern and their threats;
3 quantitative objectives for each biodiversity feature;
4 the costs of conservation in each area;
5 relative weightings of features for situations when resources are insufficient to achieve all objectives;

Box 5.1: The Islands DSS

The Gorgon Barrow Island Net Conservation Benefits Fund has funded development of the Islands DSS (2012–17), a new decision-support software system, as a contribution towards protecting the valued features of the Pilbara islands. As inputs to the Islands DSS, aside from the collation of 100 years of biological data, the funds have allowed the development of biosecurity risk models, spatially and temporally dynamic population models, cost models, action efficacy models, a feature weighting system, a new prioritisation solver (Brotankova *et al.* 2015; Urli *et al.* 2016) and a graphical user interface. The software is due to be released in 2017, with the first task to be prioritising threat mitigation actions for the Pilbara islands (Box 5.2). The software is designed to use data drawn from user-friendly spreadsheets that allow island managers to readily update input data as they gain knowledge about their environment. Managers can use model outputs to refine and support management decisions that can be implemented over both short and long timeframes. The graphical user interface includes a map of the study region highlighting islands with recommended management actions, and allows users to explore the input data, cost estimates and ecological model results that are site- and species-specific via tables and graphs.

6 a way of solving the conservation problem (i.e. identifying priorities for investment).

Areas

An area – also commonly referred to as a planning unit – is any site to be assessed for priority and compared with others in a planning region. In the case of the Pilbara islands, each island is a unique planning unit. Large islands (or the mainland) could be cut into multiple planning units when actions are unlikely to be applied across them entirely and parts of islands need to be distinguished for prioritisation. Software systems like MARXAN allow decision-makers to simply apply a grid to the landscape to create planning units, or they can apply pre-defined planning units using

Box 5.2: Prioritising invasive species management on the Pilbara islands

The Pilbara region, in the tropical-arid north-west of Australia, has ~600 islands spread over 30 000 km^2 of ocean between Exmouth Gulf and Port Hedland. Most of the islands are small, with a total area of 690 km^2 and 1490 km of coastline habitat, and the majority are conservation reserves. Seven islands have tourist accommodation and 22 islands have industrial activity (Lohr *et al.* 2016). Nine islands host assisted colonies or reintroduced populations of eight mammal and two bird species. Five groups of islands (Montebello Islands, Lowendal Islands, Barrow Island, Sunday Island and the Exmouth Gulf Islands) are listed as Important Bird Areas (Onton *et al.* 2013). Bedout Island is a seabird stronghold with thousands of nesting lesser frigatebirds (*Fregata ariel*), brown boobies (*Sula leucogaster*), masked boobies (*Sula dactylactra*), common noddies (*Anous stolidus*) and terns. Many of the other islands are nesting sites for ospreys (*Pandion cristatus*) (Dennis and Clancy 2014), white-bellied sea eagles (*Haliaeetus leucogaster*) and marine sea turtles (Pendoley *et al.* 2016). Unfortunately, at least 61 weed species (Lohr *et al.* 2015) and nine invasive animal species currently threaten biodiversity on the Pilbara islands, especially those closer to the coast. Currently, research data (2012–17) on the distribution and attributes of 57 native species and 23 threats (introduced plants and animals) are being used to develop and apply new conservation planning software using the Islands DSS (Box 5.1). The first task for the Islands DSS is to recommend a schedule for threat mitigation activities that could occur on the Pilbara islands.

GIS layers. The first version of the Islands DSS is spatially implicit with user-defined management units (Lohr *et al.* 2017).

Biodiversity features

All the conservation planning tools require decision-makers to quantify the conservation features present in each planning unit. The Islands DSS requires quantification of both the features and the threats present in each planning unit. Data that can

Fig. 5.2: North shore of Angel Island, Dampier Archipelago, Pilbara islands of Western Australia has a rapidly changing array of habitat types including mangroves (*Rhysophora stylosa* and *Avicennia marina*), red rock piles, *Acacia bivenosa* shrub land and dune vegetation. Gidley Island is visible across a narrow and shallow channel. Photo: J Brotankova.

be used to quantify features include recent survey records, estimates of species abundances and extrapolations from environmental surrogates (e.g. ecosystem types). For lesser-studied areas, like most of the Pilbara islands (Fig. 5.2; Box 5.2), quantifying the conservation features present can be surprisingly difficult as much of the data are presence–absence data. While the Islands DSS can use presence–absence data, it requires the user to make assumptions about the abundance of conservation features (e.g. long-established threats are at carrying capacity). We are in the process of databasing more than 100 years of sporadic species observations, collating expert knowledge and using high-resolution SPOT 6 and 7 satellite imagery to produce a habitat map for the islands and to quantify the threats and features on the Pilbara islands. We are concurrently engaging in similar work for the southern Great Barrier Reef islands (Box 5.3). Combining expert knowledge and habitat availability data can predict the occurrences of some species with a reliability of 95% (Pressey and Wenger 2016).

Explicit quantitative objectives

Explicit quantitative objectives are at the heart of systematic conservation planning (Margules and Pressey 2000). Importantly, objectives are typically set across the whole planning region, allowing the contributions of populations in individual areas to be compared. Non-quantitative targets increase uncertainty in the planning process and cannot be consistently reviewed for achievement. Objectives are ideally interpretations of broad conservation

Box 5.3: Islands of the southern Great Barrier Reef

Within the southern Great Barrier Reef (GBR) between Mackay and Bundaberg there are 303 islands, including 147 national park islands and four Commonwealth islands (Pressey and Wenger 2016). In parallel with similar research on the Pilbara islands, a database for the southern GBR islands was created to consolidate expert knowledge, identify survey gaps and predict species distributions on less-studied islands (Pressey and Wenger 2016) and further the development of the Islands DSS (Box 5.1). The database now contains >31 500 records from 183 sources and establishes the presence of 695 unique species, consisting of 564 native animal species (313 birds, 111 insects, 76 reptiles, 51 mammals, 10 amphibians and three others), 43 invasive animal species (12 birds, 23 mammals, five insects and three others), 252 invasive plant species and 88 species of fungi. Seventy-four unique regional ecosystems are recorded on the southern GBR islands. These regional ecosystems have been broadly categorised into four on rocky islands (coastal dunes and beaches; dry forest; wetlands, mangroves, marshes and mudflats; moist forest) and four on sand cays (coastal zone; grassland/shrubland/herbland/woodland; *Pisonia* forest; moist forest). Some 37 Priority species on the southern GBR islands (two turtles, 19 seabirds, six landbirds, four shorebirds, four raptors and two mammals) are being used to further the development of the Islands DSS in conjunction with research on the Pilbara islands.

goals previously established in policy by agencies and/or stakeholders (Pressey *et al.* 2003). Objectives need to be in the same units for which data are available and can vary within a single planning process, from number of individuals or extent of mapped ecosystems. For example, Didier *et al.* (2009) set population objectives with an aim to conserve 6800 American marten (*Martes americana*), 8000 black bear (*Ursus americanus*), 2000 common loon (*Gavis immer*) and 5000 moose (*Alces alces*) in the 19 700 km^2 of Adirondack Park. Objectives should be based on the conservation requirements of species and ecosystems, not revised downward to accommodate competing land uses (Pressey *et al.* 2003) or limited budgets.

Fig. 5.3: Crested tern (*Thalasseus bergii*) colony on Delambre Island, Pilbara, Western Australia is a feature of conservation concern. Photo: J Brotankova.

Weighting of features

Although much of the literature on conservation planning has estimated the total cost of achieving all objectives, a more common practical problem is to choose which objectives will be met when resources are insufficient to achieve all of them. This means making choices between biodiversity features (Fig. 5.3). If funds are limited, which species and ecosystems should be given precedence? This question can make managers uncomfortable but, unless it is addressed directly, some species will miss out on management actions. The weighting of species directly can dramatically alter (50%) the selection of sites in reserve design (Arponen *et al.* 2005).

Directly applying weights to dozens of conservation features is prone to personal bias and inconsistencies. A weighting system has been embedded within the Islands DSS (Fig. 5.1) based on the Analytical Hierarchy Process, a transparent process of systematic rationality which breaks down a complex task into smaller competing decision criteria (Saaty 2008), including local contextual information. The Analytical Hierarchy Process has been historically applied in conservation management for identifying priority areas (Moffett *et al.* 2006), identifying threatened species (Ou *et al.* 2008) and prioritising management techniques or tasks (Roura-Pascual *et al.* 2009; Lohr *et al.* 2014).

Conservation costs

Conservation planning requires information on the costs of conservation in each area. Previously, costs have been rough, and often partial, estimates of the full costs of applying conservation measures and then managing areas through time (Armsworth 2014; Sutton and Armsworth 2014). Estimates typically relied on simplistic proxies (e.g. island area) calculated at a low resolution and often with untested assumptions. If poorly estimated costs result in projects being underfunded, there could be an interruption in management due to insufficient funds (Simberloff 2009), which could lead to a failure to meet conservation objectives (Myers *et al.* 2000).

The Islands DSS contains a new cost model that can estimate the cost of implementing a management action at individual sites. The model uses a multi-layered cost framework that accounts for the variety of logistical issues faced by managers. At a basic level, cost estimates are the sum of costs of labour, travel, consumables and equipment. The prices of consumables and equipment do not vary greatly, depending on suppliers. Travel costs are a function of the type of transport (boat, helicopter, fixed-wing airplane), mechanical and safety limitations of transport, and distance to – and size of – islands. The cost of labour is a function of attributes of the invasive species (intrinsic rate of increase, dispersal capabilities, lifespan), the type of treatment needed, and the terrain of an island. Adjustments are used in the cost calculations for factors such as work undertaken in extreme environments (e.g. hand-spraying weeds on coastal cliffs) or time-consuming efforts at sites (e.g. searching for the last seedling of Mossman river grass, *C. echinatus*). The incorporation of these factors into the cost model increases the spatial and temporal variation in costs used in the prioritisation process, and increases the likelihood of identifying the most cost-effective management actions.

Incorporating population ecology

Prior to the Islands DSS, the integration of continuous models of management efficacy or species response to threat mitigation into systematic conservation planning was limited because it was computationally challenging (Adams and Setterfield 2014). Where prioritisation exercises have recognised dynamic costs, benefits and the efficacy of each action (Bottrill *et al.* 2008; Joseph *et al.* 2009), the species' responses to those conservation actions have been modelled external to the prioritisation process. Alternatively, it was assumed that threats were static and were eliminated instantaneously, and biodiversity features recovered immediately in response (Watts *et al.* 2009; Klein *et al.* 2010; Wilson *et al.* 2010). Adams and Setterfield (2014) demonstrated that the incorporation of continuous species response curves in systematic planning models can reduce total budget estimates by up to 20%, which could dramatically alter the allocation of resources and the decision recommendations made by a decision-support tool.

The Islands DSS contains two population ecology equations in the algorithm (Urli *et al.* 2016). The first equation describes the dynamics of a space-limited species (plant, or seabird consuming off-island resources) given competition from other space-limited species. The second equation describes the dynamics of a consumer whose resources are contained within the boundary of a given area (Lohr *et al.* 2017). Aside from standard terms for population growth and carrying capacity, the equations include a species interaction term and a management efficacy term that allows the optimiser to predict how a species might respond to the removal or suppression of a threat, and how much it might cost to mitigate the selected threat.

To use this software capability, decision-makers define conservation objectives in terms of the desired abundance of a given native species (Fig. 5.4). This will usually be defined across all islands collectively in a study region, but can also be specified for individual islands. The optimiser then seeks a suite of management actions that will max-

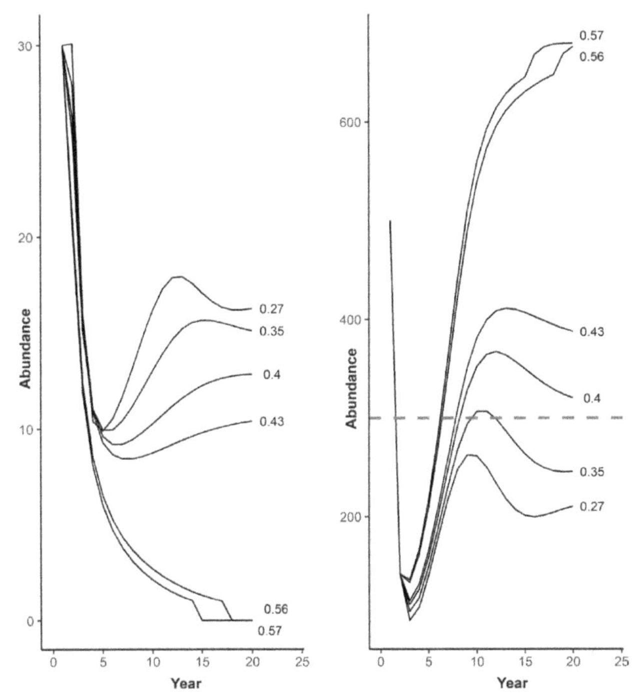

Fig. 5.4: Variation in abundance of an introduced predator (cat, left) and its prey (wallaby, right) in response to variation in the effectiveness of a management action (line labels) which is applied annually to the predator. Between 0.27 and 0.57 (27–57%) of the cat population is removed each year. Due to fluctuating predator–prey dynamics, the hypothetical island used to generate these results will support 200 wallabies if there are 16 cats (removing 27% cats each year). More than 37% of cats need to be removed each year to raise the number of wallabies above the objective. The optimiser will use the cost of implementing cat control to determine whether managers should remove more than 37% of the cat population each year.

imise the number of features that get above that objective for the least amount of money.

Using continuous species response curves gives us greater insight into lag-times between the implementation of a conservation action and the recovery of a species. Sea turtles, for example, take decades (30–50 years) to mature and breed; hence, it can take decades to observe the population-level effect of protecting sea turtle nests. Incorporating a species interaction term, a management efficacy term and associated uncertainty into continuous species response curves gives managers insight into the required quantity and duration of management actions, and associated budgets that must be implemented to achieve their objectives (Fig. 5.4). An understanding of ecosystem response rate will also help to moderate the expectations of stakeholders.

The prioritisation process

The prioritisation processes (i.e. algorithm or solver) used by conservation planning tools solve two types of problems: identifying sets of areas or tasks that achieve all conservation objectives at minimum cost, and identifying sets of areas or tasks that maximise the achievement of objectives within a budget constraint. For the Pilbara islands project, solving the second problem is more relevant to day-to-day decision-making because budgets are typically allocated in a top-down fashion.

Most conservation planning algorithms are heuristic in that they produce good solutions to problems without being able to guarantee that the solutions are mathematically optimal. Recently, however, there has been an increasing development of 'exact' methods to find optimal solutions (discussed below) that could be used in conservation (Beyer *et al.* 2016).

Genetic algorithm

The solver used in the Islands DSS is a genetic algorithm. Genetic algorithms are inspired by natural selection, hence the terms used to describe the rules to be followed by the computer in the problem-solving calculations. Within a prioritisation problem there may be hundreds of management actions that could be taken. Each potential action is synonymous with a binary (1/0) gene with two alleles. In the case of the Islands DSS, the management action options reflect 'do an action' (1) or 'don't do an action' (0) (Brotankova *et al.* 2015). As an analogy, we consider a unique combination of these genes and a suite of management actions to represent one 'chromosome' and one potential solution to the prioritisation problem.

The quality of each chromosome is determined by the fitness function, i.e. a mathematical description of the cost-effectiveness of implementing the actions included in the chromosome. This quality is considered in terms of both the number of objectives achieved and the cost incurred. The best few chromosomes found in the first iteration of the algorithm are transferred directly to the second iteration so that further selection acts on a relatively successful gene pool. Any chromosomes that violate the constraints set by the decision-maker are thrown out. Pairs of chromosomes are selected randomly from the remaining pool of chromosomes and mated, whereby they exchange genes with one another (Brotankova *et al.* 2015). The fitness value of each chromosome is calculated and the process repeats until it converges on a solution.

Temporal optimisation: optimising what occurs when, not just where

Existing systematic conservation planning tools take advantage of the fact that a conservation action (e.g. the gazetting of a protected area) can have different benefits for biodiversity depending on where it is located. This basic premise is just as true in the temporal dimension as it is in a spatial landscape. As an example, potentially valuable conservation interventions will be rendered worthless when the target biodiversity feature has been irretrievably lost (e.g. the Christmas Island pipistrelle *Pipistrellus murrayi*: Martin *et al.* 2012). Temporal variation in the efficacy of conservation actions is the basis of temporal optimisation. Temporal optimisation models are exact methods of optimisation that are currently being applied in systematic conservation planning but are yet to be incorporated into decision-support tools, including the Islands DSS, which uses a heuristic genetic algorithm.

Just as spatial conservation prioritisation requires a geographical description of the landscape, temporal prioritisation requires models that describe how the economic and ecological system evolves over time (e.g. are budgets allocated annually or biennially?). Static maps are no longer sufficient to make decisions.

Temporal optimisation, like spatial conservation planning, uses multiple variables to track threats, features and costs through time, and requires explicit objectives. However, the objectives typically state a time period within which goals should be achieved, or ask that goals be achieved in a minimum amount of time.

Once the system dynamics and management objective have been described mathematically, temporal optimisation is used to choose the best option from a set of available interventions and to create an action schedule. When the system dynamics are described by continuous time differential equations, the optimal schedule can be calculated using optimal control theory (Lenhart and Workman 2007). When the system dynamics are described by discrete time equations (particularly when they are stochastic), the optimal schedule is generally calculated using dynamic programming (Leonard and Long 1992) or stochastic dynamic programming (Marescot *et al.* 2013). Deterministic optimisation with continuous time differential equations produces a complete optimal schedule from a given initial condition, describing how much effort to allocate to different actions throughout the project period. In contrast, stochastic optimisation methods generate state-based management policies, i.e. tables that indicate the best management action to select given the current state of the conservation system.

The two methods described above use mathematically elegant theories to construct the optimal schedule over time. Where these timings cannot be applied, a final option is to use computational methods (e.g. genetic algorithms or 'brute-force' searches) to find more appropriate – but not necessarily optimal – schedules. Alternatively, users can pre-define a set of management policies that represent reasonable choices for managers and then use the system model to test which of the defined options produces the best output (often called Management Strategy Evaluation: Bunnefeld *et al.* 2011). These methods are robust and do not require sophisticated mathematical analysis, but they are unlikely to identify the true optimal schedule unless one of the pre-defined strategies is in fact optimal.

Temporal optimisation methods are computationally and mathematically challenging and are not as prevalent as spatial conservation prioritisation with heuristic algorithms. Nevertheless, these methods can provide vital insight into conservation problems and have been applied in a range of systems, including managing and removing interacting species (Bode *et al.* 2015), metapopulation management (Walker *et al.* 2015), determining the order of protected area purchase (Costello and Polasky 2004; Bode *et al.* 2008) and managing invasions through time and space (Adams and Setterfield 2015; Baker 2016; Baker and Bode 2016).

Concluding comments

The Islands DSS will be the first systematic conservation planning software to provide recommendations to conservation practitioners – initially working on islands – for management actions that are both spatially and temporally dynamic. The integration of continuous models of population ecology provides decision-makers with access to site- and species-specific population models that provide insight into lag-times between the implementation of management actions and the recovery of a species. The cost model considers logistical constraints to produce high-resolution estimates of the cost of management actions. In the future, the Islands DSS can also be applied by managers on the mainland and in the marine environment, given some assumptions regarding the dispersal of species among management areas.

The development of the Islands DSS was a collaboration with managers of the Pilbara islands and the southern Great Barrier Reef islands to ensure the system was user-friendly to practitioners, and that it generated outputs useful to managers.

In today's world, we demand efficiency, transparency and defensibility in decisions that affect our lives, especially decisions made by government agencies. The Islands DSS gives managers access to vast quantities of biological data and advanced models of island ecosystems. Collectively this information can provide a supporting basis for informed management decisions with the necessary evidence in the face of competing land uses, demands for limited funding and the need for replicability and accountability of decisions.

Acknowledgements

Funding for this research was provided by the Gorgon Barrow Island Net Conservation Benefits Fund and the Australian Research Council Centre of Excellence for Coral Reef Studies.

References

Adams VM, Setterfield SA (2014) Planning across freshwater and terrestrial realms: cobenefits and tradeoffs between conservation actions. *Conservation Letters* **7**, 425–440. doi:10.1111/conl.12080

Adams VM, Setterfield SA (2015) Optimal dynamic control of invasions: applying a systematic conservation approach. *Ecological Applications* **25**, 1131–1141. doi:10.1890/14-1062.1

Armsworth PR (2014) Inclusion of costs in conservation planning depends on limited datasets and hopeful assumptions. *Annals of the New York Academy of Sciences* **1322**, 61–76. doi:10.1111/nyas.12455

Arponen A, Heikkinen RK, Thomas CD, Moilanen A (2005) The value of biodiversity in reserve selection: representation, species weighting, and benefit functions. *Conservation Biology* **19**, 2009–2014. doi:10.1111/j.1523-1739.2005.00218.x

Baker C (2016) Target the source: optimal spatiotemporal resource allocation for invasive species control. *Conservation Letters* 10.1111/conl.12236.

Baker C, Bode M (2016) Placing invasive species management in a spatiotemporal context. *Ecological Applications* **26**, 712–725. doi:10.1890/15-0095

Ball IR, Possingham HP, Watts M (2009) Marxan and relatives: software for spatial conservation prioritisation. In *Spatial Conservation Prioritisation: Quantitative Methods and Computational Tools* (Eds A Moilanen, KA Wilson and H Possingham) pp. 185–195. Oxford University Press, Oxford.

Beyer HL, Dujardin Y, Watts M, Possingham HP (2016) Solving conservation planning problems with integer linear programming. *Ecological Modelling* **328**, 14–22. doi:10.1016/j.ecolmodel.2016.02.005

Bode M, Wilson KA, Brooks TM, Turner WR, Mittermeier RA, McBride MF, Underwood EC, Possingham HP (2008) Cost-effective global conservation spending is robust to taxonomic group. *Proceedings of the National Academy of Sciences of the United States of America* **105**, 6498–6501. doi:10.1073/pnas.0710705105

Bode M, Baker C, Plein M (2015) Eradicating down the food chain: optimal multispecies eradication schedules for a commonly encountered invaded island ecosystem. *Journal of Applied Ecology* **52**, 571–579. doi:10.1111/1365-2664.12429

Bottrill MC, Joseph LN, Carwardine J, Bode M, Cook C, Game ET, Grantham H, Kark S, Linke S, McDonald-Madden E, Pressey RL, Walker S, Wilson KA, Possingham HP (2008) Is conservation triage just smart decision-making? *Trends in Ecology and Evolution* **23**, 649–654. doi:10.1016/j.tree.2008.07.007

Brotankova J, Randall M, Lewis A, Pressey RL, Wenger A (2015) A genetic algorithm solver for pest management control in island systems. In *Artificial Life and Computational Intelligence.* (Eds T Ray, R Sarker and X Li) pp. 273–285. Springer, Berlin.

Bunnefeld N, Hoshino E, Milner-Gulland EJ (2011) Management strategy evaluation: a powerful tool for conservation? *Trends in Ecology and Evolution* **26**, 441–447. doi:10.1016/j.tree.2011.05.003

C-BIG (Conservation Biology Informatics Group) (2016) *RobOff: Robust Offsetting, Habitat Restoration, Maintenance and Management.* University of Helsinki, Department of Biosciences. <http://cbig.it.helsinki.fi/software/roboff/>

Costello C, Polasky S (2004) Dynamic reserve site selection. *Resource and Energy Economics* **26**, 157–174. doi:10.1016/j.reseneeco.2003.11.005

Dennis T, Clancy G (2014) The status of the osprey (*Pandion haliaetus cristatus*) in Australia. *Journal of Raptor Research* **48**, 408–414. doi:10.3356/JRR-OSPR-14-01.1

Didier K, Glennon MJ, Novaro A, Sanderson EW, Strindberg S, Walker S, Di Martino S (2009) The landscape species approach: spatially-explicit conservation planning applied in the Adirondacks, USA, and San Guillermo-Laguna Brava, Argentina, landscapes. *Oryx* **43**, 476–487. doi:10.1017/S0030605309000945

Januchowski-Hartley SR, Visconti P, Pressey RL (2011) A systematic approach for prioritizing multiple management actions for invasive species. *Biological Invasions* **13**, 1241–1253. doi:10.1007/s10530-011-9960-7

Joseph LN, Maloney RF, Possingham HP (2009) Optimal allocation of resources among threatened species: a project prioritization protocol. *Conservation Biology* **23**, 328–338. doi:10.1111/j.1523-1739.2008.01124.x

Klein CJ, Steinback C, Watts M, Scholz AJ, Possingham HP (2010) Spatial marine zoning for fisheries and conservation. *Frontiers in Ecology and the Environment* **8**, 349–353. doi:10.1890/090047

Kukkala AS, Moilanen A (2013) Core concepts of spatial prioritisation in systematic conservation planning. *Biological Reviews of the Cambridge Philosophical Society* **88**, 443–464. doi:10.1111/brv.12008

Lenhart S, Workman J (2007) *Optimal Control Applied to Biological Models.* Chapman and Hall/CRC, London.

Leonard D, Long N (1992) *Optimal Control Theory and Static Optimization in Economics.* Cambridge University Press, New York.

Lohr C, Lepczyk C, Cox L (2014) Identifying people's most preferred management technique for feral cats in Hawaii. *Human–Wildlife Interactions* **8**, 35–45.

Lohr C, Passeretto K, Lohr M, Keighery G (2015) Prioritising weed management activities in a data deficient environment: the Pilbara islands, Western Australia. *Heliyon* **1**, e00044. doi:10.1016/j.heliyon.2015.e00044

Lohr M, Lohr C, Keighery G, Long V (2016) The status and distribution of non-native plants on the gazetted and territorial islands off the north coast of Western Australia. *Conservation Science Western Australia* **10**, 1–74.

Lohr C, Hone J, Bode M, Dickman CR, Wenger A, Pressey RL (2017) Modeling dynamics of native and invasive species to guide prioritization of management actions. *Ecosphere* **8**, e01822. doi:10.1002/ecs21822

Marescot L, Chapron G, Chades I, Fackler PL, Duchamp C, Marboutin E, Gimenez O (2013) Complex decisions made simple: a primer on stochastic dynamic programming. *Methods in Ecology and Evolution* **4**, 872–884. doi:10.1111/2041-210X.12082

Margules C, Pressey RL (2000) Systematic conservation planning. *Nature* **405**, 243–253. doi:10.1038/35012251

Martin TG, Nally S, Burbidge AA, Arnall S, Garnett ST, Hayward MW, Lumsden LF, Menkhorst P, McDonald-Madden E, Possingham HP (2012) Acting fast helps avoid extinction. *Conservation Letters* **5**, 274–280. doi:10.1111/j.1755-263X.2012.00239.x

Moffett A, Dyer J, Sarkar S (2006) Integrating biodiversity representation with multiple criteria in North-Central Namibia using non-dominated alternatives and a modified analytic hierarchy process. *Biological Conservation* **129**, 181–191. doi:10.1016/j.biocon.2005.10.033

Moilanen A (2007) Landscape zonation, benefit functions and target-based planning: unifying reserve selection strategies. *Biological Conservation* **134**, 571–579. doi:10.1016/j.biocon.2006.09.008

Myers J, Simberloff D, Kuris A, Carey J (2000) Eradication revisited: dealing with exotic species.

Trends in Ecology and Evolution **15**, 316–320. doi:10.1016/S0169-5347(00)01914-5

Nias R, Burbidge A, Ball D, Pressey R (2010) Island arks: the need for an Australian national island biosecurity initiative. *Ecological Management and Restoration* **11**, 166–167. doi:10.1111/j.1442-8903.2010.00545.x

Onton K, Maurer G, Weller D (2013) *North West Cape Welcomes Shorebirds: Identifying Habitat and Building Conservation Capacity, Western Australia.* WA Department of Parks and Wildlife/Birdlife Australia, Perth.

Ou H, Lu C, O'Toole D (2008) A risk assessment system for alien plant bioinvasion in Xiamen, China. *Journal of Environmental Sciences (China)* **20**, 989–997. doi:10.1016/S1001-0742(08)62198-1

Pendoley KL, Whittock PA, Vitenbergs A, Bell C (2016) Twenty years of turtle tracks: marine turtle nesting activity at remote locations in the Pilbara region, Western Australia. *Australian Journal of Zoology* **64**, 217–226.

Pouzols F, Moilanen A (2013) RobOff: software for analysis of alternative land-use options and conservation actions. *Methods in Ecology and Evolution* **4**, 426–432. doi:10.1111/2041-210X.12040

Pressey R (2002) The first reserve selection algorithm: a retrospective on Jamie Kirkpatrick's 1983 paper. *Progress in Physical Geography* **26**, 434–441. doi:10.1191/0309133302pp347xx

Pressey R, Wenger A (2016) *Prioritising Management Actions for Great Barrier Reef Islands: Report to the National Environmental Research Program.* Reef and Rainforest Research Centre, Cairns.

Pressey R, Cowling R, Rouget M (2003) Formulating conservation targets for biodiversity pattern and process in the Cape Floristic Region, South Africa. *Biological Conservation* **112**, 99–127. doi:10.1016/S0006-3207(02)00424-X

Pressey R, Watts M, Barrett T, Ridges M (2009) The C-plan conservation planning system: origins, applications, and possible futures. In *Spatial Conservation Prioritization: Quantitative Methods and Computational Tools.* (Eds A Moilanen, KA Wilson and H Possingham) pp. 211–234. Oxford University Press, Oxford.

Roura-Pascual N, Richardson DM, Krug RM, Brown A, Chapman RA, Forsyth GG, Le Maitre DC, Robertson MP, Stafford L, Van Wilgen BW, Wannenburgh A, Wessels N (2009) Ecology and management of alien plant invasions in South African fynbos: accommodating key complexities in objective decision-making. *Biological Conservation* **142**, 1595–1604. doi:10.1016/j.biocon.2009.02.029

Saaty T (2008) *Decision making for Leaders: The Analytic Hierarchy Process for Decisions in a Complex World.* RWS Publications, Pittsburgh.

Simberloff D (2009) We can eliminate invasions or live with them: successful management projects. *Biological Invasions* **11**, 149–157. doi:10.1007/s10530-008-9317-z

Smith P, Theberge J (1986) A review of criteria for evaluating natural areas. *Environmental Management* **10**, 715–734. doi:10.1007/BF01867726

Sutton N, Armsworth P (2014) The grain of spatially referenced economic cost and biodiversity benefit data and the effectiveness of a cost targeting strategy. *Conservation Biology* **28**, 1451–1461. doi:10.1111/cobi.12405

UNEP-WCMC and IUCN (2016) *Protected Plant Report 2016.* UNEP-WCMC and IUCN, Cambridge, UK and Gland, Switzerland.

Urli T, Brotankova J, Kilby P, Van Hentenryck P (2016) Intelligent habitat restoration under uncertainty. *Proceedings of the 30th AAAI Conference on Artificial Intelligence*, 3908–3914.

Veitch C, Clout M (Eds) (2002) *Turning the Tide: The Eradication of Invasive Species.* IUCN SSC Invasive Species Specialist Group, Gland, Switzerland.

Visconti P, Pressey R, Segan D, Wintle B (2010) Conservation planning with dynamic threats: the role of spatial design and priority setting for species' persistence. *Biological Conservation* **143**, 756–767. doi:10.1016/j.biocon.2009.12.018

Walker A, Poos J, Groeneveld R (2015) Invasive species control in a one-dimensional metapopulation network. *Ecological Modelling* **316**, 176–184. doi:10.1016/j.ecolmodel.2015.08.015

Watts ME, Ball IR, Stewart RS, Klein CJ, Wilson K, Steinback C, Lourival R, Kircher L, Possingham

HP (2009) Marxan with zones: software for optimal conservation-based land- and sea-use zoning. *Environmental Modelling and Software* **24**, 1513–1521. doi:10.1016/j.envsoft.2009.06.005

Wilson KA, Meijaard E, Drummond S, Grantham HS, Boitani L, Catullo G, Dennis CR, Dutton I, Falcucci A, Maiorano L, Possingham HP, Rondinini C, Turner WR, Venter O, Watts M (2010) Conserving biodiversity in production landscapes. *Ecological Applications* **20**, 1721–1732. doi:10.1890/09-1051.1

6
Island pest management

Keith Springer

Summary

Island managers have grappled with issues and impacts caused by invasive species for decades, whether they are invasive invertebrates, vertebrates or plants. Unlike large mainland areas, the typically concise boundaries of islands can provide opportunities for control or eradication of pest species with greater likelihood of successful outcomes than on larger contiguous landmasses. Once pests are removed, diligent biosecurity measures are required to prevent reinvasion and lock in biodiversity gains. Australia has been at the forefront of efforts to combat the impacts of invasive species on islands and has progressively achieved widespread success in restoring natural ecosystems and protecting native species.

Introduction

Pest mammal populations have long challenged island managers in Australia, as they have globally. Typical issues include habitat alteration, competition, displacement or predation of native wildlife, or a combination of these factors. Animal species considered pests on islands are like those on mainland Australia – including foxes, feral cats, rabbits, rats and mice.

The impact of pest species varies according to local conditions and the situation of individual islands. On some islands, habitat modification by stock may be the major cause of habitation degradation and the impact of an additional grazer such as rabbits may be considered more of an economic impact on existing farming as a land use. Where islands are in a natural state, the introduction of species alien to that environment may have enormous long-term impacts on flora, fauna or both.

Islands often have a fragile ecology and rich biodiversity value due to relative isolation from other landmasses, and as a result have distinctive flora and fauna – including a high rate of endemism. Often, island species have evolved in the absence of mammalian predators and are naive to the arrival of such threats. They may also have co-evolved with other island species and suffer when codependent species are impacted by invasive ones. Partly because of these factors, the establishment of pest species on islands is a leading cause of island species' extinctions, and the local extirpation of many others. Human settlement and resulting habitat modification is another leading cause of biodiversity loss on islands.

In the global context, islands form just 5% of the Earth's landmass, yet nearly 50% of the 724 species recorded as going extinct in the last 400 years have been island species (Convention on Biological Diversity 2017). When considering avian species, some 90% of bird species that have become extinct in that period have been island species. An Australian example is Lord Howe Island (Fig. 6.1), where nine of 15 native landbirds are extinct. Of these, three were extinct by the 1860s, victims of

Fig. 6.1: A view from the summit of Mt Gower on Lord Howe Island. Pest plant and animal eradications are undertaken to enhance the island's World Heritage values. Photo: J Shick.

human collection for food or persecution as crop pests, and five became extinct after the introduction of black rats (*Rattus rattus*) from a shipwreck in 1918. In addition, 13 endemic invertebrates and two endemic plant extinctions occurred after black rats established (Lord Howe Island Board 2016). Another is Christmas Island, where invasive yellow crazy ants (*Anoplolepis gracilipes*), wolf snakes (*Lycodon aulicus capucinus*), house mice (*Mus domesticus*), black rats and feral cats (*Felis catus*) have all wreaked havoc with native bird, insect and land crab populations (BirdLife International 2008) as well as causing the extinction of two native rodent species – an endemic shrew (*Crocidura trichura*) and the Christmas Island pipistrelle (*Pipistrellus murrayi*).

Such issues have resulted in island managers often being keenly aware of the actual or potential impacts of pest species on island ecosystems. However, awareness of an issue doesn't always equate to the ability to rectify it. In recent decades, methods of managing pest animal populations have improved dramatically, especially for smaller mammals and, in particular, lagomorphs and rodents. Prior to the 1990s the main tools for managing island pest animals were poisoning, shooting and trapping; these commonly targeted larger mammals such as ungulates which were easier to detect and remove than were more cryptic species like rodents.

Another key issue for land or wildlife managers was that the most feasible objective on most islands was animal control (especially until the 1990s, for smaller mammals). Under a pest control strategy, the objective is to lower pest populations with the goal of a commensurate reduction in impacts. In such a strategy, an individual pest animal is relatively unimportant because, if impacts are reduced, a small residual population is acceptable. Animal control requires relatively constant pressure to maintain populations at a low enough level to avoid impacts from increasing again, and sometimes native species on an island can't sustain populations with even a low predator population. In addition to constant population pressure to realise outcomes, an animal control strategy is often subject to variable application of financial resources and hunting effort, and funding is vulnerable to political or management changes in funding priority. The gains made up to that point can rapidly be lost as the pest population bounces back without the hunting pressure to constrain it.

In contrast, when pursuing a pest eradication strategy, the individual is vitally important because eradication requires removal of every individual. This is rarely feasible with mainland pest populations; this strategy is pursued mostly on islands. There, the finite land margin defines the treatment area and, if reinvasion risks can be managed to near-zero and every individual can be put at risk of encountering the method used, then biodiversity gains accrued by removal of the pest population can be locked in. Pest control programs have been carried out on islands by government agencies and NGOs for decades, often as part of routine operations that are unrecorded and unreported. This chapter will focus on eradication of vertebrate pests as a management tool to enhance biodiversity and ecological restoration outcomes on islands, while acknowledging that significant impacts are caused by invasive invertebrate and plant species too.

Two developments allowed pest management to progress from control to eradication on islands, especially when targeting rodents and lagomorphs. First, the development of technology such as Global

Positioning Systems (GPS) and its application to helicopter flight management and second, the evolution of underslung buckets for agricultural applications (spreading fertiliser or seeds) to conservation applications such as accurate aerial distribution of poison bait. Utilising these techniques and developing them to enable removal of pests from islands occurred mostly in New Zealand. There, the rapid decline in native bird species due to introduced mammalian predators was causing significant concern, but there were few pest-free places to where remnant bird populations could be moved and be safe from predators. The New Zealand Wildlife Service had been clearing rats *Rattus* sp. from small islands since the early 1960s but, as more and more bird species declined on the mainland, pressure grew to clear more islands of pests so that birds unable to cope with predation could be moved there before they faced extinction (Hutton *et al.* 2007). These factors encouraged the development of aerial baiting techniques that led to demonstrated successes in removing vertebrate pests from islands. It was only logical that such techniques could be applied to Australian islands, although there were often greater challenges. New Zealand has no native land mammals other than two species of bat, but Australian islands often sustain populations of native marsupials or rodents that could be at risk from the techniques intended to remove their invasive brethren.

Australian story

The practice of pest management on islands has varied throughout Australia – logically, as some states have many islands (e.g. Western Australia, Queensland, Tasmania) and others have relatively few (e.g. South Australia, Victoria, New South Wales). All states have differing pest situations, among numerous other factors, on their islands. Most states have undertaken some degree of pest management on some of the islands under their jurisdiction. New South Wales has been successful in removing rabbits from Cabbage Tree and Broughton Islands and house mice from Montague Island, and Norfolk Island has removed rabbits from nearby Phillip Island. Western Australia's (then) Department of Conservation and Land Management has been at the forefront of island management activity. Under its Montebello Renewal project, feral cats and black rats were eradicated from the Montebello Islands – feral cats by aerial baiting and trapping, and black rats using bait stations on many small islands. New Zealand aerial baiting methodology was used after hand distribution methods failed to eradicate rats from 1020 ha Hermite Island and four nearby islands. Removal of invasive predators allowed the introduction of critically endangered native marsupial species to the islands (Burbidge 2004). Black rats have also been eradicated from 12 other Western Australian islands. Cats have been eradicated from several islands on which they were considered an invasive pest, including Rottnest and Faure Islands. An ambitious feral cat eradication is underway on Dirk Hartog Island (62 000 ha) which, if successful, will make it the largest island worldwide from which feral cats have been eradicated. House mice have been eradicated from six islands in Western Australia, and foxes controlled on several islands.

Tasmania has about 60 islands with introduced and/or pest animals. The most significant achievements to date have been the integrated pest eradication program for Macquarie Island and the eradication of rabbits from Woody Island, house mice from Fischer Island, fallow deer from Maria Island and feral cats from Tasman and Wedge Islands. A program to eradicate black rats from Big Green Island is underway.

The most comprehensive record summarising global eradications is the Database of Island Invasive Species Eradications (DIISE 2015), from which the following summary of eradication activity around Australia can be drawn. Like most databases, the picture is only as complete as the information added to it, which is a continually evolving effort. In 2016, for example, the pest species database showed a total of 179 Australian islands where 274 eradications targeting 18 species were carried out. Specifically:

- two eradications targeted bird species;
- 30 eradications targeted feral cats;
- 16 eradications targeted three species of dog or fox;
- 41 eradications targeted rabbits;
- one eradication targeted a reptile species;
- 109 eradications targeted three species of rodent;
- 75 eradications targeted seven species of ungulate.

When looking at the efficacy of these pest eradications, 22 were recorded as failures across seven species of pest, 203 were successful against 13 species of pest, one was still in progress, two were still to have an outcome determined, two were planned, and the remainder were unknown or unreported.

The removal of pest species from islands can have long-lasting positive environmental outcomes, but only if robust and diligent biosecurity procedures are in place to prevent reinvasion. Investment in biosecurity facilities and procedures is almost always cheaper than the cost of removing established pest species.

Case studies

The following case studies illustrate examples of pest management work undertaken to restore island biodiversity in Australia. All of them involve an initial program of pest animal control, which eventually changed to eradication as methodology improved and there was greater appreciation of the biodiversity benefits of eradication.

Phillip Island, Victoria

Phillip Island is a good example of a populated and heavily modified island where a pest species has been subjected to control and then eradication strategies. Located close to the Victorian mainland and connected to it by a 600 m bridge, Phillip Island is an inhabited island of 10 000 ha with a substantial area cleared for livestock and agriculture, several urban areas and some remnant woodland. The island is a significant seabird nesting site and major overwintering ground for trans-equatorial migratory shorebirds. For many years, there has been sustained impacts to stock and wildlife through predation by red foxes (*Vulpes vulpes*), which have been present since 1907. Aside from stock losses, which was the initial impact of concern to island landowners, an important economic and wildlife impact of foxes has been predation upon little penguins (*Eudyptula minor*). These birds are a major tourist attraction on the island and form an important economic nature tourism resource, attracting nearly 700 000 visitors annually (2015–16) to the nightly 'penguin parade'. This visitor activity is estimated to generate some A$400 million for the Victorian economy (2013–14) of which about A$140 million is the economic contribution to the Bass Coast Shire (unpub. data, Phillip Island Nature Parks).

Control of foxes commenced in 1918, and a bounty system introduced in 1954 was in place for nearly 30 years. A more focused effort at fox control commenced in 1980 as recognition increased of foxes' impact on the penguin population. A fox control team operated for over 20 years using poisons, shooting, snares and dogging as primary methods. This succeeded in reducing fox numbers considerably but efforts dwindled by the early 2000s as staff left and management priorities changed – a good example of the risks to a control program noted earlier in the chapter. In 2006, the strategy of fox control, in place for 88 years, was changed to eradication (McPhee and Broomfield 2004). With this stronger mandate, increased resources and focus, the number of foxes killed each year rose initially, especially after the introduction of island-wide 1080 baiting (Fig. 6.2). Greater pressure was applied, and the surviving population had dropped dramatically by the end of the 2000s (Kirkwood *et al.* 2014). The fox eradication project was reviewed in 2013 and fox detection dogs (as opposed to dogs used for driving or flushing foxes) were introduced to support the eradica-

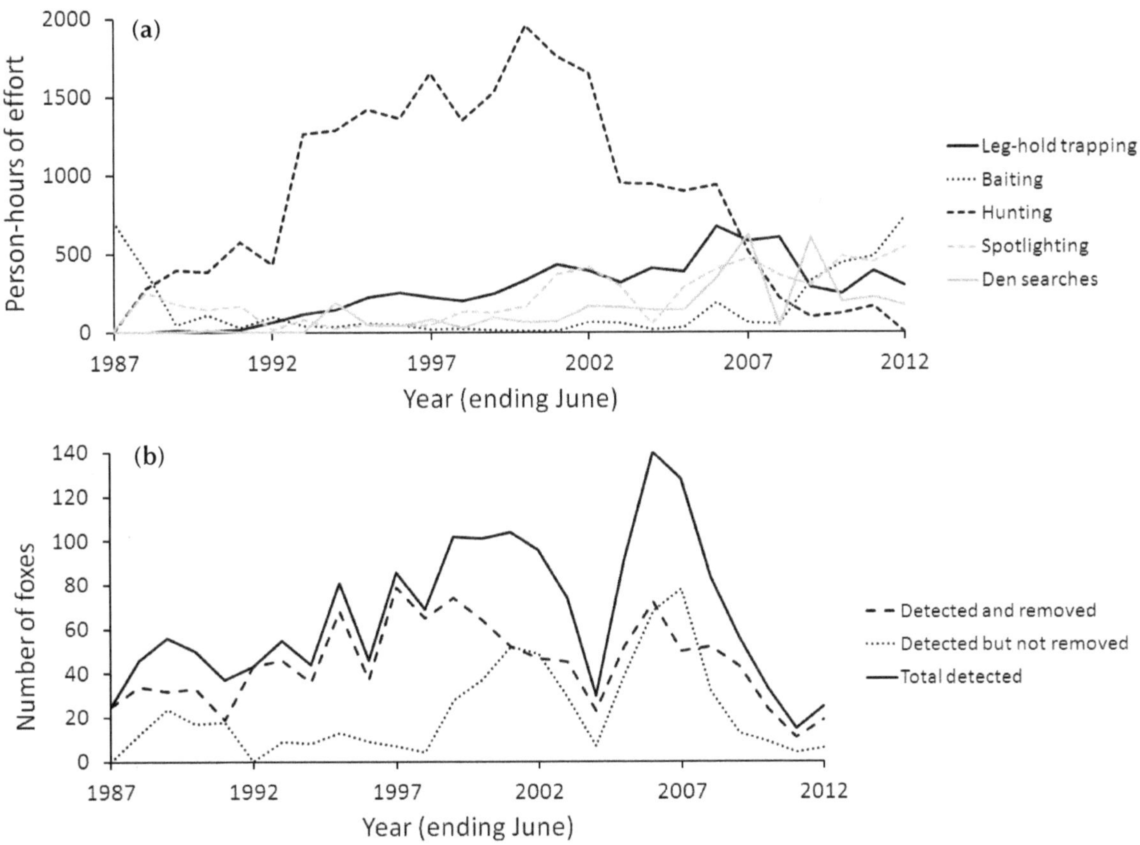

Fig. 6.2: The history of the fox eradication campaign on Phillip Island in terms of (a) person-hours of effort invested in each management action and (b) number of foxes detected (removed and sighted). All methods were used to remove foxes, and some foxes were seen but not removed when spotlighting and hunting. Source: Rout *et al.* (2014).

tion team in locating residual animals, which by then were at low density (Rout *et al.* 2014). In many pest control and some eradication programs, 'the foot is taken off the throttle' at this point. This is because managers perceive that returns (the number of dead foxes) are not worth the inputs (staff time and operating costs), especially as the last individuals in a pest population are usually the most expensive to remove. The program may be halted if it is perceived that the money can be better used elsewhere, even though success may be close.

This has not occurred on Phillip Island – the eradication effort has been maintained even though the number of foxes taken between 2010 and 2016 totalled only ~14, with no confirmed sightings since August 2015. This ability to 'stay the distance' is a critical factor in successful eradications. In the case of Phillip Island, managers recognised that even with low fox densities, significant penguin mortality could still occur because foxes often killed more birds than they would eat (surplus killing). This disproportionate impact on prey species from low predator numbers is not unusual for some species including cats and mustelids.

Fox eradication is considered feasible and if reinvasion risks can be managed (the narrow channel and the bridge being the primary reinvasion pathways) then eradication delivers better conservation, tourism and economic outcomes than long-term control. It is also cheaper, because financial resources will not be required once eradication is achieved, other than for ongoing monitoring.

Fox eradication may soon be declared successful on Phillip Island, although further monitoring work will be required to reach that point. The benefits of reducing the fox population to near-zero are discussed elsewhere in this book: eradication success appears close, but even changing the focus of resources and mindset to the goal of eradication achieved a lower fox density in 10 years than the control measures of the previous 88 years.

An interesting aspect of the Phillip Island case study is the question of what management is possible because of human attitudes to charismatic wildlife species. Luckily for penguins, they have a fascinating appeal to humans, and that appeal extends to a willingness to pay to see them – as nearly 700 000 people each year do on Phillip Island. The financial result of this appeal and the associated willingness to pay is directly responsible for the ability of Phillip Island Nature Parks to resource the eradication of foxes, as well as its other wildlife research work. If penguins did not have the ability to generate an economic return, the eradication of foxes would probably not be undertaken on Phillip Island. Given the massive penguin decline due to fox predation in the 20th century, eventually the penguin population would probably be extirpated. In this instance, their appeal is a win for penguins, for Phillip Island Nature Parks as a business entity, for other less-appealing wildlife (to humans) that are incidental beneficiaries of fox removal and for the local and state economies. It should be recognised, though, that if a less-charismatic species suffered extensive population decline through predation, then the management ability or motivation to intervene to the same extent could be much reduced.

Lord Howe Island, New South Wales

Lord Howe Island lies 760 km off the coast of New South Wales, north-east of Sydney. It covers 1450 ha with a slightly modified natural environment and a small resident population (~350). It represents a good example of an inhabited island with a range of pest species – plants, invertebrates and mammals – that have resulted in significant impacts, including multiple extinctions of endemic fauna and at least two endemic plants. Over the years, island managers have undertaken control or eradication strategies for many of these species, and others are planned.

As was typical during the European exploration of Australasia, animals such as pigs and goats were released on Lord Howe Island by the 1820s to provide food for subsequent inhabitants. After a settlement was established in 1834, feral cats and numerous invasive plant species established, notably cherry guava (*Psidium cattleianum*). These species would have had impacts immediately, although during the 19th century awareness of these was probably limited. Further damage occurred after black rats were accidentally introduced to the island when the ship *SS Makambo* ran aground in 1918. Within five years, five endemic bird species were extinct due to rodent predation. By 2003 they were joined by at least 11 species of invertebrate. Other species have been significantly impacted by rodents. Of these, the most notable was the endemic Lord Howe Island phasmid (*Dryococelus australis*), considered extinct until 2001, when a relict population was found on Balls Pyramid, 23 km south-east of Lord Howe Island.

Inscription onto the World Heritage list in 1982 brought a greater recognition of the natural values of Lord Howe, and of the threats to those values. Management of pest species – mammal, plant, invertebrate and avian – has evolved over time, with initial control actions morphing in most cases to an eradication goal.

Removal of feral goats from part of the island was attempted in 1976 and most goats were removed from the northern part of the island, which benefited vegetation and seabird habitat there. In 1979, action commenced to address wider biodiversity impacts with the eradication of cats in 1980 and pigs (*Sus scrofa*) completed in 1981. An attempt at removing all goats commenced in 1999 and was completed by 2001, although the project was not supported by all residents (Parkes *et al.* 2011). This operation left a surviving population of about five female goats to die out, with no wild goats having been seen since ~2014–15. These actions aided the recovery of the endemic and endangered Lord Howe woodhen (*Gallirallus sylvestris*) as well as benefiting other species, especially plants. Pigs were eradicated by hunting with the assistance of dogs. Feral cats were removed by using dogs, trapping and shooting. The 1976 goat control relied on ground hunting, but formidable terrain on Mt Gower in the island's north meant it was difficult to access all areas of goat habitat, and hunting pressure was variable. The eradication attempt in 1999 primarily used aerial shooting from helicopters, backed up with ground hunting methods, again using dogs.

To ensure that the species do not re-establish, a local by-law prohibits the importation or keeping of pigs, cats and goats. A 'grandfather clause' allowed domestic animals living at the time the by-law was passed to remain until they died. A small number of female domestic goats are kept on the island.

The Lord Howe Island Board commenced weed control efforts in the 1990s, which have evolved to an island-wide eradication strategy. At least 68 plant species have been declared noxious, although 40 are limited in distribution. Invasive plants are notoriously difficult to control, let alone eradicate, but the Board has displayed long-term commitment and persistence to the eradication goals and is realising significant progress. One reason for this is that many plant seeds are distributed widely by birds or wind and can remain viable in the soil for some time – regular searching is necessary to locate new individuals, which leave no sign other than their *in situ* presence. However, many formerly widespread weeds now have extremely limited distribution and over 20 species are close to being eradicated. Six species have already been officially declared as eradicated. Among the successes of the weed teams on Lord Howe Island have been the long-term involvement of volunteer programs and the development of highly accurate herbicide application by helicopter-mounted lances. This has enabled effective spot-spraying on cliff faces where the terrain is too steep to safely access or individual plants have been difficult to locate (Bower 2016).

From 2012 to 2016 an eradication program targeted African big-headed ants (*Pheidole megacephala*). This initiative appears to be on the cusp of success, with the last area of infestation having undergone treatment, from an initial infested area of around 50 ha. In common with eradication programs for many species, there will be a two-year monitoring period and if no evidence of ants is found during that time then this species will be considered removed from the island.

Among bird species, the Tasmanian masked owl (*Tyto novaehollandiae castanops*) is targeted for eradication, with localised control strategies in place for common starlings (*Sturnus vulgaris*), mallard × Pacific black duck hybrids (*Anas superciliosa* × *A. platyrhynchos*), and feral pigeon (*Columba livia*). Control strategies are in place for these species as they could recolonise the island if populations are not kept at low levels.

The two remaining mammalian pests on Lord Howe Island are black rats and house mice. Black rats are the primary suppressant of Lord Howe Island's native plants, landbirds, burrowing seabirds and invertebrates. Removing both rodent species would not only enhance the island's biodiversity values, it may also have a positive economic impact given the island's tourism marketing strategy as a 'natural paradise'. Bird populations are likely to expand significantly in the absence of rat predation and this aligns well with eco-tourism themes and specialist endemic bird-watching ventures. The technical aspects of eradicating the species are not daunting. Both species have been eradicated from a far larger island in Australia (Macquarie) and other rodent species have been removed from islands at similar latitudes and ecoregions, for example Norway (*Rattus norvegicus*) and Pacific (*R. exulans*) rats from Raoul Island (New Zealand) and from numerous islands in the Pacific. An eradication plan has been developed using a combination of standard aerial baiting methodology, hand broadcasting and bait stations to meet the criteria required to achieve success. Funding to implement the rodent eradication plan has been available since 2012 and the project is continuing to build a vital requirement – widespread community support. There has been some opposition to the rodent eradication for many years – a minority of island residents are vociferously opposed, but others are strongly supportive. This demonstrates the critical importance of a community mandate.

While universal support for eradication programs on islands can rarely be expected, such programs need to incorporate a plan demonstrating the benefits of rodent eradication to ecological restoration and improvements in island biodiversity. The Lord Howe Island Board is still motivated to eradicate rodents from the island, especially given its World Heritage status and its commitment to removing invasive species that negatively impact the island's natural values. The Board is also conscious that the amount of toxin required for rodent eradication is less than the amount required if current localised rodent control is maintained over the long term. A successful eradication would mean the cessation of future toxin use. Consultation and community engagement efforts continue, with the eradication operation scheduled for 2018.

In general terms, Lord Howe Island has achieved much in the field of pest species management. Recognition of invasive species impacts, setting clear goals, preparation of detailed management plans, regular reviews and focused application of resources at a sustainable level have meant that Lord Howe has managed to eradicate some species of plants, animals and – hopefully soon – invertebrates. Species with a less severe impact have been controlled where eradication is not desirable, feasible or cost-effective. All in all, island managers have recognised that the problems caused by pest species will only get worse if not addressed, and have decided that it is within their ability and mandate to arrest the progression of many pest species across the island.

Macquarie Island, Tasmania

Macquarie Island (54°37′S, 158°51′E, 12 865 ha) is a sub-Antarctic island lying 1500 km south-east of Hobart. It is part of Tasmania and is a Nature Reserve and World Heritage site managed by the Tasmania Parks and Wildlife Service. The Australian Antarctic Division operates a station to support scientific, island management and meteorological staff. It is a particularly important breeding site for seabirds and marine mammals in the Southern Ocean because there are so few places to breed across thousands of kilometres of ocean in that region.

Since discovery in 1810, several invasive plant and animal species have established feral populations on Macquarie Island. Feral dogs and cats were both recorded by 1820. Dogs – as a large, pack-oriented predator – were probably responsible for a rapid massive decline in the previously healthy wandering albatross population. House mice probably established by 1830, with sealer activity. Rabbits and weka (*Gallirallus australis*, a rail from New Zealand) were introduced around 1880 as food sources for workers extracting penguin oil on the island. Black rats were recorded in the early 20th century, probably established from one of the several shipwrecks occurring on the island around that time (Cumpston 1968). Of the introduced plants and invertebrates, none have dominated ecosystems or caused significant adverse impacts. The interactions and balance between native and introduced species proved complex as new species were added to the mix. For example, after 60 years co-existing with feral cats, the endemic Macquarie Island parakeet (*Cyanoramphus erythrotis*) and an endemic banded rail (*Gallirallus philippensis macquariensis*) were both extinct within 10–12 years of rabbits establishing in 1879. The increasing prey base afforded by rabbits allowed the feral cat population to expand, to the point where they soon exterminated the two endemic landbirds (Taylor 1979). Direct evidence of the impact of rodents and feral cats was absent for many years, largely due to the lack of observers on the island to evaluate the unfolding problem. However, indirect evidence can be gleaned from the numbers of seabird species that were breeding on predator-free offshore islands but not on the main island, and the number of species that could be expected to breed within the region but were not. What seems likely is that many bird species – especially burrow-breeders – were killed by feral cats and black rats before they were recorded, leading to a relatively depauperate avian fauna.

Wekas were also implicated in the reduction in bird numbers and the extinction of the two landbirds. An inquisitive and omnivorous opportunist feeder, wekas were largely restricted to the coastal margin, where they picked through piles of kelp washed up on the beaches to forage for invertebrates, a core part of their diet.

The impacts of black rats and feral cats resulted in reduced abundance of birds and species, and the impacts of rabbits were obvious. Although records of damaged vegetation ecosystems were not made until the early 1950s, impacts were well established and widespread before botanists arrived to observe and record them. A permanently staffed station was established in 1948 and from the 1950s onwards a succession of papers and reports recorded the extensive overgrazing by rabbits and the altering of vegetation communities (Taylor 1955; Costin and Moore 1960; Jenkin 1975; Jenkin *et al.* 1981). These showed that palatable species were removed across wide areas and replaced with a close-cropped sward. Numerous smaller plants were grazed to remnant populations, and dominant species such as tussock (*Poa foliosa*), Macquarie Island cabbage (*Stilbocarpa polaris*), silver-leaf daisy (*Pleurophyllum hookeri*) and prickly shield fern (*Polystichum vestitum*) were so heavily impacted that their distribution was severely restricted.

Recognition of the damage caused to the island's ecosystem by rabbit overgrazing was followed by attempts in the late 1950s and early 1960s to control rabbit numbers, first by poisoning and then by releasing the myxoma virus. Poisoning trials were not sufficiently successful to be used for wide-scale control (Johnston 1966), and the myxoma virus failed to establish. Work continued during the 1960s testing various forms of rabbit control, none of which were particularly successful. In the late 1960s a plan to introduce a different strain of myxoma virus was initiated, preceded by release of the European rabbit flea (*Spilopsyllus cuniculi*). The earlier failure of the myxoma release was probably due to the lack of a suitable vector; an island-wide distribution of the host-specific rabbit flea would be necessary for myxoma to establish successfully. Batches of fleas were distributed around the island annually for 10 years before the myxoma virus (Lausanne strain) was released in November 1978 (Brothers *et al.* 1982). The effect on the rabbit population was dramatic, with numbers dropping from an estimated 150 000 (Copson *et al.* 1981) to ~40 000 by 1982. Annual virus distribution by rangers held rabbit numbers in check through until the late 1990s, when numbers started to rise again.

The impact of feral cats on the burrowing seabird population was also causing concern among wildlife managers. With surface-nesting seabirds (e.g. Cape petrels *Daption capense*) probably extirpated before the arrival of biologists to record their existence, feral cats were known to predate heavily on burrowing seabirds, as well as on rabbits and, to a much lesser extent, rodents. A study in 1974 attempted to assess this impact and estimated that feral cats were killing some 60 000 birds and 56 000 rabbits each year: the feral cat population was estimated at 250–500 (Jones 1977). Burrowing petrels have long lifespans and low reproductive rates, making their populations sensitive to changes in adult survival (Dilley *et al.* 2017). Rabbits have short lifespans and high reproductive rates, making them better able to sustain population increases in the face of predation pressure.

Feral cat control began in 1974, intended to reduce cat numbers and therefore impacts on seabird populations. By 1982, the feral cat strategy changed from control to eradication, but with no additional planning or resources there was no real change in method or outcome. However, in 1995 a feral cat abatement plan was prepared with a clear eradication goal and, with largely federal government funding and Australian Antarctic Division logistics support, an expanded and refocused feral cat eradication commenced in 1998. This increase in hunting pressure brought feral cat numbers down rapidly and by 2000 the last known feral cat had been removed, with eradication declared successful in 2002 (Robinson and Copson 2014). When feral cat numbers were brought to near-zero, grey petrels (*Procellaria cinerea*) were discovered breeding, the first confirmed in over a century. As winter-breeding birds with large burrows, they had succumbed early to cat predation and re-established quickly when predation ceased. White-headed petrel (*Pterodroma lessonii*) and Antarctic prion abundance also increased in subsequent years, along with other burrow-nesting species expanding or re-establishing (Brothers *et al.* 2008).

Following the reduction in rabbit numbers from myxomatosis in the early 1980s, weka numbers reduced, probably because feral cats increased predation on them as rabbits became fewer. Tasmania Parks and Wildlife staff decided to increase pressure on wekas as well, shooting them on sight, until weka were eradicated by the end of 1988. This was the first of Macquarie's pest species to be successfully removed.

At about the same time as feral cat numbers diminished, by 1999 the rabbit population was increasing, with consequent impacts on vegetation. Some researchers claimed that this showed that feral cats were the primary limiting factor on rabbits, and that removing cats was the primary cause of a dramatic increase in the rabbit population and massive degradation of the Macquarie Island landscape (Bergstrom *et al.* 2009). This theory was widely accepted as it reflected a basic principle in population biology in terms of predator–prey relationships. However, it was a simplistic evaluation of the situation and failed to account for two other major influences on the rabbit population (Dowding *et al.* 2009). One was a 'double-hit' regarding myxomatosis. Production of the myxoma virus had ceased in 1999 and stocks had gradually been used up on Macquarie Island, so the amount of virus spread annually dwindled from 2001 until the last was used around 2006. More importantly, the efficacy of the virus was waning by the end of the 1990s. This mirrored the experience on the Australian mainland where myxomatosis had reduced virulence after ~20 years – the same timeframe as on Macquarie Island. Blood tests from Macquarie Island rabbits in 2003 showed a number had antibodies to the virus, inferring widespread resistance to the virus. Rabbit population increases had much more to do with resistance to and release from a biological control agent than with release from a predator – myxomatosis had killed far more rabbits than cats ever had.

The second major influence on rabbit expansion was an increasing abundance of food. This was part of the cycle of rabbit population and vegetation overgrazing that had been continuing since rabbit establishment but that was documented only after botanists arrived on the island in the mid-20th century. Myxoma release in the late 1970s was the first time a massive reduction in rabbit numbers had been achieved, and after 20 years of significantly reduced grazing pressure the dominant vegetation species recovered their former abundance, leaving vegetation cover by the late 1990s in the best condition that it had been in for at least two decades (Scott and Kirkpatrick 2013). This abundance meant that, when released from biological control and with reduced predation pressure, rabbits had ample food to support their population expanding again (Terauds *et al.* 2014). The boom in rabbit numbers therefore plateaued by ~2006 and numbers declined after that (Terauds 2009), as the food supply was largely exhausted (Fig. 6.3). Several researchers have postulated that changing climate may have favoured greater rabbit kitten survival over this period. The 'cats as the primary constraint on the rabbit population' theory failed to explain how rabbits could establish and flourish in such numbers in the presence of an established feral cat population. It also did not address the question of how feral cats failed to limit the rabbit population in the 1970s when it was greater than the mid-2000s peak (Terauds 2009).

From a management point of view, the devastation to the vegetation and landscape by the mid-2000s warranted another approach to invasive species management on Macquarie Island. Before the early 2000s, eradication of rabbits, black rats and house mice was not considered possible. Efforts were directed at rabbit control and virtually none towards rodents, although some localised rodent baiting on the mainland near offshore rock-stacks with remnant blue petrel colonies was undertaken from 1999 to 2002. However, the success of the New Zealand Department of Conservation in eradicating Norway rats from Campbell Island (one of the closest islands to Macquarie) in 2001 demonstrated that eradication of rats (at

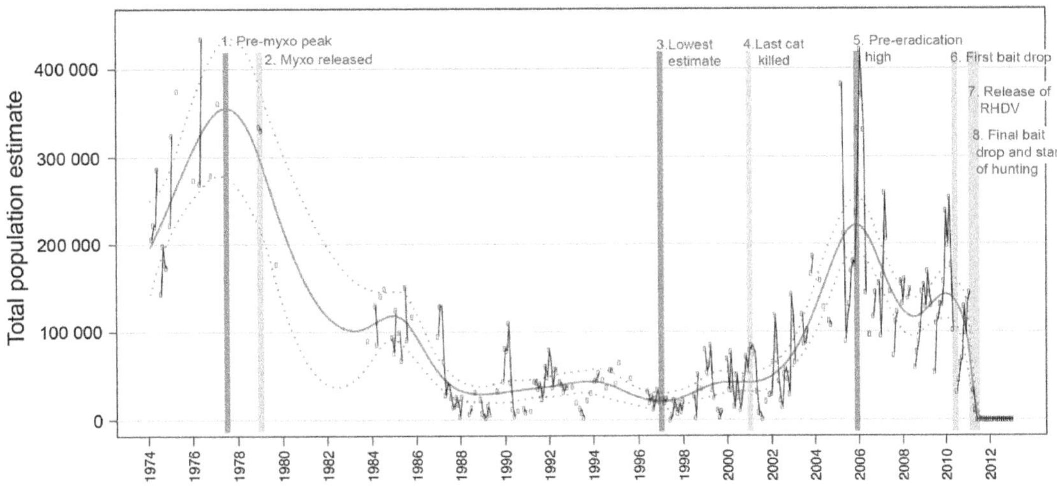

Corrected veg	Approximate date	Population estimate from model	Notes
1. Pre-myxo peak	June 1977	355 222 ± 78 556	Date from model
2. Myxo introduced	Dec 1978	297 287 ± 81 095	Copson (1981)
3. Trough after myxo	Dec 1996	20 669 ± 9211	Date from model
4. Last cat killed	Dec 2000	41 996 ± 10 319	TASPAWS
5. Pre-eradication high	Nov 2005	221 600 ± 31 680	Date from model
6. Preliminary bait drop	May 2010	135 707 ± 25 995	10% of island baited
7. RHDV released	February 2011	50 653 ± 23 879	Hand baited
8. Successful bait drop	May 2011	7825 ± 11 090	Whole island baited

Fig. 6.3: Population trends of whole-island rabbit numbers on Macquarie Island, based on thermal imaging correction and extrapolated from vegetation class areas. Points (o) indicate monthly island-wide estimates, with non-missing adjacent months joined with solid black lines. The solid red line is the smoothed estimate and red dotted lines indicate the 95% confidence interval. Source: Terauds *et al.* (2014).

least) from remote sub-Antarctic islands was indeed possible. This awareness, coupled with public concern as the massive environmental degradation on Macquarie Island became known, encouraged the Tasmania Parks and Wildlife Service to propose concurrent eradication of all three remaining invasive species from the island. With funding eventually secured from the Tasmanian and Australian governments in June 2007, planning commenced for a project that drew on experience of rodent and rabbit eradications from islands elsewhere (Springer 2011).

The plan involved aerial distribution of toxic baits containing the anticoagulant brodifacoum, aimed at eradicating all rodents and killing most rabbits. As some rabbit survivors were expected, rabbit detection dogs were trained and field huts located around the island, anticipating that field-based teams would be required for some years to hunt surviving rabbits. A first attempt at baiting in 2010 was thwarted, due to the non-availability of the ship selected to transport the expedition to Macquarie at the planned time and then by adverse weather conditions after arrival. Wind and cloud prevented helicopters from flying and only ~8% of bait was spread before the attempt was called off in July, with the impending return of many bird species that migrated from the island over winter.

A second attempt was undertaken in May 2011, preceded in February by the release of rabbit haemorrhagic disease virus (RHDV), an initiative intended to kill as many rabbits as possible in autumn. This would reduce the number subsequently killed by aerial baiting and the potential consumption of toxic carcasses by scavenging seabirds – an issue that had caused some concern after the 2010 baiting attempt. This initiative was successful; some 80–90% of rabbits were killed by

RHDV (Springer and Carmichael 2012). Aerial baiting followed in May and June 2011, and was ultimately successful in eradicating black rats and house mice. Hunting teams commenced work immediately following the completion of baiting, and with the help of the dogs located and removed 11 rabbits in the next four months, including one litter of juveniles – the only evidence of post-baiting rabbit breeding. Monitoring by the hunting teams continued for another 2.5 years, with a rodent detection team (two handlers with three rodent-detection dogs) deployed to the island for a year in April 2013. Eradication of all remaining pest species was declared successful in April 2014 (Springer 2016).

Recovery of vegetation has been dramatic since rabbits were removed (Scott 2015), with many areas recovering their former vegetation abundance (Fig. 6.4). Distribution of species has increased, especially the most palatable. *Poa cookii*, for example, has greatly expanded its range compared to its former distribution under rabbit grazing. Growing conditions are harsher on the island plateau, where there are interesting changes that are novel to contemporary eyes, as no-one living has seen what the plateau looked like without a heavy rabbit presence. Plants like the lycopodium *Huperzia australiana* and orchid species *Nematoceras dienema* and *N. sulcatum* are among species improving their distribution and abundance. Anecdotal evidence is strong for the recovery of invertebrates, especially on the plateau where they are no longer subject to mouse predation. Moths and spiders are far more evident. There have been marked improvements in seabird breeding success across many species. Blue petrels were quick to recolonise the main island from offshore rock stacks where several colonies had persisted, and colonies are now found at numerous sites around the island. Grey petrels, white-headed petrels, Cape petrels and soft-plumaged petrels all either established or expanded breeding sites, and active diving petrel burrows observed. Many of these species were assumed to have been prospecting for breeding sites for decades, but had been unable to successfully re-establish in the face of predation by cats and/or rats. Antarctic terns were quick to re-establish breeding sites on cobblestone beaches, moving down from rock-stacks where they had attempted to find protection from predation.

Fig. 6.4: Razor Back at the base of The Isthmus, Macquarie Island. Overgrazing by rabbits removed vegetation by December 2010 (top) but tussock (*Poa foliosa*) was recovering in May 2014 (bottom), three years after aerial baiting targeting rabbits and rodents. Photos: K Springer.

It is relatively early days for a pest-free Macquarie Island and it may be decades before birds, vegetation and invertebrates reach a new equilibrium, which will be different from that of pre-European times as the demonstrated effects of a changing climate become manifest. However, only five years after the last pest mammals were removed, the ecological changes are dramatic. Two introduced bird and three introduced plant species remain on the island, but to date have not demonstrated significant adverse environmental impacts (Scott 2015).

Concluding comments

Increasing global trade and travel mean that biosecurity risks to Australia and its islands are likely to continue to increase. Management of invasive pest species will remain a challenge, and on mainland Australia control is currently the only feasible management option. Eradication of invasive species on increasingly larger and more challenging islands, including inhabited islands, is an area with great potential. While often initiated for ecological benefits, eradicating pests can also have positive outcomes in social and economic terms.

Acknowledgements

Grateful thanks are given to the many people who assisted with information and checking the case studies within this chapter. In particular, Nicolas Carlile, David Priddel, Hank Bower, Sue Bower and Andrew Walsh (Lord Howe Island); Beau Fahne, Duncan Sutherland and Stu Murphy (Phillip Island); and Keith Morris and Andrew Burbidge (Western Australia).

References

Bergstrom DM, Lucieer A, Kiefer K, Wasley J, Belbin L, Pederson TK, Chown SL (2009) Indirect effects of invasive species removal devastate world heritage island. *Journal of Applied Ecology* **46**, 73–81. doi:10.1111/j.1365-2664.2008.01601.x

BirdLife International (2008) *Endemic Birds on Christmas Island are Being Devastated by Introduced Ants*. <http://www.birdlife.org/datazone/sowb/casestudy/399>

Bower S (2016) Breaking bad: 10 years into a projected 30-year weed eradication program on World Heritage listed Lord Howe Island. In *Proceedings of the 20th Australasian Weeds Conference*. 11–15 September, Perth, Western Australia. (Eds R Randall, S Lloyd and C Borger) pp. 201–208. Weeds Society of Western Australia, Perth.

Brothers NP, Eberhard IE, Copson GR, Skira IJ (1982) Control of rabbits *Oryctolagus cuniculus* on Macquarie Island by myxomatosis. *Australian Wildlife Research* **9**, 477–485. doi:10.1071/WR9820477

Brothers N, Bone C, Wilson A (2008) The response of burrow-nesting petrels and other vulnerable bird species to vertebrate pest management and climate change on sub-Antarctic Macquarie Island. *Papers and Proceedings of the Royal Society of Tasmania* **142**, 123–148.

Burbidge AA (2004) Montebello renewal: Western Shield review – February 2003. *Conservation Science Western Australia* **5**, 194–201.

Convention on Biological Diversity (2017) *Island Biodiversity: What's the Problem?* <https://www.cbd.int/island/problem/?sec=alien>

Copson GR, Brothers NP, Skira IJ (1981) Distribution and abundance of the rabbit, *Oryctolagus cuniculus* (L.), at subantarctic Macquarie Island. *Australian Wildlife Research* **8**, 597–611. doi:10.1071/WR9810597

Costin AB, Moore DM (1960) The effects of rabbit grazing on the grasslands of Macquarie Island. *Journal of Ecology* **48**, 729–732. doi:10.2307/2257346

Cumpston JS (1968) *Macquarie Island*. ANARE Scientific Reports, Series A (1), ANARE publication No. 93. Antarctic Division, Department of External Affairs, Melbourne.

DIISE (2015) *The Database of Island Invasive Species Eradications*. Island Conservation, Coastal Conservation Action Laboratory UCSC, IUCN SSC Invasive Species Specialist Group, University of Auckland and Landcare Research New Zealand. <http://diise.islandconservation.org>

Dilley BJ, Schramm M, Ryan PG (2017) Modest increases in densities of burrow nesting petrels following the removal of cats (*Felis catus*) from Marion Island. *Polar Biology* **40**(3), 625–637. doi:10.1007/s00300-016-1985-z

Dowding JE, Murphy EC, Springer K, Peacock AJ, Krebs CJ (2009) Cats, rabbits, myxoma virus, and vegetation on Macquarie Island: a comment on Bergstrom *et al.* (2009). *Journal of Applied Ecology* **46**, 1129–1132. doi:10.1111/j.1365-2664.2009.01690.x

Hutton IN, Parkes JP, Sinclair ARE (2007) Reassembling island ecosystems: the case of Lord Howe Island. *Animal Conservation* **10**, 22–29. doi:10.1111/j.1469-1795.2006.00077.x

Jenkin JF (1975) Macquarie Island, subantarctic. In *Structure and Function of Tundra Ecosystems* (Eds T Rosswall and OW Heal). *Ecological Bulletin (Stockholm)* **20**, 375–397.

Jenkin JF, Johnstone GW, Copson GR (1981) Introduced animal and plant species on Macquarie Island. *Colloque sur les ecosystems subantarctics, Paimpont, CNFRA* **51**, 301–313.

Johnston GC (1966) Macquarie Island and its rabbits. *Tasmanian Journal of Agriculture* **37**, 277–280.

Jones E (1977) Ecology of the feral cat *Felis catus* (L.), (Carnivorus Felidae) on Macquarie Island. *Australian Wildlife Research* **4**, 249–262. doi:10.1071/WR9770249

Kirkwood R, Sutherland DR, Murphy S, Dann P (2014) Lessons from long-term predator control: a case study with the red fox. *Wildlife Research* **41**, 222–232. doi:10.1071/WR13196

Lord Howe Island Board (2016) *Lord Howe Island Rodent Eradication Project: Public Environment Report*. Lord Howe Island Board, Lord Howe Island.

McPhee S, Broomfield T (2004) Strategy for the eradication of foxes from Phillip Island. Unpublished report for Phillip Island Nature Parks, Victoria.

Parkes JP, Macdonald N, Leaman G (2011) An attempt to eradicate feral goats from Lord Howe island. In *Island Invasives: Eradication and Management* (Eds CR Veitch, MN Clout and DR Towns) pp. 233–239. IUCN, Gland, Switzerland.

Robinson SA, Copson GR (2014) Eradication of cats (*Felis catus*) from sub-Antarctic Macquarie Island. *Ecological Management and Restoration* **15**, 34–40. doi:10.1111/emr.12073

Rout TM, Kirkwood R, Sutherland DR, Murphy S, McCarthy MA (2014) When to declare successful eradication of an invasive predator? *Animal Conservation* **17**, 125–132. doi:10.1111/acv.12065

Scott J (2015) Macquarie Island: exciting changes in vegetation and landscapes after eradication of rabbits and rodents. *Australian Plants* **221**, 351–357.

Scott JJ, Kirkpatrick JB (2013) Changes in the cover of plant species associated with climate change and grazing pressure on the Macquarie Island coastal slopes, 1980–2009. *Polar Biology* **36**, 127–136. doi:10.1007/s00300-012-1243-y

Springer K (2011) Planning processes for eradication of multiple pest species on Macquarie Island: an Australian case study. In *Island Invasives: Eradication and Management* (Eds CR Veitch, MN Clout and DR Towns) pp. 228–232. IUCN, Gland, Switzerland.

Springer K (2016) Methodology and challenges of a complex multi-species eradication in the sub-Antarctic and preliminary effects of invasive species removal. *New Zealand Journal of Ecology* **40**, 273–278. doi:10.20417/nzjecol.40.30

Springer K, Carmichael N (2012) Non-target species management for the Macquarie Island Pest Eradication Project. *Proceedings of the 25th Vertebrate Pest Conference*. University of California, Davis, California.

Taylor BW (1955) *The Flora, Vegetation and Soils of Macquarie Island*. ANARE Reports. B(2) No. 19, Botany, 190.

Taylor R (1979) How the Macquarie Island parakeet became extinct. *New Zealand Journal of Ecology* **2**, 42–45.

Terauds A (2009) *Changes in Rabbit Numbers on Macquarie Island 1974–2008*. Tasmania Parks and Wildlife Service, Hobart.

Terauds A, Doube J, McKinlay J, Springer K (2014) Using long-term population trends of an invasive herbivore to quantify the impact of management actions in the sub-Antarctic. *Polar Biology* **37**, 833–843. doi:10.1007/s00300-014-1485-y

7

Australian islands as 'arks' for biodiversity

Andrew A. Burbidge, Sarah Legge and John C.Z. Woinarski

Summary

Of the 46 conservation translocations of vertebrate animals and vascular plants to Australian islands since 1980, 11 have been reintroductions and 36 assisted colonisations. In Australia, islands have been targeted for conservation translocations of vertebrates because they provide a place where major threats – particularly exotic predators (red fox *Vulpes vulpes* and feral cat *Felis catus*) but also introduced rodents and herbivores – are absent or can be eradicated. Considering mammals only, and excluding all relatively recent translocations where success has yet to be verified, these post-1980 island translocations have had a 90% success rate, which is greater than the success rate of 70% for mammal translocations to mainland fenced areas over a similar period. Conservation translocations to Australian islands must, therefore, be considered a successful conservation action. We discuss planning, monitoring, what constitutes a successful translocation and the possible effects of translocations on island biodiversity values. Most translocations have been planned on a case by case basis. As there are a limited number of islands available for candidate species in terms of suitability of habitat and island area, nation-wide planning of future translocations is recommended.

Introduction

Translocation of Australian animals as a conservation action has been a long-standing practice (Armstrong *et al.* 2015), one that has, on the whole, been remarkably successful. In this chapter we review conservation translocations of indigenous Australian animals to Australian islands, provide case studies and discuss the future of this conservation action.

Island translocations

Terminology

The IUCN/SSC Guidelines for Reintroductions and Other Conservation Translocations Version 1.0 (IUCN/SSC 2013) recommends standard terminology for translocations. Two categories are relevant here: a 'conservation translocation' of a species to where it formerly occurred is a 'reintroduction' and a 'conservation translocation' of a species to where it did not occur naturally is an 'assisted colonisation'. To qualify as an extant population that can be included in IUCN Red List assessments of the taxon's conservation status, islands hosting assisted colonisations need to be geographically close to the original distribution of the taxon, although geographically close is inexactly defined (IUCN Standards and Petitions Subcommittee 2016). Of the 46 translocations in Table 7.1, 11 were reintroductions.

Ontology of translocations to islands

Humans have been translocating animals and plants to islands for thousands of years, and there are several prehistoric examples for islands in the Australian region (Heinsohn 2002). A particularly notable example is the introduction of the dingo

Table 7.1. Conservation translocations of indigenous vertebrates and vascular plants to Australian islands, 1980 to mid-2016

Species	Status	Island	State/ territory	Area (km²)	Date	R/AC	S/F	Nat. occ.
Abutilon julianae, Norfolk Island abutilon	CR	Norfolk	Com	1.9	Late 1980s-	R	S	1
Hibiscus insularis, Phillip Island hibiscus	CR	Norfolk	Com	1.9	Late 1980s-	R	S	1
Achyranthes margaretarum, Phillip Island chaff flower	CR	Norfolk	Com	1.9	Late 1980s-	AC	S	1
Dysoxylum bijugum, sharkwood	VU	Phillip	Com	34.6	Late 1980s-	AC	S	1
Euphorbia norfolkiana, Norfolk Island euphorbia	CR	Phillip	Com	34.6	Late 1980s-	AC	S	1
Meryta latifolia, broad-leafed meryta	CR	Phillip	Com	34.6	Late 1980s-	AC	S	1
Cordyline obtecta, ti, Norfolk Island cabbage tree	VU	Phillip	Com	34.6	Late 1980s-	AC	S	1
Ctenotus lancelini, Lancelin Island skink	VU	Favourite	WA	0.03	2002	AC	S	1
Callirallus philippensis andrewsi, Cocos buff-banded rail	EN	Horsburgh	Com	1	2013	R	*	1
Atrichornis clamosus, noisy scrub-bird	EN	Bald	WA	8.09	1992, 1993, 1994	AC	S	0
Eremiornis carteri, spinifexbird	LC	Hermite	WA	9.85	2010–11	R	S	>2[a]
Malurus leucopterus edouardi, white-winged fairy-wren (Barrow Island subspecies)	VU	Hermite	WA	9.85	2010–11	R	S	1
Dasyurus hallucatus, northern quoll	EN	Astell	NT	11.43	2003	AC	S	>15
		Pobasso	NT	3.96	2003	AC	S	
Parantechinus apicalis, dibbler	EN	Escape	WA	0.27	1998–2001	AC	S	2
		Gunton	WA	0.9	2015	AC	*	
Sarcophilus harrisii, Tasmanian devil	EN	Maria	TAS	104	2012, 2013	AC	*	1
Isoodon auratus, golden bandicoot	VU	Doole	WA	2.2	2011	AC	S	7
		Guluwuru	NT	76.25	2007, 2008	AC	S	
		Hermite	WA	9.85	2009	R	S	
		Raragala	NT	92.32	2008–09	AC	S	
Perameles bougainville, western barred bandicoot	VU	Faure	WA	47.95	2005	R	S	2
Perameles gunnii, eastern barred bandicoot	VU	Churchill	VIC	0.57	2015	AC	*	2
Macrotis lagotis, bilby	VU	Thistle	SA	41	1997	AC	S	0
Bettongia lesueur, boodie	NT(CD)	Alpha	WA	1.04	2011	AC	S	3
		Boodie	WA	2.05	1993	R	S	
		Faure	WA	51.48	2002	AC	S	

Species	Status	Island	State/ territory	Area (km²)	Date	R/AC	S/F	Nat. occ.
Bettongia penicillata, woylie	CR	Baird Bay island	SA	0.16	1982	AC	F	0
		Saint Peter	SA	34.39	1989	R	S	
		Venus Bay, island A	SA	0.17	1980	AC	S	
		Wedge	SA	9.47	1983	AC	S	
Potorous gilbertii, Gilbert's potoroo, ngilkat	CR	Bald	WA	8.09	2005–07	AC	S	0
		Michaelmas	WA	0.92	2016	AC	*	
Lagorchestes conspicillatus, spectacled hare-wallaby	NT	Hermite	WA	9.85	2010	R	S	1
Lagorchestes hirsutus, mala, rufous hare-wallaby	VU	Trimouille	WA	4.99	1998	AC	S	2
Petrogale persephone, Proserpine rock-wallaby	EN	Hayman	QLD	3.51	1998 and later	AC	S	0
Petrogale rothschildi, Rothschild's rock-wallaby	LC	West Lewis	WA	20.8	1982	AC	S	3
Lagostrophus fasciatus, banded hare-wallaby	VU	Faure	WA	51.48	2004-	AC	S	2
Leggadina lakedownensis, kerakenga, northern short-tailed mouse	LC	Serrurier	WA	3	1996	AC	S	1
Leporillus conditor, greater stick-nest rat	NT(CD)	Faure	WA	51.48	2006	AC	F	1
		Reevesby	SA	3.73	1990	AC	S	
		Saint Peter	SA	34.39	1993	AC	S	
		Salutation	WA	1.66	1990	AC	S	
Pseudomys fieldi, djoongari, Shark Bay mouse	VU	Doole	WA	2.20	1993–2001	AC	F	1
		Faure	WA	51.48	2002	R	S	
		North West	WA	1.23	1999, 2000	AC	S	
Total: 30 taxa		46 island translocations				11 R, 35 AC	3 F, 38 S	

[a] Resident on Barrow and Thevenard Islands; also recorded on several other near coastal Pilbara islands.
Status: IUCN Red List status as evaluated by the Mammal Action Plan (Woinarski *et al.* 2014) or Bird Action Plan (Garnett *et al.* 2011), with the status of the single reptile and vascular plants being as shown in the EPBC Act lists. CR = Critically Endangered, EN = Endangered, VU = Vulnerable, NT(CD) = Near Threatened (Conservation Dependent).
State/territory: NT = Northern Territory, QLD = Queensland, NSW = New South Wales, VIC = Victoria, TAS = Tasmania, SA = South Australia, WA = Western Australia, Com = Commonwealth territory.
R/AC: R = reintroduction, AC = assisted colonisation.
S/F: S = successful translocation, F = failed (did not establish), * recent translocation, success yet to be established.
Nat. occ. = number of islands where the species occurred naturally and persists.

(*Canis familiaris*) to Australia some 4000 years ago (Savolainen *et al.* 2004; Smith and Savolainen 2015), and its subsequent introduction to several islands in northern Australia (Abbott 1979; Abbott and Burbidge 1995). Cook deposited pigs (*Sus scrofa*) in 'Van Diemen's Land' in 1777 (Brodie 2016). European settlers have been transporting animals to Australian islands since at least the early 1800s: some were domesticated and semi-domesticated animals valued as companions or as farm animals, some were species prized for a range of perceived amenity values, from snake-eating birds to confiding (and melodious) songbirds (Lever 1992), some were accidental releases of 'hitchhikers' such as black rats (*Rattus rattus*) and house mice (*Mus musculus*), and some were pets that escaped and became wild, as was the case with the sugar glider (*Petaurus breviceps*) in Tasmania (Gunn 1851). Many non-indigenous plants have established in Australia, some on islands (Harris and McKenney 1999; Lohr and Keighery 2016; Lohr *et al.* 2016).

The use of Australian islands as arks for the conservation of threatened species probably commenced with well-meaning releases of koalas *Phascolarctos cinereus* to French Island, Victoria in the late 1800s, and in the 1920s and 1930s to six islands in Queensland, four in Victoria and Kangaroo Island in South Australia. In the same vein, superb lyrebirds (*Menura superba*) were translocated to Tasmania in several releases during 1934–49, to safeguard them from the perceived impacts of fox predation on the mainland (Lever 1992). More recently, with an upsurge in scientific research into and management of threatened species, and increasing evidence of risks of *in situ* extinction, translocation to islands has become one of several strategies used to conserve species. Translocations have become increasingly strategic, regulated, carefully planned, documented and monitored (Armstrong *et al.* 2015).

Conservation translocations to Australian islands

In Australia, islands have been targeted for conservation translocations because they provide a place where major threats are absent or can be eradicated. For threatened vertebrates, these threats have particularly been the non-indigenous predators (red fox *Vulpes vulpes*, feral cat *Felis catus* and black rat) and introduced herbivores. Islands may offer attractive benefits for translocations: in general, ensuring biosecurity for island-translocated populations is relatively cheaper to maintain than ongoing management of introduced species in mainland conservation reserves, or maintaining the security and introduced predator-free status of fenced exclosures (mainland islands). For native species that are particularly vulnerable to the impacts of invasive species, broad-scale intensive predator control may be inadequate, and islands may provide larger areas (and hence maintain larger populations) that are free from non-indigenous species than can fenced areas on the mainland. In addition, most of Australia's >8000 islands, with some important exceptions, are uninhabited and publicly or Aboriginal-owned. Because many islands have no or relatively few competing land uses, conservation translocations to islands may also be more straightforward to plan and implement than would be the case for some comparable mainland areas.

The *Feral Animals on Offshore Islands Database* (DoE 2016), a compilation of information provided by state and territory conservation agencies, includes >2000 records of non-indigenous species on Australian islands >20 ha. Our database (contact A.A. Burbidge) summarising translocations of Australian indigenous plants and vertebrates to islands since the 1800s includes translocations of nine species of vascular plants, one frog species, seven reptile species, 30 bird species and 44 mammal species. We are not aware of any conservation translocations of threatened invertebrates to Australian islands.

At least 46 conservation translocations, involving 23 vertebrate species and seven plant species, took place to Australian islands from 1980 to mid-2016. There have been few translocations of threatened plants to islands. The only recent ones have been between Norfolk Island and nearby Phillip

Box 7.1: Exchanges of threatened plants between Norfolk Island and nearby Phillip Island

John Woinarski

There are few cases of translocations of threatened plants to Australian islands. The most notable involves a series of translocations between Norfolk Island (36 km^2) and the nearby (6 km distant) Phillip Island (195 ha). With their most recent separation occurring 10 000–6000 years ago, the two islands have shared a long geological history and hence have closely related (and highly endemic) biodiversity. Both have been buffeted by marked environmental changes associated with Norfolk's waves of human settlement. But the major threat factors have differed markedly between the two islands. Norfolk has been extensively cleared and settlers introduced many weeds and – particularly detrimentally for many native plants – the black rat and pig. Because of its small size and rugged nature, Phillip Island was not settled nor deliberately cleared, but Norfolk Islanders introduced pigs, goats and rabbits to it. These multiplied rapidly and consumed almost all plant matter, other than in a few rugged extremities: by the 1860s, they had turned the island to a desolate moonscape (Coyne 2009). The pigs and goats died or were shot out by 1915, and extensive control programs for rabbits in the 1980s finally succeeded in eradicating them. Luckily, some native plants survived on cliffs and the few pockets of Phillip Island that were least accessible to introduced mammals. These included some plant species that had been lost (or were never recorded) from Norfolk Island, including three critically endangered species – the Norfolk Island abutilon (*Abutilon julianae*), Phillip Island chaffy tree (*Achyranthes margaretarum*) and Phillip Island hibiscus (*Hibiscus insularis*). In a carefully planned conservation management program, these have been successfully reintroduced (or introduced) to Norfolk Island. Conversely, the recovery of Phillip Island has been bolstered by a series of successful recent plant introductions from Norfolk, to reintroduce plant species lost from Phillip Island, to bolster their small surviving populations or as introductions to provide a second population. These Norfolk–Phillip translocations have included the critically endangered Norfolk Island euphorbia (*Euphorbia norfolkiana*) and shade tree (*Meryta latifolia*), and the vulnerable ti (*Cordyline obtecta*) and sharkwood (*Dysoxylum bijugum*) (Mills 2009; DNP 2010).

Island, the latter having been severely degraded by rabbits (*Oryctolagus cuniculus*) until their eradication in 1988 (Mills 2009) (Box 7.1).

Translocations of threatened taxa to islands are becoming increasingly sophisticated, including planning for and monitoring the maintenance of genetic diversity, consideration of possible detrimental impacts (on source populations and their environments and on recipient environments and their species), improved policy and regulation, consideration of translocated island populations as constituents of a broader metapopulation, and better monitoring.

In this chapter, we briefly document recent (from 1980) cases of conservation translocations of native species to Australian islands, review the conservation value of these translocations (and consider the meaning of translocation 'success'), identify issues affecting success or potential collateral detriment, and consider the likely future for island translocations.

How successful are conservation translocations to Australian islands?

Of the 46 conservation translocations (Table 7.1), three failed to establish and four are recent with success yet to be established (see Fig. 7.1), leaving 39 that were 'successful' (Box 7.2). These conservation translocations have resulted in the establishment of 39 subpopulations (as defined by IUCN 2001) of 30 species, almost all of them listed as Threatened. Considering mammals only, and excluding the few relatively recent translocations where success has yet to be established, island translocations have a 90% success rate (n = 34 translocations), which is greater than the success rate for mammal translocations to mainland fenced areas (70%; n = 54 translocations) over a similar period (Table 7.1, unpub.

Fig. 7.1: Release of Cocos buff-banded rail (*Gallirallus philippensis andrewsi*) to Horsburgh Island in the southern atoll of Cocos (Keeling) Islands group. Photo: Caitlyn Pink.

data). Conservation translocations to Australian islands must therefore be considered a successful conservation action. Successful conservation actions are particularly important in a world where threats continue to increase and biodiversity continues to decline (Butchart *et al.* 2010).

The reasons for the three cases of failure to establish are sometimes unclear. For the small island in Baird Bay, South Australia, woylies *Bettongia penicillata* failed to establish because of predation by a fox that crossed to the island over a sand bar that is exposed at extreme low tide (Robinson *et al.* 1996). The failure of djoongari (*Pseudomys fieldi*) to establish on Doole Island, Western Australia, after what appeared initially to be a success, has been attributed to inundation during a cyclone, as the translocated animals appeared to prefer coastal habitats (including mangrove hollows) while they persisted (Morris *et al.* 2000). The cause of failure of greater stick-nest rats (*Leporillus conditor*) to establish on Faure Island in Shark Bay, Western Australia, was unclear (Box 7.3).

Monitoring and how to evaluate success

Some monitoring of translocations is now standard practice, including consideration of population trends, reproduction, health and genetic heterogeneity of the translocated population, and some impacts of the translocation on species native to the island. For animals, monitoring is often staged, with initial monitoring used to report on the fate of released individuals, usually by radio-tracking. Often the transmitters are fitted with a mortality mode that changes the pulse rate after a certain number of hours if the transmitter does not move. Locating an animal soon after death may allow the cause of death to be ascertained, which is particularly important if the translocation is not successful or if a proportion die. Longer-term monitoring may be via trapping, track counts or observational

Box 7.2: Montebello Islands, Western Australia

Andrew Burbidge and Keith Morris

The Montebello Islands, situated off the north-western coast of Australia, comprise ~180 islands varying in size from Hermite Island (985 ha) and Trimouille Island (522 ha) to a few square metres. Three nuclear weapons were tested by the British in the Montebello Islands in 1952 and 1956. Feral cats and black rats had established by 1910, presumably via the pearling industry. As a result of these factors, three indigenous mammals and one or two species of birds had become locally extinct. A major project from 1996 to 2001 by the then Western Australian Department of Conservation and Land Management eradicated both invasive mammals. In 1998, 30 mala (*Lagorchestes hirsutus* central Australian subspecies) were translocated from the Tanami Desert, Northern Territory to Trimouille Island (Langford and Burbidge 2001). Monitoring (Fig. 7.2) since has shown that mala occupy the whole island. Trimouille Island now protects the only wild subpopulation of the subspecies and the assisted colonisation has enabled its conservation status to be upgraded from Extinct in the Wild to Endangered. In 1999 and 2000 djoongari (*Pseudomys fieldi*) were established on North West Island (135 ha) from a captive-bred population at Perth Zoo. From 2009 to 2011, under environmental offset conditions imposed on the Gorgon LNG project on Barrow Island, spectacled hare-wallabies (*Lagorchestes conspicillatus conspicillatus*), golden bandicoots (*Isoodon auratus barrowensis*), white-winged fairy-wrens (*Malurus leucopteris edouardi*) and spinifexbirds (*Eremiornis carteri*) were reintroduced to Hermite Island from Barrow Island, and boodies (*Bettongia lesueur*) were introduced to Alpha Island (104 ha). All these translocations have been successful, contributing to the improved conservation status of the species (Parks and Wildlife 2014, 2015). There is potential to translocate other species, such as the water rat (*Hydromys chrysogaster*) to islands in the group to complete this fauna reconstruction project.

Fig. 7.2: Radio-tracking mala (*Lagorchestes hirsutus* central Australian subspecies) on Trimouille Island (Montebello Archipelago), Western Australia. Photo: A.A. Burbidge.

Box 7.3: Faure Island, Western Australia

Sarah Legge, Chantelle Jackson, Michael Smith

Faure Island (5800 ha) lies in the Shark Bay World Heritage Area, only 6 km from the mainland. The island's vegetation comprises grassy shrublands on sandplains, samphire flats and spinifex coastal dunes. The island had been an operating pastoral lease for over 100 years until its purchase by Paruna Sanctuary Ltd (which later became Australian Wildlife Conservancy) in 1999. At that time, the island supported several thousand sheep and goats, a small herd of horses, house mice and feral cats. With assistance from the then WA Department of Environment and Conservation, the cats were removed in 2001 (Algar *et al.* 2010) and the larger herbivores were removed by 2004. Records from skeletal remains and old owl pellets (Baynes 2008) indicated that the pre-European mammalian fauna probably comprised three species (woylie, western barred bandicoot *Perameles bougainville* and djoongari). A single record of a fourth species (pale field rat *Rattus tunneyi*) may have derived from a much older fauna. The translocation plan for the island was developed via expert consultation in 2001, and was based partly on knowledge of the pre-European species but also on conservation need and habitat suitability of other candidate species. Of the five species selected, two were reintroductions (djoongari, western barred bandicoot) and three were assisted colonisations (banded hare-wallaby *Lagostrophus fasciatus*, boodie and greater stick-nest rat, the last being the only unsuccessful translocation). The boodie was substituted for woylies because, at that time, it was perceived to have greater conservation need. All three assisted colonisation species were previously present on the mainland adjacent to Faure Island and on other Shark Bay islands (Baynes 1990). The initial translocations of all species were carried out between 2002 and 2006, with supplementary translocations of some species since then. Djoongari, western barred bandicoots, banded hare-wallabies and boodies are now well established on the island. To date, there has been no evidence of reincursions by any of the introduced species that were removed.

counting. There is no accepted standard measure of a 'successful' translocation; however, most translocations have been deemed successful if the target species establishes, breeds and expands to fill suitable habitat over several years. The length of time this may take will depend on the generation length and fecundity of the species concerned, but is usually less than five years. Monitoring subsequent to the establishment period has been highly variable; populations on some islands have been monitored at least annually, but in other cases the monitoring has been irregular and infrequent.

Woinarski *et al.* (2014), when evaluating the Red List conservation status of mammal taxa translocated to islands and mainland islands up to 2012, used 'wild' (not closely confined in captivity), 'self-sustaining' (a population of >50 mature individuals established and reproducing for five years) and 'large' (large enough to sustain a viable subpopulation of the translocated taxon and where management is limited to monitoring and the control or eradication of introduced animals) as criteria that all needed to be met for including a translocated subpopulation in an evaluation of the conservation status of a taxon. We have followed those guidelines in assessing whether a conservation translocation is judged to be successful (see Table 7.1) and we note their congruence with the latest IUCN guidelines for inclusion of translocated populations in status assessments (IUCN/SSC 2013). The federal Department of Sustainability, Environment, Water, Population and Communities has developed a policy statement on the translocation of *Environment Protection and Biodiversity Conservation Act 1999* (EPBC Act) listed species (DSEWPaC 2013).

In the long term, conservation translocations to Australian islands can be deemed successful only if the translocation has contributed towards preventing the extinction of the species concerned. As well, the introduced subpopulation should be self-sustaining without the need for reinforcement (terminology of IUCN/SSC 2013) or feeding, and the assisted colonisation must have no major deleterious impacts on intrinsic values at the destination island.

Choosing islands, species and individuals for translocation

To date, assisted colonisations to Australian islands have been largely planned on a case by case basis. Abbott (2000) examined Western Australian islands and found that there were 254 islands of >100 ha in size, with 90% in tropical seas. He presented a procedure for selecting appropriate islands for translocation, emphasising those islands without significant biological characteristics such as large colonies of nesting seabirds and the occurrence of native mammal species, but also considering island area, degree of isolation and the habitats present. He provided a prioritised list of candidate species of native terrestrial mammals, based on size of geographical range, natural absence from any Western Australian island, and the degree to which current threatening processes are not being addressed successfully.

There is a limited number of islands available for candidate species in terms of suitability of habitat and island area, so nation-wide planning of future translocations is desirable. A translocation of one species may make it unlikely or challenging to translocate a second species to the same island. For example, translocating a predator may impact other species and translocating a species with similar food requirements to one already there may lead to competition and loss of the introduced or resident species. On the other hand, Ostendorf *et al.* (2016) found that a prior translocation of southern hairy-nosed wombats (*Lasiorhinus latifrons*) to Wedge Island, South Australia, provided, via their large warrens, additional shelter habitat for black-footed rock-wallabies (*Petrogale lateralis*). Past translocations to islands have limited the options for the future. Choosing between islands and mainland fenced exclosures for a translocation also requires consideration of relative costs and benefits. For example, some species may be unsuitable for assisted colonisations to islands because of lack of suitable habitat on islands or the lack of suitably large islands with appropriate habitat.

Once decisions have been made about which species is proposed to be translocated to a specific island, selecting individuals to be translocated is the next critical step. The number of founder animals and the genetic status of the source population can affect the longer-term genetic viability of the translocated population. DNA samples should be collected from animals before they are released in the new site so that the genetic profile of the new subpopulation can be established in the medium to long term. Reports of the genetic status of translocated subpopulations are scarce, something that should be corrected. Ottewell *et al.* (2014) assessed the genetic diversity of golden bandicoots (*Isoodon auratus*) translocated from Barrow Island to Hermite and Doole Islands and to a mainland island, and found no significant loss of genetic diversity in any of the wild-born populations compared with the source population. Further, they found no significant increase in inbreeding or relatedness among wild-born individuals compared with founder populations two years post-translocation. Cardoso *et al.* (2009) similarly found that two translocated island subpopulations of the northern quoll (*Dasyurus hallucatus*) showed a slight but non-significant reduction in genetic diversity compared with the mainland source populations.

In both these cases the source population was genetically diverse. This high diversity may not always be the case, however, as some remnant mammal populations on smaller islands, as well as some remnant mainland populations, are genetically depauperate (Eldridge *et al.* 1999, 2004a, 2004b; Cardoso *et al.* 2009). Genetic considerations should always be a factor in choosing a source population and deciding on a founder number, and genetic diversity should be monitored as a translocated population establishes.

The rationale behind key decisions underpinning any translocation are generally formalised in a peer-reviewed Translocation Proposal. Such proposals are now a standard requirement of most conservation agencies across Australia when approvals or permits are being sought for a translocation. As well as summarising much ecological and contextual information, Translocation Proposals are primarily designed to:

- justify the translocation in terms of conservation need;
- justify the selection of the donor population and the host site, detailing the methods involved including the selection of animals (number, gender, age, genetic provenance, etc.) to be translocated;
- assess risks to the donor population (e.g. arising from the removal of several individuals) as well as risks to the host site (e.g. disease transmission, competition for *in situ* species, etc.);
- outline mitigation measures for any significant risks;
- set out the post-translocation monitoring protocols and the criteria against which the success of the translocation will be judged.

To date, most recent conservation translocations to Australian islands have involved islands that are uninhabited by people and managed by a single government agency. In the future, translocations to populated islands are more likely, and these proposals will require a greater level of community involvement, support and collaboration.

Effects of translocations on island ecosystems

Assisted colonisations (introductions) to islands need careful consideration. Islands are frequently of high intrinsic value for nature conservation as intact or near-intact ecosystems. Some islands harbour endemics and species that are extinct or threatened on the mainland. Australian offshore islands have proven crucial for vertebrate conservation, with eight species of mammals that formerly occurred on the mainland becoming restricted to Australian islands (Johnson 2006; Burbidge *et al.* 2009). Many islands retain populations of animals that have declined on the mainland (Woinarski *et al.* 2014), and many have other values such as being seabird and sea turtle breeding sites.

Island faunas and floras are particularly vulnerable to introductions of alien species (Medina *et al.* 2011). With assisted colonisations to islands, the introduction of a new species to what is often a relatively simple ecosystem must have effects on the island's species and their interrelationships. The deleterious effects of some of the koala introductions to islands are well documented. For example, Masters *et al.* (2004) found that, from a release of only 18 koalas between 1923 and 1925, the estimated 27 000 koalas on Kangaroo Island have had a dramatic effect on its vegetation, with one tree, *Eucalyptus viminalis* subsp. *cygnetensis*, suffering island-wide degradation and local extinctions. The introduction of the sugar glider to Tasmania from Victoria in ~1835 (Gunn 1851) has had significant deleterious effects on hollow-nesting species, especially the swift parrot (*Lathamus discolor*) (Stojanovic *et al.* 2014; Heinsohn *et al.* 2015). The assisted colonisation of the Tasmanian devil (*Sarcophilus harrisii*) to Maria Island, Tasmania (Box 7.4), has led to predation on the Cape Barren goose (*Cereopsis novaehollandiae*), itself an introduced species to the island, and major deleterious impacts on at least one breeding colony of the little penguin (*Eudyptula minor*) (DPIPWE 2016). Some other mammal introductions have had a major detrimental impact on vegetation, e.g. the tammar wallaby *Macropus eugenii* to Granite Island, South Australia (Robinson *et al.* 1996). It is possible that translocations could affect other fauna, e.g. introducing a predator to an island with important seabird populations, or an important turtle rookery. The effects of many assisted colonisations are either unstudied or less dramatic, but some changes will have occurred.

There is a lack of information on the effects of conservation translocations to Australian islands, something that needs attention. Comprehensive surveys of the native biota of a potential host island should precede any translocation. A risk assessment process, including possible impacts on the fauna (including invertebrates) and flora of islands, should be considered as part of an assisted coloni-

Box 7.4: Translocations to Maria Island, Tasmania

Menna Jones and Vince Scoleri

Maria Island (115 km^2), a national park since 1972, lies 3.5 km off Tasmania's east coast. Dry sclerophyll woodland and forest dominate the vegetation, with pockets of wet forest, heath, coastal shrub and pasture. Mammals present at European settlement in 1825 were Tasmanian pademelons (*Thylogale billardierii*), long-nosed potoroos (*Potorous tridactylus*), wombats (*Vombatus ursinus*) and ringtail possums (*Pseudocheirus peregrinus*), with swamp rats (*Rattus lutreolus*) described as the abundant small mammal. Europeans introduced cats, black rats and house mice. Invasive feral cats and black rats may limit swamp rats, which have declined to residual pockets in some wet forest patches on higher ground. In the hope of supporting a putative introduction of thylacines (*Thylacinus cynocephalus*), in 1971 eastern grey kangaroos (*Macropus giganteus*), Bennett's wallabies (*Macropus rufogriseus*), brushtail possums (*Trichosurus vulpecula*), bettongs (*Bettongia gaimardi*) and bandicoots (eastern barred *Perameles gunnii* and southern brown *Isoodon obesulus*) were introduced. Bettongs and eastern barred bandicoots have disappeared, perhaps because of low habitat suitability and extended periods of drought. Cape Barren geese (*Cereopsis novaehollandiae*) were introduced to mitigate species decline. Tasmanian devils (*Sarcophilus harrisii*) were translocated in 2012–13 to establish a free-ranging insurance population isolated from the devil facial tumour disease and in 2016 numbered close to 100, the expected carrying capacity (DPIPWE 2016). Devils have significantly reduced possum activity and goose breeding success, and may offset the need to cull over-abundant macropods on the pasture. Devils may alter habitat use by cats, imposing resource limitations and thus reducing feral cat population size. Feral cat eradication is feasible but may lead to predatory release of black rats, with consequent increased predation on small mammals and birds.

sation planning and approval process, and potentially affected species should be the subject of ongoing monitoring. Such risk assessment should include not only the potential direct impacts of the translocated species, but also those of any parasites or diseases that may accompany the translocated stock. Even reintroductions could carry similar risks if other components of the original ecosystem are now missing, such as its original predator(s), its key original foraging resource, the pollinator in the case of plants, and/or if novel threats, including diseases, are operating on the island that affect the way the reintroduced species may interact with the new ecosystem.

Conservation translocations to islands: the future

There have been calls that assisted colonisation should be a 'last resort' action, partly because of past ill-conceived translocations, partly because of uncertainties associated with climate change predictions and species responses to it, and partly because there is insufficient understanding of the habitat requirements of species and the effects of translocations on ecosystem function (Seddon *et al.* 2009, 2015). Seddon *et al.* (2009) concluded that these concerns will lead to a substantial risk that prematurely embracing assisted colonisation will initiate a new era of ill-conceived translocations. However, in Australia, with its continuing unacceptable decline and loss of elements of its biodiversity, conservation translocations, including to islands, will continue to be an important conservation action, noting that careful review of proposed translocations to islands is necessary before approvals are given.

The continuing nature conservation value of islands for Australian animals depends on establishing and implementing adequate biosecurity measures (Nias *et al.* 2010; Chapter 6). Since 2000, the only Australian vertebrate extinctions have been of island-endemic species: the Christmas Island pipistrelle (*Pipistrellus murrayi*), almost certainly due to the impacts of introduced species, the Christmas Island forest skink (*Emoia nativitatis*), almost certainly due to introduced predators, and the Bramble Cay melomys (*Melomys rubicola*), probably due to sea level rise and storm surge (Woinarski *et al.* 2017; Fulton 2017). There have been many cases of loss of important island mammal subpopulations (Burbidge and Manly 2002; Woinarski *et al.* 2011, 2014) as well as of seabirds (Jones *et al.* 2008) due to incursions of non-indigenous species.

Additional conservation translocations to islands are being planned and will be considered in the future, subject in some cases to the eradication of invasive species or management of other threats. Lord Howe Island (1455 ha) is proposed for reintroductions, including the Lord Howe phasmid (*Dryococelus australis*), once black rats are eradicated (Carlile *et al.* 2009). Feral cats, goats (*Capra hircus*) and sheep (*Ovis aries*) are currently being eradicated from the 62 000 ha Dirk Hartog Island and up to 12 mammal species are proposed to be reintroduced and introduced there (http://www.sharkbay.org.au/dirk-hartog-island-ecological-restoration-project.aspx). Notably, if successful, Dirk Hartog will be the largest island from which feral cats have been eradicated, with the next largest being Marion Island (South Africa) at 29 000 ha then Macquarie Island at 12 780 ha (Parkes *et al.* 2014).

Most recent conservation translocations to islands have been carried out by government conservation agencies, with one exception – the Australian Wildlife Conservancy. This non-government organisation is carrying out many translocations to fenced enclosures on the mainland and has translocated five species of mammals (one unsuccessfully) to Faure Island in Shark Bay, in collaboration with the Western Australian government (Richards 2007). The Australian Wildlife Conservancy could access Faure Island because it is a pastoral lease available for private purchase, unlike most of islands in Australia. Apart from this exception, there has been limited review or involvement by non-government organisations or the broader public in island translocations.

Concluding comments

Conservation translocations to Australian islands will remain an important conservation action but, given that one translocation affects the possibility of others, a nationally coordinated approach is strongly recommended.

Acknowledgements

We thank reviewers Cheryl Gole and Peter Copley and editor Dorian Moro for their helpful comments on a draft of this chapter.

References

Abbott I (1979) Aboriginal man as an exterminator of wallaby and kangaroo populations on islands around Australia. *Oecologia* **44**, 347–354. doi:10.1007/BF00545239

Abbott I (2000) Improving the conservation of threatened and rare mammal species through translocation to islands: case study Western Australia. *Biological Conservation* **93**, 195–201. doi:10.1016/S0006-3207(99)00144-5

Abbott I, Burbidge AA (1995) The occurrence of mammal species on the islands of Australia: a summary of existing knowledge. *CALMScience* **1**(3), 259–324.

Algar D, Angus GJ, Brazell RI, Gilbert C, Withnell GB (2010) Eradication of feral cats on Faure Island, Western Australia. *Journal of the Royal Society of Western Australia* **93**, 133–140.

Armstrong D, Hayward M, Moro D, Seddon P (Eds) (2015) *Advances in Reintroduction Biology of Australian and New Zealand Fauna*. CSIRO Publishing, Melbourne.

Baynes A (1990) The mammals of Shark Bay, Western Australia. In *Research in Shark Bay. Report of the France-Australe Bicentenary Expedition Committee* (Eds PF Berry, SD Bradshaw and BR Wilson) pp. 313–325. Western Australian Museum Press, Perth.

Baynes A (2008) The original non-volant land mammal fauna of Faure Island, Shark Bay,

Western Australia. *Records of the Western Australian Museum* (Supplement No. 75), 25–31. doi:10.18195/issn.0313-122x.75.2008.025-031

Brodie N (2016) *1787.* Hardie Grant Books, Melbourne.

Burbidge AA, Manly BFJ (2002) Mammal extinctions on Australian islands: causes and conservation implications. *Journal of Biogeography* **29**, 465–473. doi:10.1046/j.1365-2699.2002.00699.x

Burbidge AA, McKenzie NL, Brennan KEC, Woinarski JCZ, Dickman CR, Baynes A, Gordon G, Menkhorst PW, Robinson AC (2009) Conservation status and biogeography of Australia's terrestrial mammals. *Australian Journal of Zoology* **56**, 411–422. doi:10.1071/ZO08027

Butchart SHM, Walpole M, Collen B, van Strien A, Scharlemann JPW, Almond REA, Baillie JEM, Bomhard B, Brown C, Bruno J, Carpenter KE, Carr GM, Chanson J, Chenery AM, Csirke J, Davidson NC, Dentener F, Foster M, Galli A, Galloway JN, Genovesi P, Gregory RD, Hockings M, Kapos V, Lamarque J-F, Leverington F, Loh J, McGeoch MA, McRae L, Minasyan A, Morcillo MH, Oldfield TEE, Pauly D, Quader S, Revenga C, Sauer JR, Skolnik B, Spear D, Stanwell-Smith D, Stuart SN, Symes A, Tierney M, Tyrrell TD, Vié J-C, Watson R (2010) Global biodiversity: indicators of recent decline. *Science* **328**, 1164–1168. doi:10.1126/science.1187512

Cardoso MJ, Eldridge MDB, Oakwood M, Rankmore B, Sherwin WB, Firestone KB (2009) Effects of founder events on the genetic variation of translocated island populations: implications for conservation management of the northern quoll. *Conservation Genetics* **10**, 1719–1733. doi:10.1007/s10592-008-9774-z

Carlile N, Priddel D, Honan P (2009) The recovery programme for the Lord Howe Island phasmid (*Dryococelus australis*) following its rediscovery. *Ecological Management and Restoration* **10**(S1), S124–S128. doi:10.1111/j.1442-8903.2009.00450.x

Coyne P (2009) *Incredible! The Amazing Story of the Birth and Rebirth of a Natural Treasure: Phillip Island, South Pacific.* Petaurus Press, Canberra.

DNP (2010) *Norfolk Island Region Threatened Species Recovery Plan.* Director of National Parks, Department of the Environment, Water, Heritage and the Arts, Canberra.

DoE (2016) *Feral Animals on Offshore Islands Database.* Department of the Environment, Canberra. <http://www.environment.gov.au/biodiversity/invasive-species/feral-animals-australia/offshore-islands>

DPIPWE (2016) *Save the Tasmanian Devil Program: 2015 Annual Report.* Department of Primary Industries, Parks, Water and Environment, Hobart.

DSEWPaC (2013) *Environment Protection and Biodiversity Conservation Act 1999 (Cth) Policy Statement. Translocation of Listed Threatened Species: Assessment under Chapter 4 of the EPBC Act.* Department of Sustainability, Environment, Water, Population and Communities, Canberra. <http://www.environment.gov.au/system/files/resources/c0463a3b-cf06-44a7-a7c6-76b488321561/files/epbc-act-policy-translocation.pdf>

Eldridge MDB, King JM, Loupis AK, Spencer PBS, Taylor AC, Pope LC, Hall GP (1999) Unprecedented low levels of genetic variation and inbreeding depression in an island population of the black-footed rock-wallaby. *Conservation Biology* **13**, 531–541. doi:10.1046/j.1523-1739.1999.98115.x

Eldridge MDB, Rummery C, Bray C, Zenger KR, Browning TL, Close RL (2004a) Genetic analysis of a population crash in brush-tailed rock-wallabies from Jenolan Caves, south-eastern Australia. *Wildlife Research* **31**, 229–240. doi:10.1071/WR03030

Eldridge MDB, Kinnear JE, Zengler KR, McKenzie LM, Spencer PBS (2004b) Genetic diversity in remnant mainland and 'pristine' island populations of three endemic Australian macropodids (Marsupialia): *Macropus eugenii, Lagorchestes hirsutus* and *Petrogale lateralis. Conservation Genetics* **5**,325–338. doi:10.1023/B:COGE.0000031148.59923.aa

Fulton GR (2017) The Bramble Cay melomys: the first mammalian extinction due to climate change. *Pacific Conservation Biology* **23**, 1–3.

Garnett ST, Szabo JK, Dutson G (2011) *The Action Plan for Australian Birds 2010*. CSIRO Publishing, Melbourne.

Gunn RC (1851) On the introduction and naturalization of *Petaurus sciureus* in Tasmania. *Proceedings of the Royal Society of Tasmania* **1**, 253–255.

Harris S, McKenney H (1999) Preservation Island, Furneaux Group: 200 years of vegetation change. *Papers and Proceedings of the Royal Society of Tasmania* **133**, 85–101.

Heinsohn T (2002) Animal translocation: long-term human influences on the vertebrate zoogeography of Australasia (natural dispersal versus ethnophoresy). *Australian Zoologist* **32**, 351–376. doi:10.7882/AZ.2002.014

Heinsohn R, Webb M, Lacy R, Terauds A, Alderman R, Stojanovic D (2015) A severe predator-induced population decline predicted for endangered, migratory swift parrots (*Lathamus discolor*). *Biological Conservation* **186**, 75–82. doi:10.1016/j.biocon.2015.03.006

IUCN (2001) *IUCN Red List Categories and Criteria. Version 3.1*. IUCN Species Survival Commission, Gland, Switzerland.

IUCN/SSC (2013) *Guidelines for Reintroductions and Other Conservation Translocations. Version 1.0.* IUCN Species Survival Commission, Gland, Switzerland.

IUCN Standards and Petitions Subcommittee (2016) *Guidelines for Using the IUCN Red List Categories and Criteria. Version 12*. Prepared by the Standards and Petitions Subcommittee. <http://www.iucnredlist.org/documents/RedListGuidelines.pdf>

Johnson C (2006) *Australia's Mammal Extinctions: A 50,000 Year History*. Cambridge University Press, Melbourne.

Jones HP, Tershy BR, Zavaleta ES, Croll DA, Keitt BS, Finkelstein ME, Howald GR (2008) Severity of the effects of invasive rats on seabirds: a global review. *Conservation Biology* **22**, 16–26. doi:10.1111/j.1523-1739.2007.00859.x

Langford D, Burbidge AA (2001) Translocation of mala from the Tanami Desert, Northern Territory to Trimouille Island, Western Australia. *Australian Mammalogy* **23**, 37–46. doi:10.1071/AM01037

Lever C (1992) *They Dined on Eland*. Quiller Press, London.

Lohr MT, Keighery G (2016) The status and distribution of alien plants on the islands of the south coast of Western Australia. *Conservation Science Western Australia* **9**(2), 181–200.

Lohr MT, Lohr CA, Keighery G, Long V (2016) The status and distribution of non-native plants on the gazetted and territorial islands off the north coast of Western Australia. *Conservation Science Western Australia* **10**(8). https://www.dpaw.wa.gov.au/CSWAjournal

Masters P, Duka T, Berris S, Moss G (2004) Koalas on Kangaroo Island: from introduction to pest status in less than a century. *Wildlife Research* **31**, 267–272. doi:10.1071/WR03007

Medina FM, Bonnaud E, Vidal E, Tershy BR, Zavaleta ES, Donlan JC, Keitt BS, Corre M, Horwath SV, Nogales M (2011) A global review of the impacts of invasive cats on island endangered vertebrates. *Global Change Biology* **17**, 3503–3510.

Mills K (2009) *The Vegetation of Phillip Island, Norfolk Island Group*. Envirofund, Norfolk Island and Department of the Environment, Canberra.

Morris K, Speldewinde P, Orell P (2000) *The Recovery Plan for Djoongari or the Shark Bay Mouse (Pseudomys fieldi) 1992–2001*. Western Australian Wildlife Management Program No. 17. Department of Conservation and Land Management, Perth.

Nias RC, Burbidge AA, Ball D, Pressey RL (2010) Island arks: the need for an Australian national island biosecurity initiative. *Ecological Management and Restoration* **11**, 166–167. doi:10.1111/j.1442-8903.2010.00545.x

Ostendorf B, Boardman WSJ, Taggart DA (2016) Islands as refuges for threatened species: multi-species translocation and evidence of species interactions four decades on. *Australian Mammalogy* **38**, 204–212. doi:10.1071/AM15018

Ottewell K, Dunlop J, Thomas N, Morris K, Coates D, Byrne M (2014) Evaluating success of translocations in maintaining genetic diversity in a threatened mammal. *Biological Conservation* **171**, 209–219. doi:10.1016/j.biocon.2014.01.012

Parkes J, Fisher P, Robinson S, Aguirre-Muñoz A (2014) Eradication of feral cats from large islands: an assessment of the effort required for success. *New Zealand Journal of Ecology* **38**, 307–314.

Parks and Wildlife (2014) *Barrow Island Threatened and Priority Species Translocation and Reintroduction Program*. Annual Report to *CALM Act* Minister, *Barrow Island Act* Minister and Joint Venturers. WA Department of Parks and Wildlife, Perth.

Parks and Wildlife (2015) *Barrow Island Threatened and Priority Species Translocation and Reintroduction Program*. Annual Report to *CALM Act* Minister, *Barrow Island Act* Minister and Joint Venturers. WA Department of Parks and Wildlife, Perth.

Richards J (2007) Return to Faure Island. *Landscope* **22**, 10–17.

Robinson T, Canty P, Mooney T, Rudduck P (1996) *South Australia's Offshore Islands*. Department of Environment and Natural Resources, South Australia.

Savolainen P, Leitner T, Wilton AN, Matisoo-Smith E, Lundeberg J (2004) A detailed picture of the origin of the Australian dingo, obtained from the study of mitochondrial DNA. *Proceedings of the National Academy of Sciences of the United States of America* **101**, 12387–12390. doi:10.1073/pnas.0401814101

Seddon PJ, Armstrong DP, Soorae P, Launay F, Walker S, Ruiz-Miranda CR, Molur S, Koldewey H, Kleiman DG (2009) The risks of assisted colonization. *Conservation Biology* **23**(4), 788–789.

Seddon PJ, Moro D, Mitchell NJ, Chauvenet ALM, Mawson P (2015) Proactive conservation or planned invasion? Past, current and future use of assisted colonisation. In *Advances in Reintroduction Biology of Australian and New Zealand Fauna* (Eds DP Armstrong, MH Hayward, D Moro and PJ Seddon) pp. 105–126. CSIRO Publishing, Melbourne.

Smith B, Savolainen P (2015) The origin and ancestry of the dingo. In *The Dingo Debate: Origins, Behaviour and Conservation* (Ed B Smith) pp. 55–80. CSIRO Publishing, Melbourne.

Stojanovic D, Webb MH, Alderman R, Porfirio LL, Heinsohn R (2014) Discovery of a novel predator reveals extreme but highly variable mortality for an endangered migratory bird. *Diversity and Distributions* **20**, 1200–1207. doi:10.1111/ddi.12214

Woinarski JCZ, Ward S, Mahney T, Bradley J, Brennan K, Ziembicki M, Fisher A (2011) The mammal fauna of the Sir Edward Pellew Islands, Northern Territory: refuge and death-trap. *Wildlife Research* **38**, 307–322. doi:10.1071/WR10184

Woinarski JCZ, Burbidge AA, Harrison PL (2014) *The Action Plan for Australian Mammals 2012*. CSIRO Publishing, Melbourne.

Woinarski JCZ, Garnett ST, Legge SM, Lindenmayer DB (2017) The contribution of policy, law, management, research, and advocacy failings to the recent extinctions of three Australian vertebrate species. *Conservation Biology* **31**, 13–23.

8

Indigenous Protected Areas and islands

Ellie Bock, Philip Rist, Tom Vigilante and Neil Waina

Summary

Caring for country and other resource conservation approaches, including those for islands, are informed by dynamic human beliefs, knowledges and perceptions. Australia's saltwater peoples see land and sea as inseparable, as integral to their living identities as Aboriginal and Torres Strait Islander peoples and their continuing survival. There is no Indigenous notion of a maritime commons, or of marine resources open for all to exploit. It is beyond dispute that traditional estates extend beyond the low-water mark from land into marine areas, encompassing associated islands. For many of Australia's First Peoples, British colonisation involved not only the loss of inherited ancestral lands and their many biocultural resources, but also the loss of inherited sea countries together with their diverse marine resources. Since colonisation, the tenured status of protected islands in Australia has become subject to national or state/territory-based statutes and regulatory mechanisms. Accordingly, the nature of associated biocultural resource governance and conservation management-related agreement-making between Australia's First Peoples and other parties varies significantly between states and territories. More fundamentally, since the late 1990s, formal recognition of Indigenous rights and interests across all jurisdictions – through legal and other effective means, including voluntarily dedicated Indigenous Protected Areas – is exerting greater influence on island conservation across Australia.

Whilst recognising and respecting that many Aboriginal and Torres Strait Islander peoples prefer to refer to themselves as custodians, this chapter uses the term 'Traditional Owner' to emphasise occupation of Australia's islands by saltwater people over time.

Introduction: Indigenous Protected Areas

Aboriginal and Torres Strait Islander rights and interests in Australia's coastal and marine waters, and by extension in Australia's islands, have prevailed for millennia. Their assertions of these rights and interests in contemporary times pre-dates native title and, not infrequently, may also be articulated as local or collective responses to matters considered likely to negatively impact the diverse biocultural resources inherently connected to their maritime (coastal and marine) estates (Smyth 1993; NLC 1996; IMAS 2013; Dale *et al.* 2016).

Regardless of the jurisdiction within which they are now situated, an increasing number of Australia's islands are being returned to Traditional Owners via a range of common law processes, including native title and other statutes. Islands returned to their Traditional Owners are generally either directly held and managed as exclusive tenures by an incorporated Indigenous body, or managed by their Traditional Owners under collaborative (also joint or shared) management arrangements. A significant number of Australian islands are now solely or collaboratively managed by their Traditional Owners as recognised protected areas under formal management arrangements, guided by global conservation norms articulated and governed through the International Union for the Conservation of Nature (IUCN), of which Australia is a state member.

Aboriginal and Torres Strait Islander peoples' governance and management of island protected areas may be underpinned legally by Federal Court determinations of exclusive and/or non-exclusive native title or Indigenous Land Use Agreements (ILUAs), or through other effective means including formal agreements or other arrangements negotiated with government, accredited regulatory arrangements and/or the establishment of an Indigenous Protected Area (IPA).

IPAs are areas of land and/or sea that have been voluntarily dedicated by their Traditional Owners, are recognised by all tiers of Australian governments as protected areas and are managed by a combination of existing legal rights and other measures that arise from diverse sources. These include tenure regimes, fisheries regulations, cultural heritage legislation and environmental protection legislation, as well as Indigenous law and non-legal measures such as management planning and implementation by Indigenous rangers (Smyth *et al.* 2016).

IPAs are a uniquely Australian response to the holistic protection of land and sea country, including both larger and smaller islands, by Traditional Owners under set IUCN criteria (Szabo and Smyth 2003; Dodson 2007). Their existence is resulting in the increasingly culturally assured safeguarding and care of island-based biocultural resources, and the wider regional coast, land- and seascapes within which these are embedded (Girringun 2014; SVA Consulting 2016). IPAs originated from the need to recognise inherent Aboriginal and Torres Strait Islander rights, responsibilities and obligations to country – country being the spiritual, ecological, social and aesthetic realm of Australia's First Peoples – encompassing the entirety of their respective ancestral domains, at land- and seascape scale (Dodson 2007).

Dedicated IPAs represent defined connections to country, all associated biocultural resources, local avian, terrestrial and marine ecologies, flora and fauna (including endemic, rare or threatened species) and ongoing custodial obligations arising from living Indigenous laws and inherited traditions. They are cared for by Traditional Owners under holistic biocultural management principles guided by continuing Indigenous knowledge systems. IPAs are considered a highly innovative mechanism for interpreting and applying IUCN criteria and Guidelines, and for the progressive conceptualisation of protected area creation and management in Australia (Szabo and Smyth 2003). By 2017, a total of 74 dedicated IPAs were in existence across Australia, including some 25 dedicated IPAs where islets, islands and/or sea country in immediate proximity to offshore islands (and associated biocultural resources) are under the culturally assured management of Traditional Owners.

However, the complex duality of interests prevailing across Aboriginal and Torres Strait Islander sea countries has meant that several spiritually significant islands have not been included in their respective land-based IPAs. Thus, in addition to dedicated IPAs involving islands, it must be acknowledged that increasingly Australian islands are being managed by their respective Traditional Owners under other arrangements or effective means (Box 8.1 and Table 8.1).

Australia's Island IPAs

Today, IPAs inclusive of islands span the breadth and width of Australian jurisdictions (Table 8.2). Tasmania's Mount Chappell Island and Badger Island IPAs, simultaneously dedicated in 2000, were Australia's inaugural island-only IPAs: today there are five Tasmanian Aboriginal-managed island-only IPAs. The Torres Straits' first of three island-only IPAs, the Warul Kawa IPA, was declared by its Traditional Owners in 2001. More recently, South

Box 8.1: IPAs – one of many ways to care for islands

In recent years, sea country IPAs (coastal and/or marine IPAs) have gained increased acceptance as Australian marine protected areas. Many sea country IPAs involve caring for islands and their associated biocultural resources, which are often subject to multiple use regimes that at times generate significant access demands and use impacts. However, IPAs are only one of several approaches available to Traditional Owners

Table 8.1. Indigenous interests on selected Australian non-IPA islands (as at January 2017)

Island/group name	Tenure/s	Indigenous governance and/or management
Queensland		
Torres Strait islands	TSILA Freehold,[a] DOGIT[b]	Torres Strait Regional Authority and 21 RNTBCs 14 ranger groups based in 13 island communities
Wunyami (Green Island)	National Park	Gunggandji Aboriginal Corporation RNTBC, TUMRA[c]
Yunbenun (Magnetic Is.)	ALA Freehold,[d] national park	Wulgurukaba Aboriginal Corporation, ILUA
Keppel Island Group	ALA Freehold, freehold, national park	Woppaburra Land Trust, Dharumbal TUMRA
Curtis Island	Various	Port Curtis Coral Coast Traditional Owners, TUMRA
K'gari (Fraser Island)	National Park, World Heritage Area	Butchulla Aboriginal Corporation RNTBC
Minjerribah (North Stradbroke Island)	National Park and other tenures	Quandamooka Yoolooburrabee Aboriginal Corporation (QYAC) and QYAC Rangers
New South Wales		
Ukerabagh Island	Reserve	via Tweed/ Byron Local Aboriginal Land Council as a registered Aboriginal Place
Northern Territory		
Crocodile Island	Aboriginal Land	Milingimbi and Outstations Progress Resource Association Inc. and Crocodile Island Rangers
Croker Island	Aboriginal Land	Injarnyala Aboriginal Corporation RNTBC, Garngi Rangers
Goulburn Island	Aboriginal Land	Mardbalk Marine Rangers
Tasmania		
truwana (in part) (Cape Barren Island)	Aboriginal Land	Aboriginal Land Council of Tasmania, Tasmanian Aboriginal Centre (TAC) Rangers
Bruny Island (in part)	Freehold	Weetapoona Aboriginal Corporation, weetapoona (Murrayfield Station) property managers
titima (Trefoil Island)	Aboriginal Land	Aboriginal Land Council of Tasmania, TAC Rangers
Steep Island	Aboriginal Land	Aboriginal Land Council of Tasmania, TAC Rangers
Victoria		
Deen Maar Island	Wildlife Reserve	Connected to but not included in the Deen Maar IPA reserve

[a] TSILA Freehold = lands returned under the *Torres Strait Islander Land Act 1991* (Qld).
[b] DOGIT (Deed of Grant in Trust) = a common tenure in some Queensland Indigenous communities.
[c] TUMRA = Traditional Use of Marine Resources Agreement.
[d] ALA Freehold = lands returned under the *Aboriginal Land Act 1999* (Qld).

in realising their aspirations for deeper involvement in the management of their sea countries, including associated islands. Depending on the jurisdiction, selected regulatory mechanisms may also allow for defined agreement-making between saltwater peoples, and between them and other parties including statutory conservation management agencies (Table 8.1).

Statutory provisions applying in Australia's states and territories have seen Aboriginal interests in island governance and management arrangements negotiated by way of ILUAs or as part of native title determinations. Indigenous groups have also had islands returned to them through property acquisition programs (e.g. through the Indigenous Land Corporation) or through land justice/tenure resolution programs (e.g. in the Northern Territory, Queensland and Tasmania). Further, eight TUMRAs – the majority of which include islands – have been entered into by Traditional Owners for areas within the Great Barrier Reef Marine Park (GBRMP) World Heritage Area. Separate provisions apply in the Torres Strait region where sea and land management, resourced by both the Australian and Queensland governments, is largely coordinated through the Torres Strait Regional Authority (TSRA). Ailan Kastom, Aboriginal lore/law and native title rights and interests guide island, sea and land management into the present, underpinning the operations of TSRA coordinated Ranger groups and the newly established Gur A Baradharaw Kod Torres Strait Sea and Land Council (TSRA 2016).

Australia's Werdang Island IPA was dedicated in 2015, Wunambal Gaambera exclusive possession islands were added to the Uunguu IPA in 2015 and the Northern Territory's Marthakal IPA (Elcho Island) was dedicated in 2016.

Numerous islands are being included within IPAs, actively extending direct Traditional Owner management from land-based IPAs into associated sea country. Other IPAs encompassing islands were originally dedicated as combined land and sea IPAs; for example, the Girringun IPA, Australia's largest collaboratively managed IPA, includes multiple Great Barrier Reef island groups in addition to terrestrial national parks, marine parks and other IPA-included tenures.

Intent of establishing an IPA

Aboriginal and Torres Strait Islander peoples voluntarily declare or dedicate an area of land and/or sea country (including islands) to become an IPA – a clear statement of intent to affect the long-term protection of the diverse biocultural (cultural, ecological, spiritual) resources associated with that defined area (Dodson 2007; Smyth 2009). Ultimately, the degree of conservation protection afforded via an IPA will vary depending on the tenures within that IPA, the success of collaborative partnerships developed through the IPA planning process and their ongoing maintenance. Presently there are no explicit legal protections to restrict extractive resource developments on IPAs other than those generally applying to Australian land use tenures (e.g. regulations prescribing exclusive nature conservation, development planning restrictions, resource usage etc.). The strongest levels of legal conservation protection presently achievable under Australian law are either an internationally accepted World Heritage listing (enacted through environmental statute) or a ministerially determined (policy-based) protected area gazettal. However, IPAs provide the opportunity to enhance protection of significant cultural and natural values through international recognition as a protected area, as well as through community and political support associated with that recognition.

Guidelines for the development of IPA plans of management (IPA plans) were first compiled in 2011 (Hill *et al.* 2011). Local factors will influence and ultimately determine the particular planning approaches adopted by Traditional Owners, and the content and format of IPA plans. Many IPA plans are based on multi-year 'country-based' or 'Healthy Country' planning processes (Smyth 2011a; Moorcroft *et al.* 2012). Country-based planning has been applied to explore 'tenure-blind' planning where strategic objectives for the holistic management of country are able to be considered

Table 8.2. IPAs inclusive of sea countries, islands and/or islets (as at January 2017)

IPA name	Dedication year/s	Primary IPA governance entity and managers
New South Wales		
Gumma	2011	Nambucca Heads Local Aboriginal Land Council for Gumbaynggirr clans
Northern Territory		
Anindilyakwa	2006, 2016	Anindilyakwa Land Council, Anindilyakwa Rangers
Dhimurru	2000, 2013	Dhimurru Aboriginal Corporation, Dhimurru Rangers
Djelk	2009	Bawinanga Aboriginal Corporation, Djelk Rangers
Laynhapuy	2006	Laynhapuy Homelands Aboriginal Corporation, Yirralka Rangers
Marthakal	2016	Marthakal Homelands Resource Centre, Gumurr Marthakal Rangers
South East Arnhem Land	2016	Yugul Mangi Aboriginal Corporation, Yugul Mangi Rangers
Yanyuwa	2011	Mabunji Aboriginal Resource Association, li-Anthawirriyarra Sea Rangers
Queensland		
Jalunji Warra (Eastern Kuku Yalanji Stage 2)	2013	Jabalbina Yalanji Aboriginal Corporation RNTBC and statutory partners, Jabalbina Rangers
Girringun	2013	Girringun Aboriginal Corporation in association with member group RNTBCs and statutory partners, Girringun Aboriginal Rangers
Mandingalbay Yidinji	2011	Mandingalbay Yidinji Aboriginal Corporation RNTBC and statutory partners, Djunbunji Rangers
Pulu Islet	2009	Island Traditional Owners, TSRA,[a] Mabuygiw Rangers
Thuwathu Bujmulla	2013	Gangalidda people, Gangalidda Ranger groups coordinated through Carpentaria Land Council
Waarberalgahl and Porumalgal	2014	Traditional Owners, TSRA, Waarberalgahl and Poruma Rangers
Warul Kawa	2000	Traditional Owners, TSRA, Malu Ki'ai and other Traditional Owner Rangers
South Australia		
Werdang Island	2015	South Australian Aboriginal Lands Trust for Narrunga clans
Tasmania[b]		
Mount Chappell Island (Hummocky)	2000	Island Traditional Owners, TAC, TAC Rangers
Badger Island	2000	Island Traditional Owners, TAC, TAC Rangers
Babel Island	2009	Island Traditional Owners, TAC, TAC Rangers
Big Dog Island	2009	Island Traditional Owners, TAC, TAC Rangers
lungtalanana (Clarke Island)	2009	Island Traditional Owners, TAC, TAC Rangers
Western Australia[c]		
Balanggarra	2013	Balangarra Aboriginal Corporation RNTBC, Balangarra Rangers
Bardi Jawi	2013	Bardi and Jawi Niimidiman Aboriginal Corporation RNTBC,[d] Bardi Jawi Rangers, Bardi Jawi Oorany Rangers
Dambimangari	2013	Dambimangari Aboriginal Corporation, Dambimangari Rangers
Uunguu	2011, 2015	Wunambal Gaambera Aboriginal Corporation, Uunguu Rangers

[a] Torres Strait Regional Authority (TSRA) coordinates Torres Strait IPA Ranger groups (as at 2017).
[b] Tasmanian Aboriginal Centre (TAC) coordinates island IPA management in Tasmania (as at 2017).
[c] The listed WA-based IPA Ranger groups are members of the Kimberley Ranger Network.
[d] RNTBC = registered native title body corporate.

at coast-, land- and/or seascape scales. Healthy Country planning focuses on adaptive management and participatory processes. Both can enable expanded scope for collaborative partnerships, where these are desired or considered necessary by Traditional Owners.

Since the 1998 dedication of the first IPA at Nantawarrina (South Australia) as part of a federal program to consolidate the emerging terrestrial National Reserve System (NRS), IPAs have developed to become an increasingly globally recognised tool in protected area management (IUCN 2003, 2014). Increased awareness of the diverse environmental, cultural, socio-economic and transformative change benefits associated with IPAs and their implementation through locally based Ranger groups is also growing within Australia (SVA Consulting 2015). The IPA model, where supported by well-practised Aboriginal or Torres Strait Islander governance and by dedicated long-term resource commitments for local Rangers, demonstrably generates real conservation benefits for native species and cultural heritage (Grace 2016; SVA Consulting 2016). IPAs are proven to assist in addressing aspects of Indigenous disadvantage while supporting Indigenous enterprise growth, employment pathways and professional development (Commonwealth of Australia 2015).

IPAs are a preferred model for land and sea management in the Kimberley, where they extend over some 90 000 km^2, providing formal structures for recognising Indigenous rights to manage country while promoting NRS-related biodiversity and conservation outcomes (Box 8.2). The west Kimberley became a National Heritage-listed area in 2011. Larger and smaller parts of the Wet Tropics World Heritage Area, listed as National Heritage for its living Rainforest Aboriginal cultures, are collaboratively managed by Traditional Owners in partnership with Australian and Queensland government agencies through the Eastern Kuku Yalanji IPA, the Mandingalbay Yidinji IPA and the Girringun IPA. These IPAs extend into the GBRMP World Heritage Area. They thereby transcend jurisdictional boundaries governing protected area management regimes, better align related management with Traditional Owner priorities, directly employ Traditional Owners in protected area management, compliance and enforcement, and create a more holistic approach to caring for islands, lands and seas into the 21st century.

Recognition

IUCN criteria and Guidelines were developed through international participatory processes over many years and have been endorsed by Australian federal, state and territory governments. Formal recognition of protected areas in Australia by ministerial decision-making is generally subject to the meeting of explicit IUCN criteria and Guidelines and the satisfaction of all applicable national and state/territory legislative and policy requirements. This enables legal gazettal of government-dedicated Australian protected tenures. However, where Aboriginal or Torres Strait Islander peoples themselves determine long-term conservation objectives for their country, and where there are customary or other agreements in place, they may voluntarily establish and manage a protected area outside the statutory protected area framework and remain consistent with the IUCN Guidelines (Smyth 2011b).

Based on their attributed IUCN criteria, dedicated land-based IPAs are routinely added to the NRS, Australia's network of recognised protected areas, to which IPAs contribute over 40% of the total area. The Australian government's NRS guidelines recognise IPAs as essential to the establishment of a truly comprehensive NRS. Australia's *Strategy for the National Reserve System 2009–2030* states that continued investment in biodiversity conservation through IPAs and shared management arrangements are key priorities for NRS expansion, for protected area planning and management, for knowledge management including science, for performance monitoring and for stronger partnerships (NRMMC 2010). The marine equivalent of the NRS, the National Representative System of Marine Protected Areas (NRSMPA) does

Box 8.2: Uunguu IPA, Bonaparte Archipelago, Kimberley

The Uunguu IPA (Fig. 8.1) in the north Kimberley is an example of how Traditional Owners have dedicated their native title lands for the benefit of the whole world and are using best practice conservation tools to manage these areas.

In 2011 the Federal Court recognised the traditional ownership of some 2.5 million ha of land and sea country by Wunambal Gaambera people, in the Wanjina Wunggurr Uunguu Native Title Determination. At the determination ceremony, Traditional Owners declared the first stage of the Uunguu IPA, releasing their 10-year Healthy Country plan of management which outlined how the country would be managed. Wunambal Gaambera Aboriginal Corporation (WGAC) through its board of directors, staff, partners and the Uunguu Rangers are responsible for implementing the plan. Dedication of the IPA was recognised by the Australian government which supports IPA implementation through a five-year funding agreement. Bush Heritage Australia also signed a 10-year agreement with WGAC to help implement the Healthy Country plan (WGAC 2010).

Wunambal Gaambera people are saltwater people, traditionally using mangrove rafts and dugout canoes to travel between islands and the mainland to access resources such as marine turtle eggs and seabird eggs. Some clans were predominantly island-based, particularly where islands held permanent freshwater, while others lived on the mainland and made short visits to islands (Crawford 1983). Islands hold many important cultural stories and have associated cultural sites. Following World War II, Traditional Owners settled in communities like Kalumburu and it became increasingly difficult to access sea country. In implementing the Uunguu IPA, Healthy Country work supports people getting back to country and caring for country (WGAC 2010).

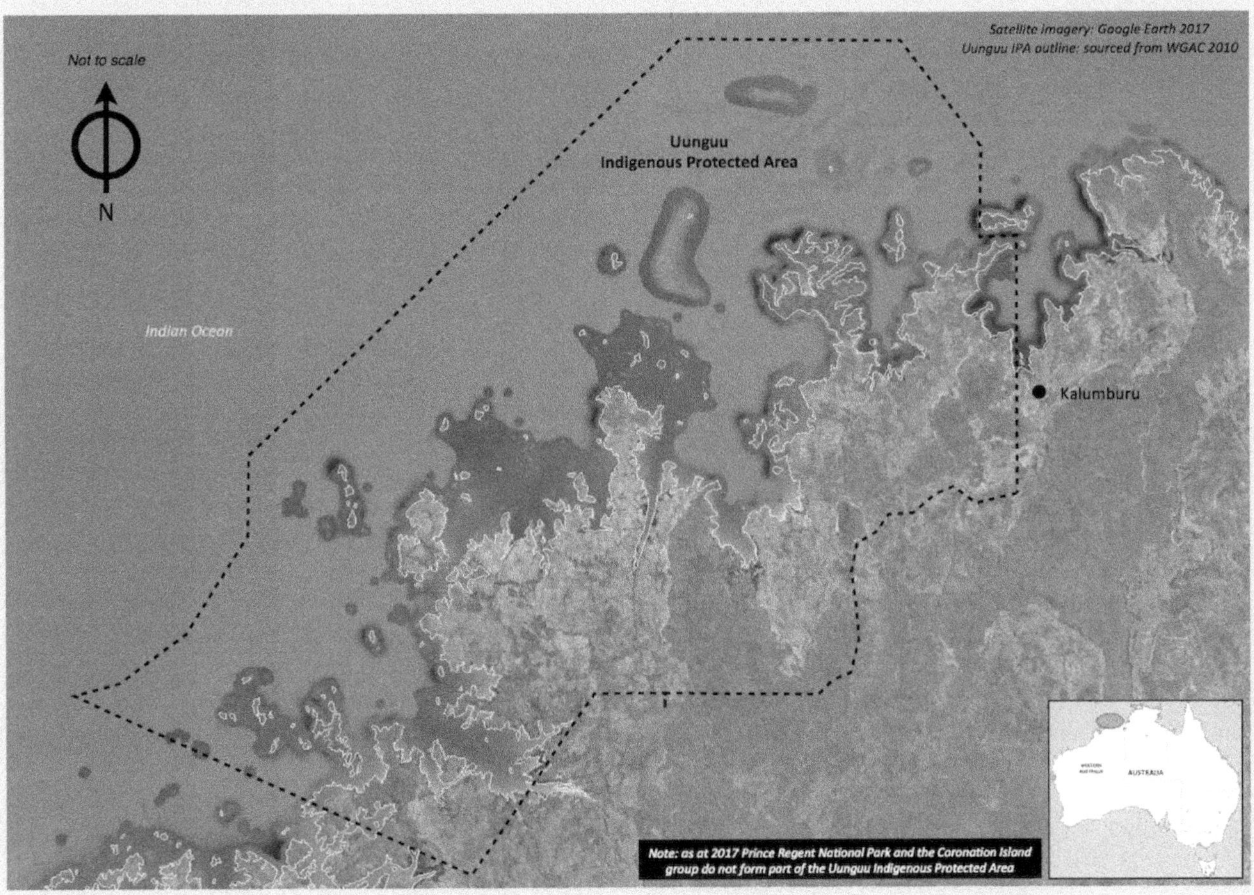

Fig. 8.1: Location of the Uunguu IPA, Western Australia.

By dedicating the Uunguu IPA, Traditional Owners have made a commitment to the whole world that they will manage their country to the IUCN standard. IUCN Guidelines prescribe that, as a category VI protected area, the majority of the protected area (typically 75%) should be for primary protection with the remainder available for sustainable and compatible enterprises and development. WGAC has undertaken planning work to designate IPA-internal areas for protection and use. Higher conservation value areas have been mapped using spatial data. The habitat of threatened species (e.g. small mammals) and culturally important species have been modelled from public datasets and Traditional Owner records with MAXENT and ZONATION software used to overlay the modelled habitats. Future land use by Traditional Owners will involve the establishment of small outstations and ranger bases with associated enterprises such as eco-tourism. In 2007, the oil and gas company INPEX undertook a feasibility study to develop a liquid natural gas-processing facility on the Maret Islands but it ultimately decided to construct a gas pipeline to Darwin (Comrie-Greig and Abdo 2012).

The conservation work undertaken by WGAC is guided by the Wunambal Gaambera Healthy Country Plan 2010–2020, developed through an exhaustive participatory planning approach with the wider Wunambal Gaambera Traditional Owner community using the Open Standards for the Practice of Conservation. This planning was facilitated by Bush Heritage Australia and funded by the Nature Conservancy and the Australian government (Moorcroft *et al.* 2012). The Healthy Country planning approach is considered best practice and has been adopted by many other Indigenous landowners across Australia, including more than 30 Indigenous groups over some 65 million ha (Godden and Cowell 2016).

WGAC established a monitoring and evaluation committee which meets annually to review the progress of the plan and to ensure that work is being done according to best practice. The committee is made up of Traditional Owners as well as external experts such as senior ecologists and anthropologists. In 2015 a mid-term evaluation of the plan was completed and a paper published on the results (Austin *et al.* 2017). An independent reviewer considered that the establishment of this monitoring and evaluation committee and its ongoing role provided an effective review function that is often lacking in most mainstream conservation organisations. The reviewer noted that 'something quite remarkable on the global scale is happening on Wunambal Gaambera country'.

The Western Australian (WA) government announced plans to declare national parks over several larger islands and much of the mainland in North Kimberley through its Kimberley Science and Conservation Strategy (Government of Western Australia 2011). Much of this area has already been declared under the Uunguu IPA. A national park can provide statutory protection and access to public funding and resources, but also presents significant challenges in that the declaration suppresses native title, requiring joint management with government agencies rather than autonomous management by Traditional Owners. ILUAs would need to be negotiated with Traditional Owners where national park declarations are future acts under the *Native Title Act*.

The expedition cruise industry has expanded on the Kimberley coast in recent decades and Traditional Owners have rapidly become concerned about visitors going to sacred sites and visiting country without permission (Scherrer *et al.* 2011a, 2011b; Scherrer and Doohan 2013). The expedition cruise industry is typically a high-end tourism product. One of the most heavily visited sites on the Kimberley coast, Bigge Island, has an important rock art site of 'Kaiara' or 'sea wanjina' ancestral spirit beings. Traditional Owners are distressed by instances where burial remains have been photographed, moved and even removed by tourists at this important site. The issue was exacerbated by the lack of legislation and regulation to support management of access to Aboriginal reserves and native title lands. This was identified as a key threat in the Healthy Country plan. There was a competing need from government and the tourism industry to maintain access to iconic tourism sites. Traditional Owners raised these issues with government and the tourism industry in forums and reports, and the idea of a permit system was discussed. The issue was not prioritised in the Kimberley Science and Conservation Strategy, however, which focused on supporting development of Indigenous tourism products. After many years of work WGAC released its Uunguu Visitor Pass, which enables visitors to access 50 permitted visitor locations on islands and coastal places (WGAC 2017; www.wunambalgaambera.org.au). The associated fee generates funds needed to manage cultural sites.

Kimberley islands are of high conservation value, containing endemic species of camaenid land snails and threatened mammal species (Gibson *et al.* 2012). Islands provide a refuge from mainland threats such as cane toads, which are establishing in the region. The WA government has published the results of a major biological survey of larger islands undertaken in collaboration with Traditional Owners (Gibson *et al.* 2012). Islands also hold unique vegetation associations, including monsoon vine thickets (Ondei *et al.* 2017). Quarantine is a priority for the management of islands. While islands are often protected from mainland threatening processes such as pest species and wildfire, regular patrols are needed to ensure that pest species do not spread from the mainland. Ocean currents and prevailing winds can carry propagules from south-east Asia while floodwaters and tides can carry pest species from the mainland. Unmanaged visitation by tourists in boats is a further potential source of pest species.

Historically, Wunambal Gaambera people burnt some islands when visiting and fire remains an important management practice today. Late dry-season wildfires started by dry lightning storms in the build-up to the wet season are a significant issue in the Kimberley, and affect islands as well as the mainland. In 2012 Bigge Island, one of the largest islands in the Kimberley, was almost completely burnt by a single lightning fire. Uunguu Rangers now undertake prescribed burning on larger islands, generally through aerial burning from a helicopter. Cultural sites are visited and burnt on foot.

not afford similar recognition to sea country IPAs, as it operates under 1998 criteria presently applicable only to legislated marine protected areas (MPAs) (Smyth *et al.* 2016). Nevertheless, the marine components of IPAs are included as a category of MPAs in the Collaborative Australian Protected Area Database (CAPAD), confirming that sea country IPAs are an integral component of Australia's MPA estate. The *Guidelines for Establishing the National Representative System of Marine Protected Areas* explicitly identify Indigenous involvement as a principle of MPA development: 'the interests of Australia's Indigenous peoples should be recognised and incorporated in decision-making'. However, providing 'for the recreational, aesthetic and cultural needs of indigenous and non-indigenous people' is ranked secondary to the primary goal of establishing and managing a 'comprehensive, adequate and representative system' of MPAs (ANZECC 1999).

The types of tenures within the scope of an IPA, and the rights and interests of an IPA's Traditional Owners at common law, will significantly influence IPA design and establishment processes and on-ground implementation of related biocultural resource management. In this respect, statutory land justice processes, native title outcomes and/or related proceedings may intersect with IPA development and formal government recognition. Traditional Owners must invest significant time, effort, goodwill and resources in IPA development and ongoing IPA implementation. They must engage with multiple statutory management agencies in the establishment of collaboratively managed IPAs where there is no legally recognised or tenured Indigenous ownership of the area/s involved. There remains a clear difference in the negotiation of sea country IPAs within what is legally considered to be the maritime commons (Smyth 2009; Smyth *et al.* 2016). Generally, most marine tenures included within IPAs will be subject to the rights and interests of non-Indigenous third parties, requiring Traditional Owners to engage with several statutory management agencies and other external stakeholders, including potential third-party partners (Box 8.3).

Resourcing

As outlined above, the Australian government initiated the IPA program in the 1990s through the then federal environment portfolio, with resourcing allocated through the Natural Heritage Trust. The IPA program is presently administered by the Department of Prime Minister and Cabinet, with complementary additional funds for many but not all IPA Ranger groups drawn from the federal Working on Country program. The program may provide limited funding over a limited number of years to Traditional Owners to develop an IPA proposal and associated plan which set out the Traditional Owners' IPA objectives, goals, strategies and actions

Box 8.3: Islands within the Girringun IPA, Great Barrier Reef

> *He rose up out of the ocean, near Hinchinbrook Island … part man, mostly eel, and gazed around … The rugged grandeur of Hinchinbrook impressed him and he rose from the water and pronounced its name loudly … [Girrugarr] then plunged into the sea and came out … at the back of Cardwell. He saw a freshwater lake surrounded by mangroves and he called it Girringun (Henry 1967).*

> *Even though we didn't stop the construction of the marina and real estate subdivision, after nearly a decade of campaigning by many, many people, many positive outcomes for Hinchinbrook were achieved … The state government also purchased property at the Cardwell Gap so the breathtaking view over the lowlands could be maintained. A proposed helicopter pad and communications towers for Mount Bowen [on the island] were also defeated (Nowakowski 2003).*

Regardless of their heritage, many Australians are familiar with the idyllic beauty of north Queensland's Dunk Island and the sheer magnificence of Hinchinbrook Island to the immediate south, one of Australia's largest all-island national parks. Many also know that these islands are surrounded by the waters of the World Heritage-listed Great Barrier Reef Marine Park. However, not many people are aware that these islands, together with the neighbouring Goold, Brook and Family group islands, are located within Australia's largest collectively governed and collaboratively managed IPA – the Girringun IPA (Fig. 8.2).

This exceptional IPA is collectively overseen by eight Traditional Owner groups – the Bandjin, Djiru, Girramay, Gugu Badhun, Gulnay, Nywaigi, Warrgamay and Warungnu peoples – through a

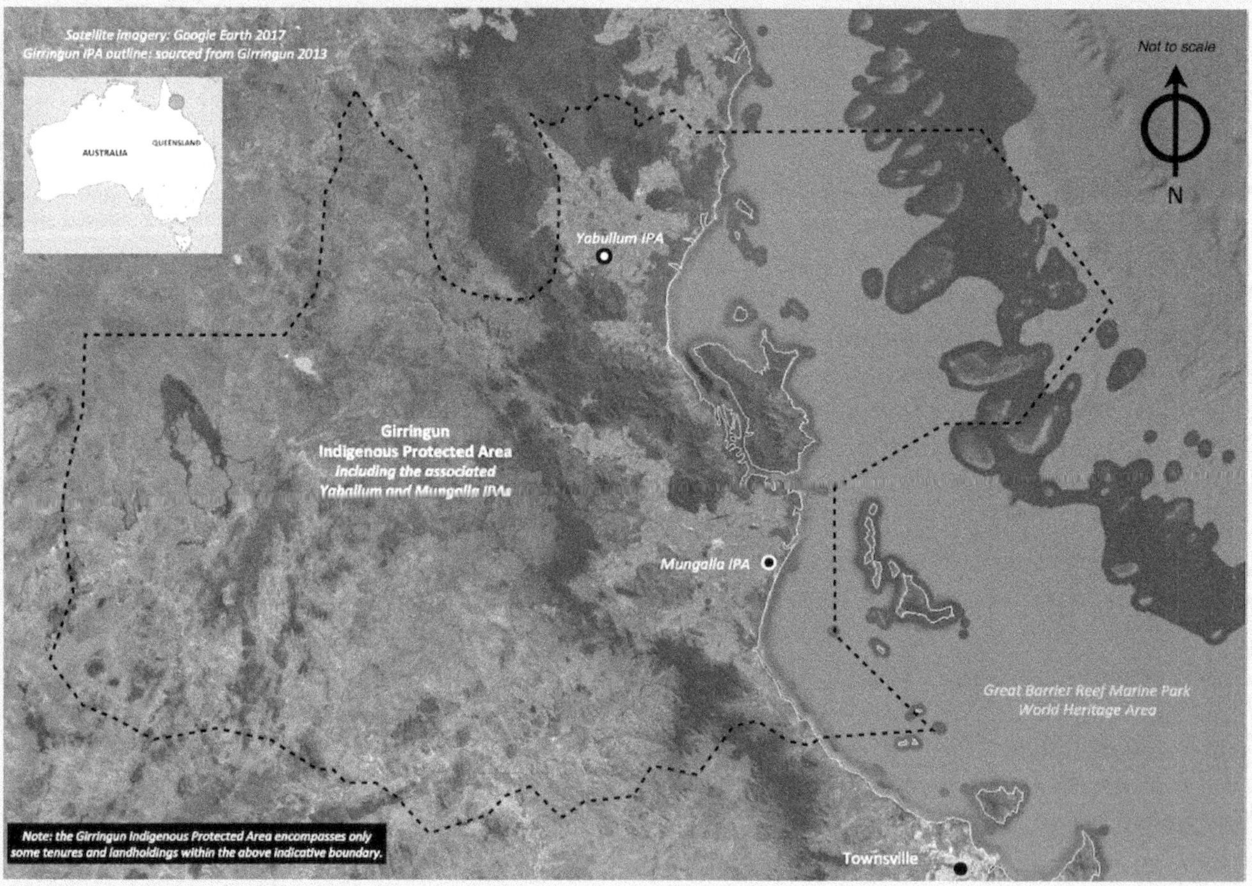

Fig. 8.2: Location of the Girringun IPA, Queensland.

regional Aboriginal alliance centred on the Girringun Aboriginal Corporation (GAC). Founded in the mid-1990s by senior Elders, GAC adopted a working group model whereby multiple Traditional Owner groups with common objectives came together at a regional community-based table to negotiate and achieve positive change. The Elders' original intent to change the earlier name of today's Girringun National Park has evolved into their lasting legacy – the 2013 declared Girringun IPA, which extends across diverse public and private tenures including significant portions of the Great Barrier Reef and the Wet Tropics World Heritage areas. Uniquely, Aboriginal and non-Indigenous collaborators have dedicated private landholdings for inclusion in this IPA. Negotiation of the 2005 Girringun Traditional Use of Marine Resources Agreement (TUMRA) involved Bandjin, Djiru, Girramay, Gulnay, Nywaigi and Warrgamay Traditional Owners engaging with neighbouring Aboriginal peoples to define an agreed TUMRA area, which today forms the sea country area of the Girringun IPA (Smyth *et al.* 2016).

Hinchinbrook, Dunk and all other islands within the Girringun IPA are named places embedded within the dynamic and continuing traditions of the region's saltwater Rainforest Aboriginal peoples, providing impressive stages for Story into the present, connecting ancient cultural land- and seascapes through tens of thousands of years of human occupation and changing sea levels (Campbell 1979; Girringun 2013). Saltwater Traditional Owners know, through their living traditions, that areas which today form part of the Great Barrier Reef were a coastal plain where kangaroos flourished some 8000 years ago. The story of Girugarr, ancestral surveyor of the Girringun region and adjacent areas, says that in his day people could walk between the mainland and these islands (Dixon 1972; Girringun 2013). The Story quotation recounted at the beginning of this section provides a brief insight into the prodigious naming of places undertaken by Girugarr in his legendary travels across the southern Wet Tropics.

Named places, together with their associated stories and customary Aboriginal law, are passed on to today's generations by their Elders. The Girringun IPA's many islands hold numerous Aboriginal sites of significance recorded by Traditional Owners including sacred places, camping and hunting sites, middens and stone fish traps. Cultural data are securely stored on Girringun's in-house restricted access database: Hinchinbrook Island alone contains over 50 recorded Aboriginal cultural sites (Girringun 2007). Hinchinbrook's Muhr-Amalee Restricted Access Area was established under state regulations. Dunk Island holds numerous sites of Aboriginal cultural significance with Story connecting that island to adjacent islands and the mainland, and to the outer Reef. Over the years Hinchinbrook Island, its channel and the adjacent mainland have been in the spotlight of conservation concerns regarding controversial coastal developments (Nowakowski 2003).

Traditional Owners engage in collective governance and priority-setting through entities such as Elders, registered native title bodies corporate (RNTBCs) or other representative organisations. Collaborative management is undertaken by the Girringun Aboriginal Rangers, who work with statutory, non-government and private managers to care for IPA-included lands, islands, seas and freshwaters (Zurba *et al.* 2012). Agencies managing the island under statute include the GBRMP Authority and multiple state agencies responsible for national parks, environment and heritage, fisheries, biosecurity, natural resources (water) and development (tourism). Rangers are gaining formal compliance and enforcement powers under regulatory mechanisms required for effective island protected area management in this region. Aboriginal and non-Indigenous partners work together on governance and on-ground management through the Girringun IPA Co-operations and Compliance Steering Committee.

As this IPA includes protected areas falling within both Commonwealth and state jurisdictions, related collaborative management involves multiple statutory partners. Hinchinbrook Island is an example in terms of these complexities: it is a national park (gazetted 1932 and 1934) situated within the GBRMP (gazetted 1975) now within the GBRMP World Heritage Area (listed 1981) and a wild river (declared 2007) situated within the Wet Tropics bioregion. None of these statutory conservation actions involved seeking free, prior and informed consent from the island's Traditional Owners.

Today, through the Girringun IPA, the strategic and operational involvement of Traditional Owners is confirmed via processes routinely engaging Elders and representative bodies, with priorities for IPA governance and on-ground management implementation negotiated within the context of a long-standing regional alliance. All Girringun IPA partners must operate within a highly complex cross-cultural and multi-jurisdictional alliance, frequently challenged by a multitude of external pressures arising for Traditional Owners, a proliferation of short-term policy-constrained public investment programs, radical changes of public administration and related machinery of government upheavals.

(Dodson 2007). IPA planning is undertaken independently by Traditional Owners or with the assistance of facilitation and planning professionals. External professionals are generally directly engaged by Traditional Owners or hosted by them as employees of Traditional Owner-endorsed IPA partners. Limited public funding for ongoing IPA coordination, governance and management may be provided by the Australian government once an IPA plan has been endorsed and an IPA is formally dedicated (Dodson 2007). IPA plans are generally integrated with the operational work plans of allied Ranger groups and may also be linked to strategic or corporate plans underpinning local or regional IPA governance. At the discretion of Traditional Owners, privately managed funds may be committed to IPAs by non-government conservation organisations. Third parties with direct interests in an IPA consultation area will generally be approached and engaged early in any planning processes.

Recent 'social return on investment' analysis concluded that the A$35.2 million investment in IPAs from government and a range of third parties has generated social, economic, cultural and environmental outcomes with an adjusted value of A$96.5 million (SVA Consulting 2016). Similarly, there are multiple benefits from the ongoing conservation of the many bioculturally diverse and unique island IPAs dedicated by their Traditional Owners.

There remains significant variance in the recognition of Indigenous peoples' sea rights across legal and policy jurisdictions, with change in this respect generally occurring later and to a lesser extent than those dealing with land (Smyth *et al.* 2016). At the conclusion of the 2012 National Indigenous Sea Country Workshop, delegates issued the National Indigenous Sea Country Statement, calling for government to provide sufficient resources to enable equitable participation by Traditional Owners in MPA policy development, allocation setting and management of MPA regulatory schemes in Australia. Delegates also issued the following MPA-specific statement:

> *The delegates are aware of the bioregional marine planning process being conducted by the Australian Government and the pending declarations of MPAs by the Australian Government in Commonwealth waters.*
>
> *According to customary law, these MPAs will include Indigenous sea country and/or the species, communities and ecosystem processes in the MPAs will be contiguous with Indigenous sea country.*
>
> *MPAs will require management plans and implementation of these plans hence significant investment in funds and resources. Indigenous land and sea managers have demonstrated capacity to develop and implement plans for sea country and adjacent lands.*
>
> *We the delegates of the National Indigenous Sea Country Workshop request that Indigenous Traditional Owners and their representative organisations are given first option to undertake the development and implementation of MPA management plans.*

2012 National Workshop delegates also issued the following statement, specific to IPAs (NAILSMA 2013):

> *Recognising that saltwater people all around Australia see land and sea as indivisible parts of their traditional estate (country), the National Indigenous Sea Country Workshop urges all levels of government to recognise and support the inclusion of sea country into coastal and island Indigenous Protected Areas (IPAs), consistent with international best practice protected area guidelines in collaboration with other users and managers.*

Concluding comments

In supporting Aboriginal and Torres Strait Islander peoples as professional managers of their land and sea countries in the contemporary context, the proven benefits of IPAs and dedicated Aboriginal and Torres Strait Islander Ranger programs deserve and require long-term commitments in maintaining and sustaining Australia's unique island, marine and terrestrial environments.

Acknowledgements

The authors and co-authors wish to acknowledge the ongoing efforts of Aboriginal and Torres Strait Islander peoples in progressing, governing and managing IPAs and island IPAs across Australia, and their integral role in the conservation of these significant places. We also acknowledge the expertise of this chapter's peer reviewers and thank them sincerely.

References

ANZECC (1999) *Strategic Action Plan for the National Representative System of Marine Protected Areas: A Guide for Action by Australian Governments.* Australian and New Zealand Environment and Conservation Council (ANZECC) Task Force on Marine Protected Areas, Environment Australia, Canberra.

Austin BJ, Vigilante T, Cowell S, Dutton IM, Djanghara D, Mangolomara S, Puermora B, Bundamurra A, Clement Z (2017) The Uunguu Monitoring and Evaluation Committee: intercultural governance of a land and sea management program in the Kimberley, Australia. *Ecological Management and Restoration* **18**(2), 124–133. doi:10.1111/emr.12257

Campbell JB (1979) Settlement patterns on offshore islands in northeastern Queensland. *Australian Archaeology* **9**, 18–32.

Commonwealth of Australia (2015) *Reporting Back 2013/14: Working on Country and Indigenous Protected Areas Programmes.* Australian Government, Canberra.

Comrie-Grieg J, Abdo L (Eds) (2012) *Ecological Studies of the Bonaparte Archipelago and Browse Basin.* INPEX Operations Australia, Perth.

Crawford IM (1983) Aboriginal exploitation of resources, Admiralty Gulf. In *Archaeology at ANZAAS* (Ed. M Smith) pp. 46–51. Western Australian Museum Press, Perth.

Dale A, George M, Hill R, Fraser D (2016) *Traditional Owners and Sea Country in the Southern Great Barrier Reef: Which Way Forward?* Report to the National Environmental Sciences Programme. Reef and Rainforest Research Centre, Cairns.

Dixon RMW (1972) *The Dyirbal Language of North Queensland.* Cambridge University Press, Cambridge.

Dodson M (2007) Indigenous Protected Areas in Australia. Paper presented to International Expert Group Meeting on Indigenous Peoples and Protection of the Environment. 27–29 August. Department of Economic and Social Affairs Division for Social Policy and Development, Khabarovsk, Russian Federation.

Gibson LA, Yates S, Doughty P (2012) Biodiversity values on selected Kimberley islands, Australia. *Records of the Western Australian Museum* **81**(Supplement), 1–14.

Girringun (2007) *The Girringun Way: The Integration of Aboriginal Cultural and Environmental Knowledge Systems with GIS for NRM Application and Other Cultural Projects.* Girringun Aboriginal Corporation, Cardwell.

Girringun (2013) *Girringun Region Indigenous Protected Areas Plan of Management 2013–2023.* Girringun Aboriginal Corporation, Cardwell.

Girringun (2014) *Girringun's Culturally Assured Process.* Video interview with Phil Rist of the Girringun Aboriginal Corporation for Indigenous Governance Awards. <http://toolkit.aigi.com.au/resource/video-girringuns-culturally-assured-process>

Godden L, Cowell S (2016) Conservation planning and Indigenous governance in Australia's Indigenous Protected Areas. *Restoration Ecology* **24**, 692–697. doi:10.1111/rec.12394

Government of Western Australia (2011) *Kimberley Science and Conservation Strategy.* Government of Western Australia, Perth.

Grace P (2016) Indigenous-led conservation: experiences from the Kimberley. *Australian Environmental Law Digest* **December**, 21–26.

Henry CJ (1967) *Girroo Gurrll The First Surveyor and Other Aboriginal Legends.* Fortitude Valley Press, Brisbane.

Hill R, Walsh F, Davies J, Sanford M (2011) *Our Country Our Way: Guidelines for Australian Indigenous Protected Area Management Plans.* CSIRO Ecosystem Sciences and Department of Sustainability, Environment, Water, Population and Communities, Canberra.

IMAS (2013) *Your Marine Values Public Report*. Institute for Marine and Antarctic Studies, Hobart.

IUCN (2003) *Proceedings of the World Parks Congress*. Durban, 8–17 September. IUCN, Gland, Switzerland. <http://worldparkscongress.org/iucn_archives.html>

IUCN (2014) *Proceedings of the World Parks Congress*. Sydney, 12–19 November. IUCN, Gland, Switzerland. <http://worldparkscongress.org/drupal/node/146>

Moorcroft H, Ignjic E, Cowell S, Goonack J, Mangolomara S, Oobagooma J, Karadada R, Williams D, Waina N (2012) Conservation planning in a cross-cultural context: the Wunambal Gaambera Healthy Country project in the Kimberley, Western Australia. *Ecological Management and Restoration: Special Issue* **13**, 16–25. doi:10.1111/j.1442-8903.2011.00629.x

NAILSMA (2013) *National Indigenous Sea Country Workshop Report*. NAILSMA Knowledge Series 014/2013. Darwin.

NLC (1996) *Ecopolitics IX: Perspectives on Indigenous Peoples' Management of Environmental Resources*. Northern Territory University. Northern Land Council, Darwin.

Nowakowski S (2003) *Hinchinbrook: A Sacred Wilderness*. Little Ramsey Press, Cairns.

NRMMC (2010) *Australia's Strategy for the National Reserve System 2009–2030*. Natural Resource Management Ministerial Council, Canberra.

Ondei S, Prior LD, Williamson GJ, Vigilante T, Bowman DMJS (2017) Water, land, fire, and forest: multi-scale determinants of rainforest in the Australian monsoon tropics. *Ecology and Evolution* **7**(5), 1592–1604. doi:10.1002/ece3.2734

Scherrer P, Doohan K (2013) Taming wicked problems: towards a resolution of tourism access to Traditional Owner lands in the Kimberley region, Australia. *Journal of Sustainable Tourism* **22**, 1003–1022. doi:10.1080/09669582.2013.847943

Scherrer P, Smith A, Dowling R (2011a) Visitor management practices and operational sustainability: expedition cruising in the Kimberley, Australia. *Tourism Management* **32**, 1218–1222. doi:10.1016/j.tourman.2010.09.007

Scherrer P, Smith A, Randall M, Dowling R (2011b) Environmental and cultural implications of visitor access in the Kimberley Region, Australia. *Australian Geographer* **42**, 257–271. doi:10.1080/00049182.2011.595766

Smyth D (1993) *A Voice in All Places: Aboriginal and Torres Strait Islander Interests in Australia's Coastal Zone*. Resource Assessment Commission, Canberra.

Smyth D (2009) Just add water? Taking Indigenous Protected Areas into Sea Country. In *Indigenous Governance and Management of Protected Areas in Australia* (Eds D Smyth and G Ward) pp. 95–110. Australian Institute of Aboriginal and Torres Strait Islander Studies, Canberra.

Smyth D (2011a) *Guidelines for Country-based Planning*. <http://sbconsultants.com.au/index.php/country-based-planning>

Smyth D (2011b) Country-based Indigenous Protected Areas: a new approach to integrated coastal land and sea management. In *Queensland Coastal Conference*. 19–21 October. (Ed. D Smyth). Workshop at the 2011 Queensland Coastal Conference, Cairns.

Smyth D, Gould J, Ayre M, Bock E, Duffer-Hyams M, Vernes T (2016) Indigenous-driven co-governance of sea country through collaborative planning and Indigenous Protected Areas. *Indigenous Law Bulletin* **8**, 15–20.

SVA Consulting (2015) *Girringun Indigenous Protected Area: Social Return on Investment Analysis*. Social Ventures Australia for the Department of Prime Minister and Cabinet, Canberra. <https://www.dpmc.gov.au/sites/default/files/publications/Girringun-SROI-Summary.pdf>

SVA Consulting (2016) *Consolidated Report on Indigenous Protected Areas following Social Return on Investment Analyses*. Social Ventures Australia for the Department of Prime Minister and Cabinet, Canberra. <https://www.dpmc.gov.au/sites/default/files/publications/SROI-Consolidated-Report-IPA_1.pdf>

Szabo S, Smyth D (2003) Indigenous Protected Areas in Australia. In *Innovative Governance: Indigenous Peoples, Local Communities and Protected Areas* (Eds H Jaireth and D Smyth) pp. 145–164. Ane Books, New Delhi.

TSRA (2016) *Land and Sea Management Strategy for Torres Strait 2016–2036.* Report prepared by Land and Sea Management Unit, Torres Strait Regional Authority, Queensland.

WGAC (2010) *Wunambal Gaambera Healthy Country Plan: Looking after Wunambal Gaambera Country 2010–2020.* Wunambal Gaambera Aboriginal Corporation, Kalumburu.

WGAC (2017) *Uunguu Visitor Pass.* Wunambal Gaambera Aboriginal Corporation. <http://wunambalgaaambera.businesscatalyst.com/what-is-uvp.htm>

Zurba M, Ross H, Izurieta A, Rist P, Bock E, Berkes F (2012) Building co-management as a process: problem-solving through partnerships in Aboriginal Country, Australia. *Environmental Management* **10**, 10.1007/s00267-012-9845-2.

9

Integrating resource development with island conservation: Barrow Island as a model for conservation and development

Dorian Moro, Johann van der Merwe, Melissa Thomas, Andrew Smith and Russell Lagdon

Summary

While at face value resource development on islands can be viewed as problematic and at odds with conservation values, there are exceptions where development has shifted from legislated environmental management actions to the management of an environmental stewardship paradigm. This chapter discusses the development of an environmental ethos – or culture – among an oil and gas workforce on Barrow Island, Western Australia. We provide examples from an oil and gas company operating on Barrow Island that demonstrates how correct environmental designs in the engineering phase of a project, together with positive environmental behaviours by personnel during the construction and operations phase of a project, can shape an environmental stewardship in the workforce, leading to a well-managed environment from a nature conservation perspective. The Barrow Island example can be used as a model for establishing environmental cultures in teams working within the wider resource sector, particularly where new developments are being planned in sensitive areas.

Introduction

Islands around Australia have many conservation and cultural values: as refuges to fauna and flora, as important geological remnants offering healthy ecosystems relative to mainland equivalents, as nesting sites for marine species such as seabirds and sea turtles, and as important archaeological places with evidence of early occupation and exploration. As an offshore landmass, islands can also act as platforms to develop or extract natural resources, or as intermediary landmasses used for processing resources present in deeper waters offshore.

Historically, some islands were viewed by the explorers as remote locations offering a wealth of resources ready for exploitation. For example, Macquarie Island was visited by vessels from many countries across the northern and southern hemispheres to exploit its rich marine fauna and take penguin and seal oil back to the home colonies (Cumpston 1968). Christmas Island was used in the late 1880s to exploit its phosphate-bearing minerals in rock deposits, and presently provides an economic return based on phosphate (Adams 1993).

Table 9.1. Australian islands with resource developments active during 2011–16

Island	Jurisdiction	Resource	Activity
Christmas Island	Commonwealth	Phosphate	1891–present
Thevenard Island	Western Australia	Oil	1989–2014
Barrow Island	Western Australia	Oil and gas	Oil: 1967–present Gas: 2016–present
Varanus Island	Western Australia	Oil and gas	1986–present
Cockatoo Island	Western Australia	Iron ore Iron	1951–1984 1990–present
Koolan Island	Western Australia	Iron ore Iron	1959–1993 2007–2014
Groote Eylandt	Northern Territory	Manganese	1960–present
North Stradbroke Island	Queensland	Zircon, rutile, ilmenite	1949–present
Curtis Island	Queensland	Gas	2015–present

The exploitation of natural resources on islands continues today, albeit within a more legislated framework.

There are nine islands around Australia which in the past five years have supported or currently support resource developments (Table 9.1). In addition to managing their business needs, each company must adhere to environmental regulations under Australian and state legislation (e.g. the *Environment Protection Act 1986* in Western Australia), and any environmental commitments unique to their project. While this may appear to be incongruous, with business and conservation representing competing priorities, Barrow Island (Western Australia) is one example which demonstrates that, with careful integrated management, business and conservation objectives can be achieved. In this chapter, we explore how Chevron Australia has managed to integrate running an oil and gas development and production operation on Barrow Island, while maintaining and protecting the island's conservation values.

Barrow Island: context and environment

Barrow Island, ~23 400 ha and some 70 km off the north-west coast of Western Australia (WA) (Fig. 9.1), is considered one of the world's most significant conservation areas (DPW 2015). It supports 13 species of resident terrestrial mammals, some of which have become extinct or threatened on mainland Australia (Moro and MacAulay 2010). It is an important staging site for many international migratory shorebirds (Bamford and Moro 2011), and a regionally important rookery for green (*Chelonia mydas*) and flatback (*Natator depressus*) turtles (Pendoley Environmental 2005). A rich diversity of invertebrates is also present on the island (Callan *et al.* 2011).

Barrow Island is managed by the Department of Parks and Wildlife (DPW 2015). Since 1967, Barrow Island has supported Australia's largest onshore oilfield. Both private sector and government organisations have staff on the island, who work collaboratively to maintain its nature conservation values. Since commencement of the oilfield on Barrow Island, the fauna and flora have functioned amid a network of managed roads, flow pipelines, storage tanks, accommodation, over 900 cleared well pads (with ~460 producing wells), water injection wells, a marine port and loading facilities, airport and other infrastructure (Fig. 9.1).

In 2009, following a rigorous environmental assessment process (Chevron Australia 2005; Lagdon and Moro 2013), the Gorgon Project was approved (Commonwealth Government of Australia 2009; Government of Western Australia 2009). The Gorgon Project includes the construction of a

Fig. 9.1: Location of Barrow Island and the network of roads (lines), oil well pads (dots) and infrastructure (shaded areas) currently established. The dark shaded area on the east coast is the location of the Gorgon Gas Development and associated offshore infrastructure.

15.6 million tonne per annum liquefied natural gas processing plant (Chevron Australia 2005), undersea pipeline easements and associated offshore marine offloading facility and jetty. It is subject to some of the most stringent conditions imposed on a major resource project anywhere in the world (Commonwealth Government of Australia 2009; Government of Western Australia 2009). More than 400 environmental conditions apply to the Gorgon Project, involving ~1200 commitments. This has required the compilation and implementation of 33 government-approved Environmental Management Plans.

Environmental monitoring

During the environmental impact assessment process, several areas of the island ecosystem were assessed as at risk from the construction and operations of the LNG plant and associated facilities (Chevron Australia 2005, 2008). Examples include potential impacts to the flatback turtle population from dredging and construction activities and from light emissions from the LNG plant and accommodation areas (Chevron Australia 2015a), and risks to terrestrial mammal species from noise, vibration and an increase in vehicle numbers (Chevron Australia 2015b). Subject matter experts have been involved in the planning, development, execution and diagnostic phases of the monitoring programs on Barrow Island. Collaboration between the company and external scientists has resulted in robust data that are scientifically defensible, thus ensuring that the public, government and scientific community have confidence in Chevron's commitment to the environment and meeting its environmental obligations. Monitoring results have been communicated widely at scientific conferences (Moro 2012; Smith 2014) and through peer-reviewed publications (Gunawardene *et al.* 2013; Pendoley *et al.* 2014).

The monitoring programs were designed to be fit-for-purpose and not to collect a 'laundry list' of unstructured information with no purpose other than data-gathering (Lindenmayer and Likens 2010). For example, the flatback turtle monitoring program was designed using a conceptual model that focused monitoring efforts in relation to the turtle's life-history and the likely stressors that are known to affect this species (Fig. 9.2).

Baseline data proved to be an invaluable resource for the monitoring programs. For example, the baseline data collected over five years for the flatback turtle provided historic population

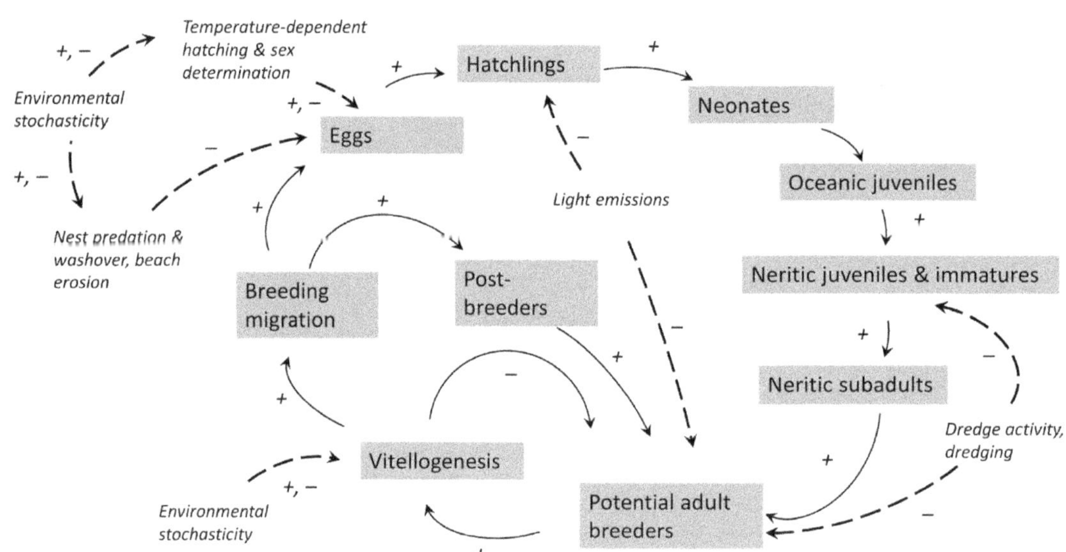

Fig. 9.2: Conceptual framework illustrating the flatback turtle lifecycle and the main natural and anthropogenic stressors that may have positive (+) or negative (–) effects on each life-stage. Environmental impact monitoring for the Gorgon Project on Barrow Island focuses on many of these life-stages and stressors. Source: modified from Chaloupka (2004).

data (e.g. nester abundance, clutch frequency, egg hatching probability, hatchling misorientation and disorientation) against which to measure potential changes in nester abundance during the construction (and future operations) of the development. It also provided a measure of natural variation of nesting females in the population (Chaloupka *et al.* 2012; Chevron Australia 2015a). The implication of this baseline data for industry (and here, Chevron) is that monitoring for environmental impact can be made over time with confidence given the science used to collect and interpret the data.

Biosecurity

Effective biosecurity management of Barrow Island is viewed as critical to operating on the island. Managing the risk of introducing non-indigenous species was first implemented by operators of the oilfield in the 1960s, and through continuous improvement has provided an effective layer of protection to the conservation values of the island by reducing the introduction of non-indigenous species (Butler 1987, 1989). For example, all goods and cargo shipped to Barrow Island were subject to biosecurity procedures, and a weed control program was maintained annually to manage the presence of buffel grass (*Cenchrus ciliaris*) on the island.

The construction and operation of a gas-processing facility on Barrow Island was recognised by the Western Australian Environment Protection Authority as a substantially higher risk to biodiversity conservation than the operation of an oilfield. To address concerns regarding biosecurity, Chevron Australia, together with the then Department of Conservation and Environment, developed a Quarantine Advisory Committee to advise Chevron on planning its heightened biosecurity for Barrow Island as part of the impact assessment process. Following environmental approvals, an independent Quarantine Expert Panel (QEP) was required as a ministerial condition to provide advice to Chevron Australia and the WA Minister for Environment. The focus of the QEP was on upgrading biosecurity measures in relation to the increase in freight profile during construction of the gas plant. To achieve this, baseline environmental data were collected to support future biosecurity surveillance programs. These baseline surveys commenced in early 2005.

In addition to the establishment of the QEP, community engagement was undertaken. The community involvement process for the Gorgon Project began during the early impact assessment period, with the identification of relevant community groups, scientists, specialist environmental consultants, government and non-government organisations, sections of private industry and members of the public who had a potential interest in the biosecurity management of Barrow Island. These organisations were invited to participate in the community involvement process to address the biosecurity issue for the Gorgon Development on Barrow Island. This resulted in the development of a pioneering risk-based Quarantine Management System (QMS). The overarching aim of the QMS is to facilitate the construction and operation of a gas-processing facility on Barrow Island while simultaneously protecting the conservation values of the island (Chevron Australia 2014). The QMS remains the largest non-government biosecurity initiative not just in Australia, but globally.

The QMS approach is to focus on pre-border prevention of non-indigenous species introductions, supported by contingencies for border and post-border biosecurity controls (Chevron Australia 2014). A systematic hazard evaluation of potential invasion pathways was developed in collaboration with 45 independent technical subject matter specialists (Stoklosa 2005). This process identified 13 pathways by which non-indigenous species could be transported to Barrow Island (Stoklosa 2005). A preliminary list of conceptual barriers was then identified for each pathway, to reduce the likelihood of introductions. These included design criteria that eliminated biosecurity risks, that were in turn supported by procedural requirements. The preliminary barriers were subject to scrutiny in risk assessment workshops by independent specialists in their respective fields. Once deemed effective in systematically reducing biosecurity risks to an acceptably low level, the barriers were incorporated into procedures, specifications and other administrative processes (e.g. training, audits, performance monitoring). The barriers include inspections and specific design modifications, and were implemented globally for material destined for Barrow Island.

Design modifications were undertaken in most instances during the planning stages of the project, and were aimed at reducing biosecurity risk from material hitchhiking on cargo. For example, large modules and pipe racks were designed to prevent the pooling of water that can harbour fugitive species such as mosquitos. The kitchen on Barrow Island was designed as a no-egress facility with complex air filters, airlocks and double doors to prevent non-indigenous species associated with food – that may have avoided detection during pre-border inspection – from escaping the controlled confines of the kitchen. Similarly, wood was excluded from accommodation areas for the construction of 5000 rooms, and each of the large accommodation buildings was sealed by a double door that prevents anthropogenic commensal species (e.g. cockroaches, spiders) from escaping to the environment. These designs were maintained through a process of audits to ensure compliance with project commitments, and reinvigorated through an education program both on and off the island.

Border and post-border controls focused on a mixture of inspections and surveillance in addition to observations made by personnel on the island. This active participation of the workforce, including local contractors and the overseas workforces, and the development of a positive work ethic of shared values and behaviours around biosecurity, has been central to the successful implementation of the QMS. One of the strengths of biosecurity awareness in the large workforce was the reporting of observations of unusual seed, invertebrate or vertebrate material (Box 9.1). The baseline environmental data collected before the commencement of the Gorgon Project provided essential information for taxonomists to identify whether species detected during the non-indigenous species surveillance programs were native or non-indigenous, and whether any non-indigenous species detected were a consequence of the Gorgon Project or present before development (Callan *et al.* 2011; Majer *et al.* 2013). The ongoing design of the biosecurity surveillance programs aim to detect non-indigenous species if they are present on Barrow Island (Jarrad *et al.* 2011a, 2011b; Whittle *et al.* 2013).

Developing an environmental culture

Prior to 2009, the island oilfield workforce comprised some 300 shift workers who remained on Barrow Island on a two-week rotation: a 10–15 year association with the island was not uncommon. This degree of consistent presence by a regular crew instilled a level of ownership, with personnel treating Barrow Island like a second home. Environmental technicians became the focal points for injured wildlife, and were trained in skills such as basic animal handling and care. Over time these people became viewed by their colleagues as focal points, or champions of the environment, when environmental concerns arose. Importantly, people could engage directly with the champions in the field rather than remotely by phone or email. Work crew often voluntarily committed their own time to participate in environmental activities which regularly took place outside work hours (Box 9.2). Among these role models there was one 'champion' whose impact on the workforce was both profound and lasting since the early years of exploration – Dr Harry Butler.

Dr Butler, a renowned Australian conservationist, valued face-to-face engagement with the island workforce. As an external consultant, he shared his extensive knowledge of the island biota. He also advised on how development can proceed in a manner that does not threaten the island's biodiversity, reinforcing the message that 'with management you can have conservation with development' (Butler 1989). With time, he was respected as an island champion for his ability to engage with people in a way that was educational and passionate yet directive.

Box 9.1: Effective biosecurity using work personnel

Scientific monitoring and surveillance is undergoing a dramatic change in the manner in which data are collected. Classic monitoring and surveillance methodologies implemented by scientists provide reliable data, but are often resource-intensive. For surveillance and monitoring programs that require extensive effort or are resource-scarce, program managers are required to think beyond classic methods and call on additional resources, namely citizens (Conrad and Hilchey 2011). However, questions remain as to the integrity of the data that citizens collect (Lewandowski and Specht 2015). On Barrow Island, data collected by biologists using classic data collection techniques have been augmented with data collected by the workforce to complete a non-indigenous species surveillance program for Barrow Island (Thomas *et al.* 2017). Quality control was achieved by using external taxonomic specialists for verification of species identification.

As part of the Gorgon Gas Development, there have been more than 13 000 flights transporting 700 000 passengers to Barrow Island. Every passenger has their shoes and luggage externals inspected before departure from the mainland. These pre-border biosecurity inspections have detected organisms (mostly invertebrates and seed material) on 6992 occasions (Scott *et al.* 2017; Fig. 9.3). The success of the pre-border biosecurity barrier is demonstrated in the low (n = 144) number of post-border non-indigenous species detected over a six-year period (September 2009 to September 2015) as part of the non-indigenous species surveillance effort on Barrow Island.

To meet Chevron's regulatory requirements of detecting a non-indigenous species if present on Barrow Island with 80% probability, a terrestrial surveillance system model was developed (Davidovitch *et al.* 2009; Jarrad *et al.* 2011a, 2011b; Whittle *et al.* 2013). The two

Fig. 9.3: Inspections of large machinery by quarantine personnel are one of many pre-border surveillance tools that reduce the risk of non-indigenous species entering Barrow Island with cargo. Photo: Chevron Australia.

main components of the surveillance model are active surveillance conducted by trained biologists implementing structured trapping and baiting, and passive surveillance conducted by direct observation and reporting by personnel working on the island.

Prior to commencing work on Barrow Island, all personnel undertake an induction which includes training on biosecurity awareness. Since 2009, over 35 000 personnel have been inducted. These personnel have contributed significantly to the biosecurity program on Barrow Island through submissions of suspect non-indigenous species. As a result, most post-border detections of terrestrial non-indigenous vertebrate and invertebrate species have been made by personnel (Thomas *et al.* 2017).

Given the scale of effort required to implement an effective biosecurity system, future developments can benefit from improved biosecurity practices by engaging the workforce in surveillance and monitoring programs. On Barrow Island, this method improved the spatial and temporal elements of the formal scientific surveillance effort, thus augmenting the classic trapping methodology. The Barrow Island example demonstrates that pre-border barriers, in combination with an effective post-border surveillance effort, are a highly effective biosecurity system.

Since the commencement of construction of the Gorgon Gas Development in 2009, personnel transfers to the island increased and the workforce peaked at more than 8000. This significantly larger workforce was more culturally diverse in origin than the historically local workforce, because of the expertise required for the complex nature of the infrastructure being built in Australia and globally. Workers included construction-oriented people with diverse skill sets (e.g. engineers, catering staff, mechanics, pipe fitters, environmental advisors, project managers, operators of large plant and mobile equipment). Their tenure on the island was relatively shorter than that of the previous workforce. In addition, this demographic group did not necessarily have an environment background or the correct mind-set required to work on an environmentally protected area such as Barrow Island. There was a need for alternative approaches to engage this more itinerant workforce, especially because of the large number of environmental commitments associated with the Project and the need to manage people's behaviours and mind-set.

Awareness of the environment became a deliberate part of the induction process for workers on Barrow Island. It highlighted the valued asset that Barrow Island represented to the community. This early engagement of new workers aimed to invigorate an expectation of the importance of the island on which they were employed. Senior managers in the environment area spoke to the workforce about the values of Barrow Island and the commitment from Chevron, and about the behaviours expected onsite. Senior construction managers felt ownership and responsibility for maintaining the QMS and other environmental programs.

With time, an environmental skill set was developed by the people working on Barrow Island. Environmental commitments were showcased to the large numbers of people on the island through educational campaigns that specifically aimed to avoid legislated directives, instead focusing on addressing the reasons for the environmental concerns; for example, switch off room lights because they affect marine turtle navigation.

Skill sets extended to managing biosecurity practices both offshore and onshore for large amounts of cargo, people and modules. Biosecurity training modules were developed for people working in critical interfaces associated with cargo preparation, disassembly of complex equipment, fabrication yards, seaports, airports, freight forwarding operations, biosecurity-approved premises, laydown and storage facilities and on-island project areas.

The involvement of senior managers was critical to maintaining the momentum and shared environmental values for progressing environmental stewardship. Managers were sensitised to the high environmental expectations and hurdles during

Box 9.2: Barrow Island Environment Month

Workforce engagement has been one of the key environmental protection strategies utilised for more than 50 years, since Chevron Australia has been operating the Barrow Island oilfield. From the beginning, it was recognised that the company was operating in a special location and it needed to do things differently from standard oilfield practices. Harry Butler was asked for advice on how the company could keep the island as it first found it. Two of his recommendations were to initiate a biosecurity system for the island, and to enrol and educate the workforce (Butler 1989).

Environment Month, a set of environment-focused activities at a particular time of the year, is an example of how workforce engagement can aid environmental protection through education and participation. It was initially set up to support 'Clean Up Australia Day'. Workgroups were encouraged to spend the last hour of the day patrolling their allocated section of road, work area or camp to pick up any litter or rubbish that had blown off work vehicles or had escaped from worksites. From worksites and roads, the activity grew into beach clean-ups where workforce volunteers would walk the beaches after hours, picking up any rubbish washed ashore. The island Environment Team would then put on a beach BBQ to thank the volunteers. Perth-based environmental staff started participating in the activities when onsite, and were invited to talk about the results of monitoring programs and other environmental topics of interest to the workforce.

As momentum grew, and as the activities diversified and others from Chevron Australia wanted to participate, the date was changed from Clean Up Australia Day to align with World Environment Day. Volunteers were sought from office-based staff to help deploy the program across different projects and contractors.

The program of activities and learning was not about compliance or regulation. It was more focused on conservation values and the environment, aiming to increase the workforce's sense of ownership and responsibility for the special place where they work.

From modest beginnings, the field program now regularly includes 'toolbox talks' to small workgroups, site-wide talks with up to 500 participants, island tours for people who champion the environment (Fig. 9.4), recycling waste into art competitions, and the

Fig. 9.4: Island tours have been an important way to share environmental experiences and educate the industry workforce about the values of the environment in which they work. Photo: Russell Lagdon.

opportunity for workers and their families to participate in office-based activities during their time off. The activities run in Perth include lunchtime talks, lunchtime demonstrations (e.g. fauna handling, marine monitoring, biosecurity inspector dogs, GIS for environmental data), 'speed presentations' on technical environmental subjects, competitions for children, and a family day where employees and contractors can bring their family into a venue, hear some talks and have a hands-on experience with interactive displays.

In 2016, the program transitioned from a focus on Gorgon construction, during which there were up to 8000 people on the island, to building and supporting a lower number of long-term operational workforce. Over the years there has been significant change out of the original champions – the program is self-sustaining and the workforce continues to come up with new ideas.

the approvals process, and so recognised the need to factor environmental risk into the engineering design and decision-making stages of construction (Box 9.3).

Ownership of conservation values

Understanding – and managing – the risks to biodiversity on Barrow Island has been central to the ongoing presence of Chevron on Barrow Island. These collective efforts have ensured that the island's full suite of native species has continued to thrive alongside development.

Skill sets alone could not, however, be relied upon to ensure the implementation of conservation values. The discipline of implementation lay with the workforce and the development of a mind-set. This took time to develop, with lessons learnt from the application of environmental and biosecurity commitments then adaptively implemented back into the design and execution decision cycles of the Project. While workforce safety was embedded in a commitment to zero injuries, environmental management across the island required more than environmental impact (scientific) monitoring with such a large workforce. All workers were granted a stop-work authority. That authority was a safety message consistently used by Chevron's CEO to empower people to reduce risk of injury in the workforce. This message was also applied to the environment, which calibrated a mind-set of zero tolerance to environmental harm beyond that required for the development. This expectation laid the foundation for an environmental commitment that historically had been evident only among the more stable workforce associated with the oilfield operations.

Construction people came up with their own initiatives of environmental management practices, not dictated solely by their environmental commitments but by their own values for the environment where they worked (Box 9.2). Work teams actively took part in environment competitions, and individuals were encouraged by line managers to innovate ideas or practices that could balance site construction activities while meeting environmental obligations and addressing environmental concerns. These ideas were encouraged through an award system and island environmental champions were recognised, to encourage participation among other workforce teams. This mind-set was viewed by Chevron as shared values akin to an environmental culture: that is, a belief, practice and often unarticulated assumption that underlies human relations with their environments (Head *et al.* 2005). Other initiatives and activities implemented on the island, that were not required by legislated environmental commitments, included:

- coastal clean-up (which won a Certificate of Merit in the Golden Gecko Environment Awards 2000), managed by the workforce on the island and aimed at removing debris washed up around Barrow Island's coastline;
- family days, which rewarded workforce members who demonstrated leadership and commitment to protecting the environment. Members were nominated and had the opportunity to invite a family member or friend to Barrow Island for the day to

Box 9.3: Managing artificial light emissions for marine turtles

Flatback turtle (*Natator depressus*) nesting adults (Fig. 9.5) are sensitive to the effects of artificial light and it can deter them from selecting suitable nesting beaches (Limpus 2009). Additionally, artificial lighting close to a nesting beach can mask a hatchling's ability to see natural light horizons and hence disrupt sea-finding behaviour, reducing hatchling survival (Tuxbury and Salmon 2005; Kamrowski *et al.* 2014). Hatchling behaviour for some turtle species has been shown to be less influenced by wavelengths in the yellow to red colour spectrum (Witherington and Bjorndal 1991). However, lights in this spectrum are not suitable for people to work under as they generally have poor colour rendition, meaning that some colours become less clear or can disappear completely as the light source becomes monochromatic. This results in a potentially significant safety risk when working in hazardous areas.

In designing the LNG plant to have minimal impact on marine turtles, several factors were considered during the early stages of the engineering phases with the primary aim being to minimise light glow and eliminate direct light spill on nesting beaches. Design features or engineering controls to manage lighting included directional or shielded lighting (Fig. 9.6), the mounting of light fittings as low as practical, louvred lighting on low-level bollards in pedestrian areas, screening of luminaires, and recessed lighting. In other areas, automatic timers or photovoltaic switches were installed to lights, black-out blinds were fitted to windows and there is a curfew on the use of outdoor sports facilities that require lighting. All areas within 500 m of the coast are kept dark, with task lighting manually activated on an as-needed basis.

Architecturally, accommodation buildings were oriented so that a minimal number of windows faced the beaches, and parking areas were located such that the illumination of vehicle headlights onto dunes could be reduced. Areas that did not require night work, including some of the road networks, security fences and holding ponds, were left unlit. The flare within the LNG plant was designed as a ground-box flare rather than the more conventional stack-flare to reduce light glow, with the flare positioned low in the landscape

Fig. 9.5: Flatback turtle. Photo: Dorian Moro.

Fig. 9.6: Plant lighting showing downward-facing and shielded luminaires to direct light spill onto work areas and reduce light glow. Photo: Dorian Moro.

and as far as practical from the coast. A 14 m high louvred shielding wall further reduced the effects of light from the flare.

Lighting management along the LNG jetty and causeway (Fig. 9.1) adopted many of the design features used for the plant and accommodation areas. As it is located perpendicular to turtle nesting beaches, additional light management measures included the use of photovoltaics to automatically turn lights on and off as vehicles pass, and the use of non-reflective matte paint on the superstructure to reduce glare. LNG loading activity is supported by a fleet of tugs that have been custom-built to minimise external light spill, and LNG carriers are specifically requested to minimise non-essential lighting while moored at the loading jetty.

Administrative controls include a lighting-specific audit program that assesses the effectiveness of guidelines and procedures for minimising light spill. The focus of the audit program coincides with the turtle breeding season (October to March) but it is conducted beforehand with ample time to implement corrective actions if needed, although light reduction techniques apply across the year.

Educational programs are conducted at the start of each turtle breeding season to raise awareness among the workforce and further engage them in the effort to reduce lighting wherever possible. Importantly, an ongoing turtle research and monitoring program is in place to assess light levels during the breeding season as well as potential impacts on adult or hatchling turtle behaviour (Chevron Australia 2015a; Pendoley and Kamrowski 2016).

experience life on the island and to view the special environment where the family member worked;

- heavy vehicles were driven only in convoy at dawn and dusk, when it is dark and mammals are active on the roads. Vehicle speed was already reduced at night to 40 km h^{-1} as a self-imposed commitment, and the practice of driving in convoy was suggested by site personnel as a means of further reducing the risk of animal strike;
- the reinforcement of environmental messages to the island workforce was an important approach by Chevron to ensure continued environmental behaviours, and a level of sustainability of the expected environmental mind-set. This is vital, because cultural shifts can affect people over time if their mind-set and practices are not reinvigorated (Head *et al.* 2005).

Sustaining the environmental culture: leaving a legacy

The management of oilfield operations on Barrow Island is widely recognised as an industry benchmark among resource sector industries and local conservation authorities. There are some key lessons.

- Workforce engagement encourages a sense of ownership and responsibility for the place where they work and for its environmental management. The ownership of conservation values must come from the people working on the island. This requires early engagement with the workforce and leadership from senior management that empowers them to act and to drive initiatives. Importantly, senior management must be seen to be authentic and trustworthy in their commitment to safeguarding the conservation values. Companies should not be complacent about senior managers leading initiatives which require technical oversight and the reinforcement of key messages.
- Fostering an environmental mind-set among an itinerant workforce can lead to an environmental culture.
- Environmental risk can be balanced with risk mitigation. The development of a gas plant on a nature reserve may – if carefully planned and managed – proceed in the face of perceived environmental risks by implementing engineering design controls to reduce the likelihood of known or predicted risks, by monitoring for environmental and biosecurity impacts using projects grounded in science, and by using scientific approaches and engineering designs to address knowledge gaps to reduce unknown consequences.

These lessons have the potential to be applied across the wider range of developments and activities that occur elsewhere on Australian islands.

Acknowledgements

We are grateful to Eugene Carmody (Department of Lands) for assistance with Table 9.1, and to Andrew Burbidge for review comments to improve this chapter content. Finally, we dedicate this chapter to the late Dr Harry Butler AO CBE (1930–2015), who mentored us, who inspired us to keep caring, and who championed and challenged Chevron and its staff to look after Barrow Island. This culture of a workforce caring for the environment in which they work carries on today.

References

Adams J (1993) *Christmas Island. The Early Years 1888 to 1958: Historic Photographs with Many Untold Tales from the Early Years of Christmas Island, an Isolated Island in the Indian Ocean*. Chapman, Canberra.

Bamford M, Moro D (2011) Barrow Island as an important bird area for migratory waders in the east Asian–Australasian flyway. *Stilt* **60**, 46–55.

Butler WH (1987) Management of disturbance in an arid remnant: the Barrow Island experience. In *Nature Conservation: The Role of Remnants of Native Vegetation* (Eds A Saunders, G Arnold, A Burbidge and A Hopkins) pp. 279–285. Surrey Beatty & Sons, Sydney.

Butler WH (1989) Management of Barrow Island. In *Australian and New Zealand Islands: Nature Conservation Values and Management* (Ed. A. Burbidge) pp. 193–199. Occasional Paper 2/89. Department of Conservation and Land Management, Perth.

Callan SK, Majer JD, Edwards K, Moro D (2011) Documenting the terrestrial invertebrate fauna of Barrow Island, Western Australia. *Australian Journal of Entomology* **50**, 323–343. doi:10.1111/j.1440-6055.2011.00818.x

Chaloupka M (2004) *Stochastic Simulation Model of Southwestern Pacific Hawksbill Sea Turtle Population Dynamics: Users Guide*. Ecological Modelling Services, University of Queensland, Brisbane.

Chaloupka M, Pendoley K, Moro D (2012) Control charts: a robust approach for monitoring endangered species exposure to a major construction project. *SPE/APPEA International Conference on Health, Safety and Environment in Oil and Gas Exploration and Production*. Perth, 11-13 September.

Chevron Australia (2005) *Draft Gorgon Environmental Impact Statement/Environmental Review and Management Programme for the Proposed Gorgon Development. 8. Existing Environment*. Chevron Australia, Perth.

Chevron Australia (2008) *Gorgon Gas Development Revised and Expanded Proposal Public Environmental Review.* Chevron Australia, Perth.

Chevron Australia (2014) *Barrow Island Quarantine: Terrestrial and Marine Quarantine Management System.* Chevron Australia, Perth.

Chevron Australia (2015a) *Gorgon Gas Development and Jansz Feed Gas Pipeline: Long-term Marine Turtle Management Plan.* Chevron Australia, Perth.

Chevron Australia (2015b) *Gorgon Gas Development and Jansz Feed Gas Pipeline: Terrestrial and Subterranean Environment Monitoring Program.* Chevron Australia, Perth.

Commonwealth Government of Australia (2009) *Approval: Gorgon Gas Development (EPBC Reference: 2008/4178), 26 August 2009.* Canberra.

Conrad CC, Hilchey KG (2011) A review of citizen science and community-based environmental monitoring: issues and opportunities. *Environmental Monitoring and Assessment* **176**, 273–291. doi:10.1007/s10661-010-1582-5

Cumpston JS (1968) *Macquarie Island.* Publication No 93. ANARE Scientific Reports, Series A. Department of External Affairs, Antarctic Division, Melbourne.

Davidovitch L, Stoklosa R, Majer J, Nietrzeba A, Whittle P, Mengersen K, Ben-Haim Y (2009) Info-Gap theory and robust design of surveillance for invasive species: the case study of Barrow Island. *Journal of Environmental Management* **90**, 2785–2793. doi:10.1016/j.jenvman.2009.03.011

DPW (2015) *Barrow Group Nature Reserves Management Plan 82.* Department of Parks and Wildlife, Perth.

Government of Western Australia (2009) *Statement that a Proposal may be Implemented: Gorgon Gas Development Revised and Expanded Proposal: Barrow Island Nature Reserve.* Ministerial Statement No. 800, 10 August 2009. Government of Western Australia, Perth.

Gunawardene NR, Majer JD, Taylor CK, Harvey MS (Eds) (2013) *The Terrestrial Invertebrate Fauna of Barrow Island, Western Australia.* Records of the Western Australian Museum Supplement 83. Western Australian Museum, Perth.

Head L, Trigger D, Mulcock J (2005) Culture as concept and influence in environmental research and management. *Conservation and Society* **3**, 251–264.

Jarrad FC, Barratt S, Murray J, Mengersen K (2011a) Ecological aspects of biosecurity surveillance design for the detection of multiple invasive animal species. *Biological Invasions* **13**, 803–818. doi:10.1007/s10530-010-9870-0

Jarrad FCS, Barrett J, Murray J, Parkes J, Stoklosa R, Mengersen K, Whittle P (2011b) Improved design method for the biosecurity surveillance and early detection of non-indigenous rats. *New Zealand Journal of Ecology* **35**, 132–144.

Kamrowski RL, Limpus C, Pendoley K, Hamann M (2014) Influence of industrial light pollution on the sea-finding behaviour of flatback turtle hatchlings. *Wildlife Research* **41**, 421–434. doi:10.1071/WR14155

Lagdon R, Moro D (2013) The Gorgon Gas Development and its environmental commitments. *Records of the Western Australian Museum* **83**, 9–11. doi:10.18195/issn.0313-122x.83.2013.009-011

Lewandowski E, Specht H (2015) Influence of volunteer and project characteristics on data quality of biological surveys. *Conservation Biology* **29**, 713–723. doi:10.1111/cobi.12481

Limpus C (2009) *A Biological Review of Australian Marine Turtles.* Environmental Protection Agency, Brisbane.

Lindenmayer DB, Likens GE (2010) The science and application of ecological monitoring. *Biological Conservation* **143**, 1317–1328. doi:10.1016/j.biocon.2010.02.013

Majer JD, Callan SK, Edwards K, Gunawardene NR, Taylor CK (2013) Baseline survey of the terrestrial invertebrate fauna. *Records of the Western Australian Museum* **83**(Supplement), 13–112. doi:10.18195/issn.0313-122x.83.2013.013-112

Moro D (2012) Maximising scientific value on Barrow Island: a Class A nature reserve. Abstract. *Island Arks Symposium III.* University House, Canberra, 7–9 February.

Moro D, MacAulay I (2010) *A Guide to the Mammals of Barrow Island*. Chevron Australia, Perth.

Pendoley K, Kamrowski RL (2016) Sea-finding in marine turtle hatchlings: what is an appropriate exclusion zone to limit disruptive impacts of industrial light at night? *Journal for Nature Conservation* **30**, 1–11. doi:10.1016/j.jnc.2015.12.005

Pendoley KL, Bell CD, McCracken R, Ball KR, Sherborne J, Oates JE, Becker P, Vitenbergs A, Whittock PA (2014) Reproductive biology of the flatback turtle *Natator depressus* in Western Australia. *Endangered Species Research* **23**, 115–123. doi:10.3354/esr00569

Pendoley Environmental (2005) Sea turtles. In *Draft Gorgon Environmental Impact Statement/Environmental Review and Management Programme for the Proposed Gorgon Development* (Ed. RPS BBG) Technical Appendix C7. Chevron Australia, Perth.

Scott JK, McKirdy SJ, van der Merwe J, Green R, Burbidge AA, Pickles G, Hardie D, Morris K, Kendrick P, Thomas ML, Horton K, O'Connor SM, Downs J, Stoklosa R, Lagdon R, Marks B, Nairn M, Mengersen K (2017) Zero-tolerance biosecurity protects high-conservation-value island nature reserve. *Scientific Reports* doi:10.1038/s41598-017-00450-y

Smith A (2014) Managing biodiversity on Barrow Island during infrastructure development. Abstract. *27th Australasian Wildlife Management Society Annual Conference.* 2–4 December, Brisbane.

Stoklosa R (2005) How-to guide for conducting risk-based assessments of quarantine threats to Barrow Island. In *Draft Gorgon Environmental Impact Statement/Environmental Review and Management Programme for the Proposed Gorgon Development* (Ed. RPS BBG) Technical Appendix D2. Chevron Australia, Perth.

Thomas ML, Gunawardene N, Horton K, Williams A, O'Connor S, McKirdy S, van der Merwe J (2017) Many eyes on the ground: citizen science is an effective early detection tool for biosecurity. *Biological Invasions* **19**, 2751–2765.

Tuxbury SM, Salmon M (2005) Competitive interactions between artificial lighting and natural cues during seafinding by hatchling marine turtles. *Biological Conservation* **121**, 311–316. doi:10.1016/j.biocon.2004.04.022

Whittle PJ, Jarrad FC, Barrett S, Martin P, Majer JD, Stoklosa R, Mengersen K (2013) A method for designing complex biosecurity surveillance systems: detecting non-indigenous species of invertebrates on Barrow Island. *Diversity and Distributions* **19**, 629–639. doi:10.1111/ddi.12056

Witherington BE, Bjorndal KA (1991) Influences of wavelength and intensity on hatchling sea turtle phototaxis: implications for sea-finding behaviour. *Copeia* **4**, 1060–1069.

10

Australia's World Heritage islands

Jennie Whinam and Justine D. Shaw

Summary

Australia has four islands that are listed as World Heritage properties. In addition, there are several other World Heritage properties (including the Great Barrier Reef) with islands significantly contributing to their Outstanding Universal Value. The four World Heritage islands have different geologic, climatic and biological values; three of the islands formed as isolated landmasses in vast oceans. Lord Howe, Macquarie and Heard-McDonald have endemic biotas due to long periods of evolutionary isolation. These three islands have only recent histories of human habitation, that have allowed evolutionary processes to occur largely free from human influence. In contrast, Fraser Island is a sand island close to the Australian mainland and has a comparatively long history of human occupation. The number of established non-native species varies between the islands yet their impacts have been substantial across all islands. The geographical setting of each island has led to recognition of their aesthetic values on a global scale except for the Heard and McDonald Island group, which includes Australia's only active volcano and largest mountain. The threats to the natural values of these World Heritage properties are like those faced by many Australian islands – climate change, biosecurity, feral species and human usage, including tourism. The isolation of these islands makes management and mitigation of the threats difficult, but their isolation has also saved them. It has limited the degree of degradation to date and offers opportunities to protect many threatened species. The demand for access by researchers and tourists to these World Heritage-listed islands is likely to increase as the number of wilderness areas with high integrity declines; this is a challenge to be managed into the future.

Introduction

The World Heritage Convention is a key international agreement that underpins conservation globally. The Convention was created in 1972, with the main aim of identifying and protecting the world's natural and cultural heritage considered to be of Outstanding Universal Value (OUV). It is based on the principle that protecting places of OUV is the responsibility of the whole international community, regardless of where the place is located (UNESCO 1972).

World Heritage properties are special areas with outstanding universal values of great significance to society, including future generations (Rao 2010). The World Heritage List is governed by the World Heritage Committee supported by the UN Educational, Scientific and Cultural Organisation (UNESCO) World Heritage Centre, the secretariat for the Convention and three technical advisory bodies. The International Union for Conservation of Nature (IUCN) is the advisory body on natural heritage. The IUCN's World Heritage Programme supports and promotes the World Heritage Convention in protecting biodiversity and aims to strengthen the conservation and management of natural World Heritage sites (IUCN 2016). The

World Heritage Programme works in close collaboration with IUCN's World Commission on Protected Areas (WCPA) and Global Protected Areas Programme (GPAP).

IUCN evaluates sites that are nominated using the relevant World Heritage Convention criteria vii–x (UNESCO 2005):

- vii to contain superlative natural phenomena or areas of exceptional natural beauty and aesthetic importance;
- viii to be outstanding examples representing major stages of Earth's history, including the record of life, significant ongoing geological processes in the development of landforms, or significant geomorphic or physiographic features;
- ix to be outstanding examples representing significant ongoing ecological and biological processes in the evolution and development of terrestrial, freshwater, coastal and marine ecosystems and communities of plants and animals;
- x to contain the most important and significant natural habitats for *in situ* conservation of biological diversity, including those containing threatened species of outstanding universal value from the point of view of science or conservation. The criteria listed here are the current unified criteria rather than the criteria in use at the time of listing.

The World Heritage List contains 1031 properties of OUV; natural sites represent 22% of the list, including 32 mixed (both cultural and natural) sites and 197 natural sites in 97 countries protecting over 280 million ha of land and sea (UNESCO 2017). Of these properties, globally only 17 are composed entirely of islands or archipelagos. Four of these island properties are in Australia: Fraser Island, Lord Howe Island, Macquarie Island and Heard and McDonald Islands (Fig. 10.1). Each island has several properties that make it special (Tables 10.1–10.5, Box 10.1).

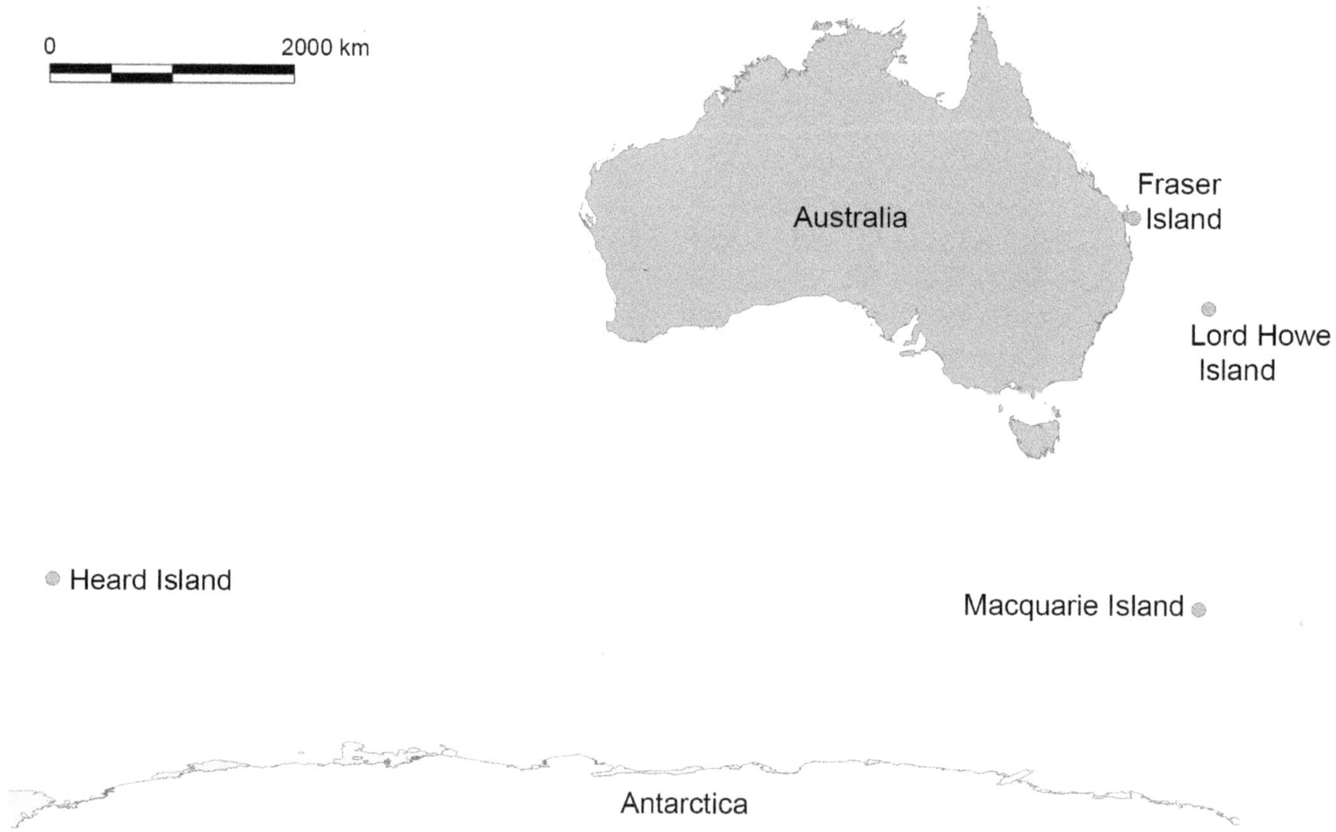

Fig. 10.1: The geographic location of Australia's World Heritage islands including remote sub-Antarctic Macquarie and Heard (and McDonald) Islands in the Southern Ocean.

Box 10.1: Biodiversity and conservation values of Australia's major WHA islands

Tables 10.1–10.5 show the range of natural values found on Australia's major World Heritage Area islands, their conservation status and the non-native vertebrate and plant species that have established.

Table 10.1. Number of native species found on Australia's major WHA islands (extant species)

	Mammals	Birds	Plants	Invertebrates
Fraser	33	330[a]	860	ND
Lord Howe	1	15	241	1600
Macquarie	3	24	42	350
Heard-McDonald	2	19	11	166

[a] Includes non-breeding species.
ND = no comprehensive data available.

Table 10.2. Number of threatened species found on Australia's major WHA islands (as listed under the *Environment Protection and Biodiversity Conservation Act 1999*)

	Mammals	Birds	Reptiles	Frogs	Plants	Invertebrates
Fraser	0	0	1	1	4	0
Lord Howe	0	9	2	0	2	6
Macquarie	2	10	0	0	2	0
Heard-McDonald	2	5	0	0	0	0

Table 10.3. Number of species known to be extinct on Australia's major WHA islands

	Mammals	Birds	Reptiles	Frogs	Plants	Invertebrates
Fraser	0	0	0	0	1	0
Lord Howe	1	9	9	0	2	4
Macquarie	0	2	2	0	0	0
Heard-McDonald	0	0	0	0	0	0

Table 10.4. Number of endemic species found on Australia's major WHA islands

	Mammals	Birds	Reptiles	Frogs	Plants	Invertebrates
Fraser	0	0	0	0	1[a]	0
Lord Howe	2[b]	13[d]	2	0	113	6
Macquarie	0	3[c]	0	0	3	0
Heard-McDonald	0	1	0	0	0	0

Several of these endemic species have become extinct.
[a] Species now extinct.
[b] One species now extinct.
[c] Two species now extinct.
[d] Nine species now extinct.

Table 10.5. Number of non-native vertebrates and plant species found on Australia's major WHA islands

	Vertebrates	Plants
Fraser	24	ND
Lord Howe	17	271
Macquarie	2	3
Heard-McDonald	0	1

Vertebrates include reptiles, birds and mammals. Only vascular plants were included.
ND = no comprehensive data available.

In Australia, the overarching management of the World Heritage values occurs under the national *Environment Protection and Biodiversity Conservation Act 1999*. The EPBC Act is the statutory instrument for implementing Australia's obligations under several multi-lateral environmental agreements, including the World Heritage Convention. Management Plans that list the significant heritage aspects of a place and how the values of a site will be managed are required by the EPBC Act (DEE 2017).

K'gari-Fraser Island

K'gari-Fraser Island is situated ~20 km off the coast of Hervey Bay, Queensland, Australia. It is the largest sand island in the world. Lying at 25°S it is a subtropical island, 35 km from the town of Hervey Bay on the Queensland coast (Fig. 10.1, Fig. 10.2a). Archaeological evidence suggests that Aboriginal Butchulla people have lived in the Great Sandy area for at least 5000 years, including Fraser Island (K'gari) and the adjacent mainland. Captain Cook first sighted Fraser Island in May 1770 and named the island 'Great Sandy Peninsula', thinking it was part of the mainland. Matthew Flinders proved this to be incorrect when he explored parts of Hervey Bay in 1799. When Europeans settled the island in the 1840s they caused great disruption to Indigenous people. The impact of the settlers increased in the 1860s as they began clearing land for pasture and commenced the harvesting of timber. Conflict and disease took their toll on the Butchulla, with only 230 recorded as remaining in 1880. The last of the Butchulla left the island in 1904 when the mission was closed, as the island was declared a Forestry Reserve (Brown *et al.* 2015). Despite this, the Butchulla continue to have cultural affiliations with the island. After some unsuccessful attempts, a native title claim was lodged on behalf of the Butchulla people in November 2009 over most of the island and some marine and adjacent land. The claim was successfully determined in October 2014 (Brown *et al.* 2015).

Logging commenced on Fraser Island in 1863, the main target species being kauri pine (*Agathis robusta*), hoop pine (*Araucaria cunninhamii*) and cypress pine (*Callitris columellaris*). In 1925, this changed when loggers realised that satinay (*Syncarpia hilli*) was resistant to marine borers and the focus changed to that species. The first mining leases on Fraser Island were granted in 1949 for the extraction of heavy minerals, mainly rutile and zircon. Sand mining exploration increased in the 1960s, attracting strong opposition from conservationists and community groups. This led to the cessation of sand mining in Great Sandy in 1976; logging continued until 1991. In 1971, the first national parks were declared in the northern part of Fraser Island, with other areas added over time (DNPSR 2011).

Fraser Island has been used for recreation since the 1870s, but the first commercial tours began in the 1930s. Visitor numbers dramatically increased during the controversies surrounding sand mining in the 1970s and logging in the early 1990s. There are currently around 350 000–400 000 visitors per year (Wardell-Johnson *et al.* 2015). Management of dingo–visitor interactions is a major issue on the island (Burns and Howard 2003; Allen *et al.* 2015).

Fraser Island is 122 km long. It has over 250 km of clear sandy beaches with long, uninterrupted sweeps of ocean beach, strikingly coloured sand cliffs and spectacular blowouts. There are 16 discrete habitats and vegetation types defined for Fraser Island. Remnants of tall rainforest grow on coastal dune systems. The island includes the world's largest unconfined aquifer on a sand island, with half the world's perched freshwater dune lakes located inland from the beach. These physical features combine to make Fraser Island an outstanding example of ongoing biological, hydrological and geomorphological processes (DNPSR 2011).

Fraser Island is a particularly important site for migratory waders and the eastern ground parrot (*Pezoporus wallicus*). There are only a few mammal species present on the island. One of the best

Fig. 10.2: The four distinct World Heritage island properties ecosystems. (a) Fraser Island, the world's largest sand island. Photo: Kate Kiefer. (b) Heard Island, with southern elephant seals lying in the foreground of the active volcano Big Ben. Photo: Kate Kiefer. (c) Macquarie Island, where rolling green hills provide a backdrop to beach-dwelling king penguins. Photo: Justine Shaw. (d) Lord Howe Island, which supports a diverse array of vegetation communities and over 100 endemic plant species. Photo: Jessica Peatey.

known is the dingo (*Canis familiaris*), which arrived on Fraser Island between 3000–8000 years ago (Burns and Howard 2003). The island supports a relatively large population of dingoes ('wangari') and, as domestic dogs are now prohibited, the conservation significance of Fraser Island dingoes is well recognised and widely accepted (Thompson *et al.* 2003; Allen *et al.* 2012). Dingoes on Fraser Island are one of the purest strains of dingo in Australia, as they have not crossbred with domestic or feral dogs to the same extent as most mainland populations (Allen *et al.* 2015). Dingoes are considered an important element of the island's cultural, ecological and conservation values.

Fraser Island was inscribed on the World Heritage List in 1992 for criteria vii, viii and ix, following protracted negotiations with stakeholders (Commission of Inquiry 1991) and was added to the National Heritage List in 2007.

Integrity

The property includes all of Fraser Island and several small adjacent islands off the west coast including Stewart and Dream Islands, and covers an area of 181 851 ha. Most of Fraser Island is national park but there is a small area of private land, with several small settlements and a tourist resort. All marine area within the property lies within the Great Sandy Marine Park. The property is managed by the Queensland government (DNPSR 2011).

Key threats addressed in the management plan include degradation due to high visitor numbers, inappropriate fire regimes, pest plants and animals, and climate change. Intensive recreational use, especially four-wheel driving, has had major impacts. Localised degradation has occurred from excessive numbers of visitors. Lake water quality has suffered serious impacts (DNPSR 2011). A recent study by Wardell-Johnson *et al.* (2015) highlights that the island is at the crossroads of World Heritage status, due to drastically increased visitation since inscription and destructive environmental use, all in combination with climatic change.

Lord Howe Island group

The first reported sighting of Lord Howe Island was in 1788 when HMS *Supply* sailed from Botany Bay to establish a penal settlement on Norfolk Island. On the return journey, Lieutenant Henry Lidgbird Ball sent a party ashore to claim Lord Howe Island as a British territory. It was initially used as a port to provision the whaling industry, and was permanently settled in June 1834. Whaling declined in the 1880s when an international export trade in the endemic kentia palms (*Howea foresteriana*) began, and this remains a key component of the island's economy. Tourism began in 1945 and is now the major industry (Hutton *et al.* 2007).

The Lord Howe Island group – Lord Howe Island, Admiralty Islands, Mutton Bird Islands, Ball's Pyramid, and associated coral reefs and marine environments – covers a total of 146 300 ha (including a terrestrial area of 1540 ha) and lies 5760 km east of Port Macquarie, off the coast of New South Wales (Fig. 10.1, Fig. 10.2d). Lord Howe Island has spectacular landscapes, including volcanic mountains, and diverse low-lying rainforests, palm forests and grasslands.

The long isolation of this island group has resulted in a diverse range of ecosystems exhibiting evolutionary processes and high endemism, with human disturbance being recent. The island has high floral endemicity and there are four endemic birds found on the island today. High concentrations of seabirds occur on the island, including the providence petrel (*Pterodroma solandri*) and red-tailed tropicbird (*Phaethon rubricauda*). Notable iconic species of the island include the endemic flightless Lord Howe woodhen (*Gallirallus sylvestris*), once regarded as one of the rarest birds in the world, and the endemic Lord Howe Island phasmid (*Dryococelus australis*), the world's largest stick insect, thought to be extinct until its rediscovery on Ball's Pyramid in 2001 (Priddel *et al.* 2003).

The property was inscribed on the World Heritage List in 1982 for criteria vii and x and was added to the National Heritage List in 2007. About 75% of the terrestrial part of the Lord Howe Island property is managed as a permanent park preserve. The property is governed the New South Wales statutory authority established under the *Lord Howe Island Act 1953* (DECC NSW 2007).

Integrity

The integrity of the property is at risk from development pressures, introduced plants and animals (particularly rodents) and high tourist numbers. However, implementation of a management program has improved the conservation status of the Lord Howe woodhen and biodiversity values have been improved through the successful eradication of feral pigs and cats as well as a marked reduction in goats. Weed control and eradication programs have been undertaken on the island over the years. These programs targeted numerous species and utilised different technologies depending on available funding and distribution of the target weeds. Recent initiatives have included helicopter delivery of herbicide to steep, remote infested cliff faces (Lord Howe Island Board 2016). Non-native ants have also been targeted by island management authorities (DECC NSW 2007).

Heard and McDonald Islands

The first confirmed sighting of Heard Island was in November 1853 by Captain John Heard on the merchant vessel *Oriental*. The McDonald Islands, which lie ~40 km further west, were discovered by Captain William McDonald in January 1854, but poor weather prevented his crew from landing.

Heard and McDonald Islands have been an Australian external territory since 1947. The islands are situated on the Kerguelen Plateau in the Southern Indian Ocean (Fig. 10.1, Fig. 10.2b). They are the only two active volcanoes in Australian territory; Mawson Peak on Heard Island is Australia's highest mountain (2745 m). There are significant marine reserves surrounding the islands.

Heard and McDonald Islands have a relatively short human history. Sealing gangs arrived in the 19th century and were followed by research expeditions between 1947 and 1955. Early visits were undertaken as part of the Australian National Antarctic Research Expeditions and associated expeditions (Brown 1964; Gressitt 1970). The last year-long research expedition was in 1992, which was the first all-winter occupation since 1954. Since then there have been a few research expeditions to Heard Island of several months duration, by the Australian government. There are only two documented brief scientific visits recorded for McDonald Island. Private expeditions and tourist ships occasionally visit Heard Island. Commercial fishing occurs outside the reserve (McIvor 2007).

Heard and McDonald Islands are among some of the most remote, least visited and dynamic islands on Earth and are volcanically active. McDonald Island has doubled in size since the 1990s through volcanic eruptions (Australian Antarctic Division website). The glaciers on Heard Island are relatively shallow and fast-flowing, which makes them particularly responsive to climate change. Significant glacial retreat has been recorded in recent decades (Thost and Truffer 2008).

The island group's physical processes provide valuable indicators of the role of crustal plates in the formation of ocean basins and continents, of dynamic glacial changes in the coastal and submarine environment, and of atmospheric and oceanic warming. The islands have large concentrations of seabirds (petrels and albatrosses), penguins and marine mammals (seals) but low species diversity. Heard Island has several endemic species including the Heard Island cormorant (*Phalacrocorax atriceps nivalis*), Heard Island sheathbill (*Chionis minor nasicornis*) and several endemic invertebrates (Green and Woehler 2006).

Heard and McDonald Islands cover a total area of 658 903 ha (~37 000 ha is terrestrial). They were inscribed on the World Heritage List in 1997 for criteria viii and ix and listed on the National Heritage List in 2007. The area is managed as a strict nature reserve (IUCN Category 1a). It is the only island WHA property managed solely by the Australian government, through the Australian Antarctic Division (Australian Antarctic Division 2005).

Integrity

Heard and McDonald Islands are the only sub-Antarctic island group that has an intact ecosystem, where the ongoing evolution of plants and animals occurs in a natural state, with few introduced species (Green and Woehler 2006).

The introduction of non-native species is the main threat to the islands' ecosystems. To date, the low temperatures (Chown *et al.* 2005) and especially low visitor numbers to the islands have controlled the number of non-native species (Chown *et al.* 1998). Any changes to human visitation through increased visits by scientists, mountaineers, tourists or increased fishery activities increases the likelihood of invasion.

Macquarie Island

Macquarie Island was discovered in 1810 when Frederick Hasselborough was exploring the Southern Ocean for new sealing grounds. Sealing gangs were present on the island for much of the 1800s, and Joseph Hatch was granted a lease for an oil industry based on harvesting penguins and seals between 1902 and 1920 (Cumpston 1968).

Between 1911 and 1914, the island became a base for the Australasian Antarctic Expedition under Sir Douglas Mawson, with an operational meteorological station between 1911 and 1915. In 1933, the island was declared a wildlife sanctuary under the Tasmanian *Animals and Birds Protection Act 1928*, and in 1972 it was made a state reserve under the Tasmanian *National Parks and Wildlife Act 1970*. The Australian National Antarctic Research Expeditions established a research station on Macquarie Island on 25 May 1948 (Selkirk *et al.* 1990).

Macquarie Island (34 km long and 5 km wide) is located in the Southern Ocean, almost 1500 km to the south-east of Tasmania and about half-way between Australia and Antarctica (Fig. 10.1, Fig. 10.2c). The property includes Macquarie Island, Judge and Clerk Islets, the Bishop and Clerk Islets, rocks and reefs, and covers a total area of 557 280 ha, including ~12 785 terrestrial hectares and all surrounding waters out to three nautical miles. It is managed as a nature reserve by the Tasmanian Parks and Wildlife Service (PWS) (TPWS 2006). Most of the waters out to 200 nautical miles to the east of the reserve are within the Macquarie Island Commonwealth Marine Reserve, managed by the Australian government in cooperation with the Tasmanian PWS.

Macquarie Island was inscribed on the World Heritage List in 1997 for criteria vii and viii and listed on the National Heritage List in 2007. Macquarie Island is of outstanding geoconservation significance as it is the only place on Earth where rocks from the Earth's mantle (6 km below the ocean floor) are exposed above sea-level (Selkirk *et al.* 1990), providing a unique opportunity to study geological features and processes of oceanic crust formation and plate boundary dynamics. These unique exposures include excellent examples of pillow basalts and other extrusive rocks. The island's remote and windswept landscape of steep escarpments, lakes and dramatic changes in vegetation creates an outstanding wild natural beauty with huge concentrations of wildlife including penguins and seals (TPWS 2006).

Integrity

Commercial exploitation of seals and penguins, together with the introduction of non-native species, has resulted in the extinction of some native species and major population declines in others. Active management programs aimed to reverse these trends began in the 1960s. They included the eradication of wekas (*Gallirallus australis scotti*) by 1988 and feral cats by 2002. A world-first multi-vertebrate species (rabbit, black rat and house mouse) eradication program was implemented in 2010 and was declared a success in 2014 (Terauds *et al.* 2014; Springer 2016; Chapter 6). While the recovery of natural ecosystem processes because of these eradication programs is likely to take decades (Whinam *et al.* 2014a) some recovery in palatable vegetation and vulnerable seabirds is already apparent (Shaw *et al.* 2011; Springer 2016). Biosecurity and climate change pose the greatest threats to the integrity of the island, with major decline in the keystone endemic cushion plant (*Azorella macquariensis*) (Box 10.2).

Other islands in Australian World Heritage properties

Australia's first World Heritage properties were listed in 1981. As of 2017 Australia has a total of 19 listed properties – 12 natural, four mixed (natural and cultural) and three cultural. In addition to the four islands/island groups already discussed, several other properties contain islands within their boundaries. The Great Barrier Reef includes over 900 islands, which contribute significantly to the OUV of this property. The Great Barrier Reef property stretches over 2000 km adjacent to the coast of Queensland and includes some of the most spectacular scenery on Earth. The vast mosaic patterns of reefs, islands and coral cays create an extraordinary aerial panorama and contain an exceptional array of both marine and terrestrial biodiversity.

The Shark Bay WHA in Western Australia includes Bernier, Dorr, Dirk Hartog and Faure Islands, which have high conservation values. This property was inscribed because of its exceptional natural features, including its large diverse seagrass beds. However, it is most renowned for its stromatolites, which are colonies of microbial mats that form hard dome-shaped deposits thought to be the oldest life forms on Earth (DEC

Box 10.2: Decline of an endemic plant on Macquarie Island

Islands present many challenges for managers due to their isolation and associated logistic limitations. Their isolation drives the biogeography of islands and, in particular, island endemism. On islands, we can investigate ecological processes and ecosystem function without many of the confounding influences of densely inhabited mainland areas. Islands present unique opportunities to study the impacts of climate change on biodiversity, including species extinction (Watson 2016). Despite reduced human pressures, the extreme isolation and high conservation values can present challenges for conservation management.

A decline, or dieback, in the sub-Antarctic Macquarie Island endemic cushion (*Azorella macquariensis*) was observed during the summer of 2008–09. The decline was rapid and extensive and resulted in the species being listed as Critically Endangered in August 2010 under the EPBC Act and as Endangered under Tasmania's *Threatened Species Protection Act 1995*. This dramatic decline in conservation status prompted a major collaborative research and conservation management project involving botanists, pathologists, horticulturalists, eco-physiologists, ecosystem modellers and land managers. Dieback has continued across the entire range of this keystone species and has been recorded in bryophytes at affected sites.

Measurements from photo plots of *A. macquariensis* from 2009 to 2013 found that the most common type of damage occurred on the most wind-exposed parts of cushions at the most wind-exposed sites (Whinam *et al.* 2014b). The association between wind exposure and damage has been reported previously in *Azorella* cushion plants (Taylor 1955; Nyakatya and McGeoch 2008). However, areas of dieback associated with surface soils dominated by small particles showed symptoms that were consistent with those of a pathogen. To date, there has been no conclusive evidence that the widespread death of *A. macquariensis* was caused by a pathogen. Some samples contain bacterial, fungal and oomycete taxa, some potentially pathogenic, which suggests that stressed cushions may become susceptible to infection.

Plant functional morphology and long-term climate record of climate parameters relevant to plants were examined to determine whether environmental conditions had dramatically changed (Bergstrom *et al.* 2015). The study found the major cause of the dieback of *A. macquariensis* and other fellfield species is likely to be associated with climatic change, specifically changes in summer water availability. Increased wind speed, sunshine hours and evapotranspiration resulted in accumulated deficits of plant available water spanning 17 years (1992–2008). The morphology of *A. macquariensis* means that it is unlikely to be able to adapt to extreme and repeated moisture deficits in an environment that has changed from constantly moist to periods of drying (Bergstrom *et al.* 2015). This is not surprising for an endemic plant that has evolved in a sub-Antarctic cool, wet, windy environment. Managers now face the dilemma of managing this species with ongoing climatic change.

Seeds from healthy populations of *A. macquariensis* have been collected and lodged with the Tasmanian Seedbank, although the germination requirements of the species are not yet fully understood. Small healthy cushion plants were brought back for horticultural studies, but plant morphology and performance is different under nursery conditions. An *ex situ* population has been established on Macquarie Island with the intention of developing a seed orchard as an insurance population (Bergstrom *et al.* 2015).

2012). Shark Bay has a rich marine life, including a large population of dugongs. The islands have high numbers of endemic species and support many threatened plants, reptiles and five species of mammals. Several of the islands have been used for conserving populations of other threatened marsupials (DEH 2003). The fringing reefs of the Muiron Islands are part of the Ningaloo Coast property and the islands themselves are an extension of the Cape Range.

The Tasmanian Wilderness property includes the Maatsuyker Island group which consists of six islands and two groups of rocks, which are designated as an Important Bird Area, providing habitat for the threatened northern soft-plumaged petrel (*Pterodroma mollis*) along with large populations of short-tailed shearwaters (*Ardenna tenuirostris*), fairy prions (*Pachyptila turtur*) and black-faced cormorants (*Phalacrocorax fuscescens*). Other island groups in the Tasmanian Wilderness property are the Pedra Branca, Trumpeter Islet, Breaksea Island and Swainson groups (Brothers *et al.* 2001).

Management of properties to protect OUV

Many of the conservation issues relating to World Heritage island properties are similar, with active management actions required to protect their OUV. The major threats include climate change, biosecurity, weed/pest management and issues associated with resource extraction, tourist visitation and scientific research.

Climate change

Current predictions are that islands around the world are likely to be affected by the impacts of climate change, especially through rising sea-levels and increased storm events (Chapter 15). IUCN has identified climate change as the biggest potential threat to natural World Heritage sites (Osipova *et al.* 2014). Climate change has been identified as a threat to many of the values of Australia's World Heritage-listed properties. Management of other threats to these values, to build resilience, is the most likely successful mitigation strategy for individual properties (Dunlop *et al.* 2012). Recently, Australia recorded its first mammal extinction from an island adjacent to the Great Barrier Reef WHA, with the critically endangered Bramble Cay melomys (*Melomys rubicola*) declared extinct from the island in 2016 (Gynther *et al.* 2016; Watson 2016) due to sea-level inundation. Different components of climate change are likely to be important on the various World Heritage islands, and often several components will combine to increase the threat to the OUV, such as increasing temperatures and changes in patterns of precipitation. Climatic change has been identified as a driver of species conservation status on Macquarie Island (Box 10.2).

The IUCN has assigned a high threat level to the impacts of climate change on Fraser Island, with irreversible change to some physical properties of the site already apparent (Osipova *et al.* 2014). Ocean acidification and increases in temperature in the marine environment have been identified as the main threats to Shark Bay (DEC 2012). Increases in seawater temperatures have already had some impact on the Ningaloo Coast and some fish species might be under pressure from recreational fishing. Oceanic warming has been identified as a threat to Lord Howe Island. While Heard and McDonald Islands are assessed at a low threat level, the issue of accelerating glacial retreat providing increased areas of potential habitat for the cosmopolitan weed *Poa annua* has been highlighted (Scott and Kirkpatrick 2005). Macquarie Island has been classified at a high threat level, with the potential impacts of climate change identified as a risk to its OUV (Osipova *et al.* 2014).

Non-native species

The isolation of these islands means that species have unique assemblages and the island ecosystems are extremely vulnerable to invasion (Simberloff 1995; Courchamp *et al.* 2003). The land use and histories of Australia's World Heritage islands are varied, although there are similarities between the history and human impacts on all the sub-Antarctic islands (Convey and Lebouvier 2009). There is a huge range in the biodiversity, remoteness and size of Australia's World Heritage islands, yet all of them have been invaded by non-native species to some extent. All four of the listed World Heritage islands support non-native plants and invertebrates. Heard and McDonald Islands are the only islands of the four groups not to be invaded by rodents, cats and birds. Once species become pests in an island ecosystem, they are typically difficult and expensive to manage and/or eradicate (Martins *et al.* 2006).

Fraser Island currently has many pest vertebrates and plant pests. Feral rodents, cats, fish and birds occur on the island and affect its biodiversity. Feral pigs are transient visitors to the island. Cane toads are considered the worst invaders, negatively affecting Fraser Island's biodiversity, particularly the reptiles. Cattle and feral horses have been removed from the island.

Lowe Howe Island supports numerous feral birds (Hutton 1990), rodents, weeds and invertebrates. Feral pigs, cats and goats have been successfully eradicated from the island (Hutton *et al.* 2007). There are ongoing plans to eradicate rats, masked owls, weeds and African big-headed ants, all of which affect island biodiversity (Lord Howe Island Board 2017). For further discussion on the role of community engagement in these initiatives, see Chapter 12.

Macquarie Island currently has at least two species of non-native landbird, at least three established non-native plants and several species of non-native invertebrates (Selkirk *et al.* 1990). Two non-native plant species have been removed from Macquarie Island (Copson and Whinam 2001), with two others under active management (Pertierra *et al.* 2016). In recent years, the island has been subject to multi-million-dollar eradication programs to remove cats (Robinson and Copson 2014), rodents and rabbits (Springer 2016).

Two non-native species have been detected on Heard Island, a plant (grass) and an invertebrate (thrip) (Scott 1989; Green and Mound 1994). To date, no direct on-ground management has occurred for these species. However, in recent years biosecurity measures have been put in place to minimise the risk of visiting researchers transporting seeds of the grass *Poa annua* (Shaw 2014). The spread of this grass (Scott and Kirkpatrick 2005) has been identified as one of the few threats to the OUV of the near-pristine island, especially as rapid glacial retreat is providing suitable habitat for this highly invasive species.

Non-native species management is critical across all the islands to protect their individual OUV. There is a fundamental difference between 'management' of weeds and pests and 'eradication' – the appropriate management response depends on several factors, such as whether the organism(s) are invasive, the extent of the infestation, resources and the availability of appropriate tools to undertake management or eradication (Helmstedt *et al.* 2016; Chapter 6).

Biosecurity

The most effective and efficient management action to protect island biodiversity is to prevent the introduction of pests and diseases. Island biosecurity is critical for ensuring the conservation of island biodiversity and ecosystem processes. It is often the major management issue for islands.

The best strategy for managing biosecurity for islands is the establishment and implementation of biosecurity protocols and controls at pre-departure points (e.g. seaports and airports on the mainland) and entry points (immediately before/upon arrival at the island) (Houghton *et al.* 2016). Biosecurity is extremely difficult to manage for islands that have high levels of visitation, receive large volumes of cargo and have resident populations, such as Fraser and Lord Howe Islands.

The IUCN has identified biosecurity as the single greatest threat to the natural values of Heard and McDonald Islands. Biosecurity has been a major focus of management of these islands, particularly as they are recognised as one of the world's rare pristine island ecosystems, with few non-native species (Box 10.1). There has been recognition of the importance of biosecurity for the sub-Antarctic islands more broadly through international collaboration between nations, scientists, managers and tourism operators (Box 10.3).

Tourism/resource extraction/expeditioners

The OUV associated with World Heritage islands infers certain biodiversity values that are of considerable interest to those wishing to utilise specific resources or visit to experience the islands' biodiversity, geodiversity and aesthetic values. The history of European colonisation of these islands demonstrates that both visitation and resource extraction have resulted in negative outcomes for some of the special values for which the properties are listed.

Developments on the mainland Queensland coast have had major impacts on catchment runoff and resulted in poor water quality for the Great Barrier Reef. Developments have included port and shipping infrastructure, urban settlements,

Box 10.3: Biosecurity for sub-Antarctic islands

Non-native plants, invertebrates, fungi and microbes occur on all sub-Antarctic islands, as well as on the Antarctic continent. They have arrived there in the last two centuries in conjunction with human activity. Most non-native plants and vertebrates are European in origin (Frenot *et al.* 2005). These alien species have direct and indirect impacts on these species-poor ecosystems. Non-native species have changed ecosystem processes and vertebrates have caused substantial loss of biodiversity. Some 108 non-native vascular species have been recorded for the sub-Antarctic, with the cosmopolitan weed *Poa annua* present on all sub-Antarctic islands (Frenot *et al.* 2005; McGeoch *et al.* 2015).

To improve biosecurity practices, an investigation was undertaken to determine whether Australian expeditioners were inadvertently acting as vectors for alien species. The study found 981 propagules (both reproductive and vegetative) from 90 species on the clothing and equipment of 64 expeditioners travelling from Hobart to Macquarie Island (Whinam *et al.* 2005). Equipment cases, daypacks and the cuffs and velcro closures of outer clothing were identified as high-risk vectors. Many of the seeds were viable (Whinam *et al.* 2005).

This Australian study springboarded an international program 'Aliens in Antarctica' as part of the International Polar Year involving researchers from nine nations (Chown *et al.* 2012; Huiskes *et al.* 2014). The project was initiated to quantify the propagule pressure of humans travelling to Antarctica and assess the invasion risk at a continental scale. Personal clothing, field equipment, fresh food, cargo and more than 27 ships and aircraft were examined. The likelihood of introduction of non-native species by national Antarctic programs and tourist operations was determined.

Chown *et al.* (2012) found that scientists are more likely to be carrying seeds on their person or in their gear than are tourists or ships' crews. However, it is worth noting that thousands more tourists than scientists visit the region annually. Visitors unintentionally carrying seeds were likely to have on average 9.5 seeds per person (Chown *et al.* 2012). Fresh fruit and vegetables are also known vectors of fungi (Whinam *et al.* 2005) and live invertebrates to sub-Antarctic islands (Houghton *et al.* 2016). It was determined that 53% of people travelling to the islands had visited a cold temperate, alpine or polar environment in the prior year (Chown *et al.* 2012). Thus, the origin of transported seeds may well have been a similar environment with suitable conditions to the sub-Antarctic.

People movement to the region is solely by ships, which can also act as vectors to the islands. Research has shown that non-native mussels (*Mytilus galloprovincialis*) can be transported live to sub-Antarctic islands aboard resupply ships (Lee and Chown 2007). These highly invasive organisms could survive the journey through the Southern Ocean and Antarctic waters. Ships are also vectors of live terrestrial invertebrates to Macquarie Island (Houghton *et al.* 2016). This is concerning for managers, particularly as many of the invertebrates are flying insects which can self-disperse from ship to shore upon arrival at the island.

The number of tourists visiting the sub-Antarctic and Antarctica has increased steadily over recent decades (Chown *et al.* 2012). Increased visitation increases the risk of alien species being introduced to sub-Antarctic islands (Chown *et al.* 1998). At the same time, ameliorating conditions associated with climate change are likely to increase the risk of pest species colonising sub-Antarctic islands and parts of Antarctica (Chown *et al.* 2012).

Mitigation measures must focus on reducing propagule loads on humans and on their food, cargo and transport vessels (Frenot *et al.* 2005; Whinam *et al.* 2005; Houghton *et al.* 2016). Fumigation, cargo cleaning, cargo checks and rodent detection dogs are now standard procedures for Macquarie Island annual resupply. Strict biosecurity measures at both pre-embarkation ports and before disembarkation at sub-Antarctic islands are likely to be the most successful management measure to minimise further introductions of non-native species.

coal mining and farming (Commonwealth of Australia and Queensland Government 2015). On Lord Howe Island, the impacts of infrastructure associated with settlement and tourism activity, such as the proposed expansion of the airport runway, are considered threats to the natural values. Increased recreational fishing is a threat both directly and indirectly (through disruption to the food chain) to the Great Barrier Reef and Ningaloo Coast. The risk of oil spills is identified as a major threat to the Great Barrier Reef and Ningaloo Coast.

Perhaps one of the best examples of contemporary challenges in balancing the maintenance of World Heritage values with socio-economic pressures, in particular eco-tourism, occurs on Fraser Island. Much has changed since the island was listed, with a notable increase in visitors and associated vehicle traffic. The increased visitation has led to increased pollution, siltation, disturbance and the introduction of pest species. Managers of the property are faced with the challenge of implementing innovative, adaptive management quickly – or risking the World Heritage property becoming just another degraded recreational facility (Wardell-Johnson *et al.* 2015).

Tourists and expeditioners visit both of Australia's sub-Antarctic World Heritage islands. Tourist visits to Heard Island are relatively few (they are not permitted on McDonald Island). Geographic isolation helps limit visitor numbers and ship visits. However, tourist ships visit Macquarie Island each summer, most en route to the Antarctic and as part of sub-Antarctic itineraries that include the New Zealand sub-Antarctic islands. Managing visitor numbers and the risks of inter-island biosecurity is a priority for both Australia and New Zealand (Box 10.3).

Most sub-Antarctic expeditioners provide support services to keep these remote stations operating, such as doctors, diesel mechanics, electricians etc. The sub-Antarctic islands offer unique opportunities to study aspects of the terrestrial and marine environment on isolated specks of land situated in vast expanses of ocean. The number of expeditioners to the sub-Antarctic is limited by the respective management plans (de Villiers *et al.* 2006) – demand to visit and work on these islands exceeds the number of people able to secure ship berths and places on the islands. The science programs also pose a biosecurity risk (Chown *et al.* 2012) and highlight the need for stringent biosecurity measures for researchers and their equipment.

Management requirements

Both the range and intensity of threats to the OUV of World Heritage islands is increasing globally. While site managers have little direct control over the threats posed by climate change, they do have some capacity to mitigate against detrimental human activities. Increased management is required to maintain the various outstanding values of Australia's World Heritage islands and to protect their integrity. This will require strategic policy frameworks relating to sustainable development and environmental protection, suitable management capacity and adequate resources, as well as collaboration and coordination between government agencies and user groups.

Concluding comments: key messages

Australia's World Heritage-listed island properties play a critical role in the protection of Australia's biodiversity, geodiversity, cultural and historic heritage values. Despite being isolated discrete landmasses these islands face multiple threats, highlighting the need for ongoing conservation management. Island management must be integrated across the many tiers of government and multiple agencies, especially in terms of biosecurity. Pest species have invaded all islands, even those that are remote and cold and have low human visitation. Pest control and eradication programs have been successfully implemented on some islands, but they are costly. The most cost-effective way to protect the islands' conservation values is through rigorous biosecurity before embarkation and upon arrival. This should be supported by ongoing pest surveillance. Demand to visit these universally outstanding places is likely to rise, resulting in increased pressures from tourists, developers and researchers. Policy-makers and management authorities need to ensure that adequate resources are focused on maintaining and protecting the identified conservation values of these properties. It is critical to keep these outstanding places exceptional, so that future generations can continue to enjoy visiting them places without damaging the values that they come to see.

Acknowledgements

We thank Veronica Blazeley and Aleks Terauds for comments on a draft version. Perpetua Turner provided helpful feedback through the review process. We thank Rod Fensham, April Reside and John Olds for clarification of species records. This study was partly supported funding from the Australian government's National Environmental Science Program through the Threatened Species Recovery Hub.

References

Allen BL, Boswell J, Higginbottom K (2012) *Fraser Island Dingo Management Strategy Review*. Report to the Department of Environment and Heritage Protection. Ecosure, Brisbane.

Allen BL, Higginbottom K, Bracks JH, Davies N, Baxter GS (2015) Balancing dingo conservation with human safety on Fraser Island: the numerical and demographic effects of humane destruction on dingoes. *Australasian Journal of Environmental Management* **22**, 197–215. doi:10.1080/14486563.2014.999134

Australian Antarctic Division (2005) *Heard and McDonald Islands Marine Reserve*. Department of National Parks, Canberra.

Bergstrom DM, Bricher PK, Raymond B, Terauds A, Doley D, McGeoch MA, Whinam J, Glen M, Ziqing Y, Kiefer K, Shaw JD, Bramely-Alves J, Rudman T, Mohammed C, Lucieer A, Visoiu M, van Vuuren BJ, Ball MC (2015) Rapid collapse of a sub-Antarctic alpine ecosystem: the role of climate and pathogens. *Journal of Applied Ecology* **52**, 774–783. doi:10.1111/1365-2664.12436

Brothers N, Pemberton D, Pryor H, Lucieer V (2001) *Tasmania's Offshore Islands: Seabirds and Other Natural Features*. Tasmanian Museum and Art Gallery, Hobart.

Brown KG (1964) The insects of Heard Island. *Australian National Antarctic Research Expeditions Report* **1**, 1–39.

Brown S, Baldwin C, Chandler L (2015) Representation of Butchulla cultural heritage values in communication of K'gari (Fraser Island) as a tourism destination. *Australasian Journal of Environmental Management* **22**, 163–180. doi:10.1080/14486563.2014.985266

Burns GL, Howard P (2003) When wildlife tourism goes wrong: a case study of stakeholder and management issues regarding dingoes on Fraser Island, Australia. *Tourism Management* **24**, 699–712. doi:10.1016/S0261-5177(03)00146-8

Chown SL, Gremmen NJM, Gaston KJ (1998) Ecological biogeography of Southern Ocean islands: species–area relationships, human impacts, and conservation. *American Naturalist* **152**, 562–575.

Chown SL, Hull B, Gaston KJ (2005) Human impacts, energy availability and invasion across Southern Ocean islands. *Global Ecology and Biogeography* **14**, 521–528. doi:10.1111/j.1466-822x.2005.00173.x

Chown SL, Huiskes AHL, Gremmen NJM, Lee JE, Terauds A, Crosbie K, Frenot Y, Hughes KA, Imura S, Kiefer K, Lebouvier M, Raymond B, Tsujimoto M, Ware C, Van de Vijver B, Bergstrom DM (2012) Continent-wide risk assessment for the establishment of non-indigenous species in Antarctica. *Proceedings of the National Academy of Sciences of the United States of America* **109**, 4938–4943. doi:10.1073/pnas.1119787109

Commission of Inquiry (1991) *Commission of Inquiry into the Conservation, Management and Use of Fraser Island and the Greater Sandy Region*. The Commission, Brisbane.

Commonwealth of Australia and Queensland Government (2015) *Reef 2050 Long-Term Sustainability Plan*. Canberra.

Convey P, Lebouvier M (2009) Environmental change and human impacts on terrestrial ecosystems of the sub-Antarctic islands between their discovery and the mid-twentieth century. *Papers and Proceedings of the Royal Society of Tasmania* **143**, 33–44.

Copson GR, Whinam J (2001) Review of ecological restoration programme on subantarctic Macquarie Island: pest management progress and future directions. *Ecological Management and*

Restoration **2**, 129–138. doi:10.1046/j.1442-8903.2001.00076.x

Courchamp F, Chapuis JL, Pascal M (2003) Mammal invaders on islands: impact, control and control impact. *Biological Reviews of the Cambridge Philosophical Society* **78**, 347–383. doi:10.1017/S1464793102006061

Cumpston JS (1968) *Macquarie Island*. Department of External Affairs, Antarctic Division, Melbourne.

DEC (2012) *Shark Bay Terrestrial Reserves and Proposed Reserve Additions*. Management Plan No. 75. Department of the Environment and Conservation, Conservation Commission of Western Australia, Perth.

DECC (2007) *Lord Howe Island Biodiversity Management Plan*. NSW Department of Environment and Climate Change, Sydney.

DEE (2017) *Managing World Heritage in Australia*. Department of the Environment and Energy, Canberra. <http://www.environment.gov.au/heritage/about/world/management-australias-world-heritage-listed/managing-world-heritage-australia>

DEH (2003) *Report on Threatened Shark Bay Marsupials, western barred bandicoot Perameles bougainville bougainville, burrowing bettong Bettongia lesueur lesueur, banded hare-wallaby Lagostrophus fasciatus fasciatus, and rufous hare-wallabies Lagorchestes hirsutus bernieri and Lagorchestes hirsutus dorreae*. Department of the Environment and Heritage, Canberra.

de Villiers MS, Cooper J, Carmichael N, Glass JP, Liddle GM, McIvor E, Micol T, Roberts A (2006) Conservation management at Southern Ocean islands: towards the development of best-practice guidelines. *Polarforschung* **75**, 113–131.

DNPSR (2011) *Great Sandy Management Plan 1994*. Department of National Parks, Sport and Racing, Brisbane. <http://www.nprsr.qld.gov.au/managing/plans-strategies/great_sandy_region.html>

Dunlop M, Hilbert DW, Ferrier S, House A, Liedlo A, Prober SM, Smyth A, Martin TG, Harwood T, Williams KJ, Fletcher C, Murphy H (2012) *The Implications of Climate Change for Biodiversity Conservation and the National Reserve System: Final Synthesis*. Report prepared for the Department of Sustainability, Environment, Water, Population and Communities, and the Department of Climate Change and Energy Efficiency. CSIRO Climate Adaptation Flagship, Canberra.

Frenot Y, Chown SL, Whinam J, Selkirk PM, Convey P, Skotnicki M, Bergstrom DM (2005) Biological invasions in the Antarctic: extent, impacts and implications. *Biological Reviews of the Cambridge Philosophical Society* **80**, 45–72. doi:10.1017/S1464793104006542

Green K, Mound LA (1994) An extension to the insect fauna of Heard Island. *Polar Record* **30**, 131–132. doi:10.1017/S0032247400021343

Green K, Woehler E (2006) *Heard Island: Southern Ocean Sentinel*. Surrey Beatty & Sons, Sydney.

Gressitt JL (1970) Subantarctic entomology and biogeography. *Pacific Insects Monograph* **23**, 295–374.

Gynther I, Waller N, Leung LK-P (2016) *Confirmation of the Extinction of the Bramble Cay Melomys rubicola on Bramble Cay, Torres Strait: Results and Conclusions from a Comprehensive Survey in August–September 2014*. Brisbane.

Helmstedt K, Shaw JD, Bode M, Terauds A, Springer K, Robinson S, Possingham HP (2016) Prioritizing eradication actions on islands: it's not all or nothing. *Journal of Applied Ecology* **53**, 733–741. doi:10.1111/1365-2664.12599

Houghton M, McQuillan PB, Bergstrom DM, Frost L, van den Hoff J, Shaw J (2016) Pathways of alien invertebrate transfer to the Antarctic region. *Polar Biology* **39**, 23–33. doi:10.1007/s00300-014-1599-2

Huiskes AHL, Gremmen NJM, Bergstrom DM, Frenot Y, Hughes KA, Imura S, Kiefer K, Lebouvier M, Lee JE, Tsujimoto M, Ware C, van de Vijver B, Chown SL (2014) Aliens in Antarctica: assessing transfer of plant propagules by human visitors to reduce invasion risk. *Biological Conservation* **171**, 278–284. doi:10.1016/j.biocon.2014.01.038

Hutton I (1990) *Birds of Lord Howe Island: Past and Present*. Ian Hutton, Coffs Harbour.

Hutton I, Parkes JA, Sinclair ARE (2007) Reassembling island ecosystems: the case of Lord Howe

Island. *Animal Conservation* **10**, 22–29. doi:10.1111/j.1469-1795.2006.00077.x

IUCN (2016) *IUCN and World Heritage.* <https://www.iucn.org/theme/world-heritage/about/iucn-and-world-heritage>

Lee JE, Chown SL (2007) *Mytilus* on the move: transport of an invasive bivalve to the Antarctic. *Marine Ecology Progress Series* **339**, 307–310. doi:10.3354/meps339307

Lord Howe Island Board (2016) *Draft Lord Howe Island Weed Management Strategy 2016.* Lord Howe Island Board, Lord Howe Island.

Lord Howe Island Board (2017) *Environmental Programs.* <http://www.lhib.nsw.gov.au/environment/environmental-programs>

Martins TLF, Brooke MDL, Hilton GM, Farnsworth S, Gould J, Pain DJ (2006) Costing eradications of alien mammals from islands. *Animal Conservation* **9**, 439–444. doi:10.1111/j.1469-1795.2006.00058.x

McGeoch MA, Shaw JD, Terauds A, Lee JE, Chown SL (2015) Monitoring biological invasion across the broader Antarctic: a baseline and indicator framework. *Global Environmental Change* **32**, 108–125. doi:10.1016/j.gloenvcha.2014.12.012

McIvor E (2007) Heard Island and McDonald Islands. *Papers and Proceedings of the Royal Society of Tasmania* **141**, 7–10.

Nyakatya MJ, McGeoch MA (2008) Temperature variation across Marion Island associated with a keystone plant species (*Azorella selago* Hook. (Apiaceae)). *Polar Biology* **31**, 139–151. doi:10.1007/s00300-007-0341-8

Osipova E, Shi Y, Kormos C, Shadie P, Zwablen C, Badman T (2014) *IUCN World Heritage Outlook 2014: A Conservation Assessment of All Natural World Heritage Sites*. IUCN, Gland, Switzerland.

Pertierra LR, Baker M, Howard C, Vega GC, Olalla-Tarraga MA, Scott JJ (2016) Assessing the invasive risk of two non-native *Agrostis* species on sub-Antarctic Macquarie Island. *Polar Biology* doi:10.1007/s00300-016-1912-3

Priddel D, Carlile N, Humphrey M, Fellenberg S, Hiscox D (2003) Rediscovery of the 'extinct' Lord Howe Island stick-insect (*Dryococelus australis* (Montrouzier)) (Phasmatodea) and recommendations for its conservation. *Biodiversity and Conservation* **12**, 1391–1403. doi:10.1023/A:1023625710011

Rao K (2010) A new paradigm for the identification, nomination and inscription of World Heritage sites. *International Journal of Heritage Studies* **16**, 161–172. doi:10.1080/13527251003620594

Robinson S, Copson GR (2014) Eradications of cats (*Felis catus*) from subantarctic Macquarie Island. *Ecological Management and Restoration* **15**, 34–40. doi:10.1111/emr.12073

Scott JJ (1989) New records of vascular plants from Heard Island. *Polar Record* **25**, 37–42. doi:10.1017/S0032247400009967

Scott JJ, Kirkpatrick JB (2005) Changes in subantarctic Heard Island vegetation at sites occupied by *Poa annua*. *Arctic, Antarctic, and Alpine Research* **37**, 366–371. doi:10.1657/1523-0430(2005)037[0366:CISHIV]2.0.CO;2

Selkirk PM, Seppelt RD, Selkirk DR (1990) *Subantarctic Macquarie Island. Environment and Biology.* Cambridge University Press, Cambridge.

Shaw JD (2014) Southern Ocean islands invaded: conserving biodiversity in the world's last wilderness. In *Plant Invasions in Protected Areas* (Eds L Foxcroft, DM Richardson, P Pysek and P Genovesi) pp. 449–470. Springer, New York.

Shaw JD, Terauds A, Bergstrom DM (2011) Rapid commencement of ecosystem recovery following aerial baiting on sub-Antarctic Macquarie Island. *Ecological Management and Restoration* **12**, 241–244. doi:10.1111/j.1442-8903.2011.00611.x

Simberloff D (1995) Why introduced species appear to devastate islands more than mainland areas. *Pacific Science* **49**, 87–97.

Springer K (2016) Methodology and challenges of a complex multi-species eradication in the sub-Antarctic and immediate effects of invasive species removal. *New Zealand Journal of Ecology* **40**, 273–278. doi:10.20417/nzjecol.40.30

Taylor BW (1955) *The Flora, Vegetation and Soils of Macquarie Island.* ANARE Reports B(2) No. 19, Botany. Antarctic Division, Department of External Affairs, Melbourne.

Terauds A, Doube J, McKinlay J, Springer K (2014) Using long-term population trends of an invasive herbivore to quantify the impact of management actions in the sub-Antarctic. *Polar Biology* **37**, 833–843. doi:10.1007/s00300-014-1485-y

Thompson J, Shirreffs L, McPhail I (2003) Dingoes on Fraser Island: tourism dream or management nightmare. *Human Dimensions of Wildlife* **8**, 37–47. doi:10.1080/10871200390180136

Thost D, Truffer M (2008) Glacier recession on Heard Island, Southern Indian Ocean. *Arctic, Antarctic, and Alpine Research* **40**, 199–214. doi:10.1657/1523-0430(06-084)[THOST]2.0.CO;2

TPWS (2006) *Macquarie Island Nature Reserve and World Heritage Management Plan.* Tasmanian Parks and Wildlife Service, Department of Tourism, Heritage and Arts, Hobart.

UNESCO (1972) *The World Heritage Convention.* UN Educational Scientific and Cultural Organisation. <http://whc.unesco.org/en/conventiontext>

UNESCO (2005) *The Criteria for* Selection. UN Educational Scientific and Cultural Organisation. <http://whc.unesco.org/en/criteria>

UNESCO (2017) *World Heritage List.* UN Educational Scientific and Cultural Organisation. <http://whc.unesco.org/en/list>

Wardell-Johnson G, Schoeman D, Schlacher T, Wardell-Johnson A, Weston MA, Shimizu Y, Conroy G (2015) Re-framing values for a World Heritage future: what type of icon will K'gari-Fraser Island become? *Australasian Journal of Environmental Management* **22**, 124–148. doi:10.1080/14486563.2014.985267

Watson J (2016) Bring climate change back from the future. *Nature* **534**, 437. doi:10.1038/534437a

Whinam J, Chilcott N, Bergstrom DM (2005) Subantarctic hitchhikers: expeditioners as vectors for the introduction of alien organisms. *Biological Conservation* **121**, 207–219. doi:10.1016/j.biocon.2004.04.020

Whinam J, Fitzgerald N, Visoiu M, Copson GR (2014a) Thirty years of vegetation dynamics in response to a fluctuating rabbit population on sub-Antarctic Macquarie Island. *Ecological Management and Restoration* **15**, 41–51. doi:10.1111/emr.12076

Whinam J, Abdul-Rahman JA, Visoiu M, di Folco M-BF, Kirkpatrick JB (2014b) Spatial and temporal variation in dieback in a threatened subantarctic cushion species. *Australian Journal of Botany* **62**, 10–21. doi:10.1071/BT13207

11

The interplay between tourism and conservation on islands

Derek Ball, Sally L. Bryant and Roland Mau

Summary

Australia's islands form the foundation of an estimated A$13 billion/annum nature-based tourism industry that has evolved from traditional tourism, to eco-tourism, to thematic-based and hands-on experiential programs. This reimagining of the industry offers great opportunity to focus on lower-volume higher-value enterprises in unique niche markets that can play an active role in helping conserve the natural values upon which they depend. There can be risks when the pressure from tourism leads to degradation of values or conflict within communities facing growth, change and mass homogeneity with little control or share in the benefits. This chapter discusses the potential role of different types of eco-tourism and provides several case studies of operations that contribute to or detract from island conservation efforts.

Introduction

For Australians, thoughts of islands conjure visions of holiday getaways, the sun and the sea, relaxing in tranquil tropical settings with family and friends. Indeed, it is undeniable that many of Australia's iconic holiday destinations are islands. Australia's islands are also focal points for international visitors and many such destinations lie in the common imagination of travellers world-wide – Rottnest, the Whitsundays and others in the Great Barrier Reef region, K'gari-Fraser Island, Bruny and Kangaroo Islands all receive large numbers of international tourists. The oceanic islands of Norfolk and Lord Howe offer unique experiences well away from the Australian continent, and opportunities to visit sub-Antarctic Macquarie Island now attract visitors from around the globe. Tourism is often a major, or sometimes the only, foundation of island economies and thus is an essential component in supporting island communities. However, uncontrolled or poorly managed tourism also has the potential to degrade the natural environment and consequently disrupt economic returns. Tourism has evolved on Australian islands from traditional tourism, to eco-tourism, to that which is more thematic-based such as bird-watching tours and to not-for-profit (more accurately, profit-for-purpose) tourism enterprises that seek to help manage and conserve the natural values upon which the industry depends.

In this chapter, the evolution of tourism is discussed in terms of how it might better aspire to, and achieve, true economic and ecological sustainability. This is supported by four case studies of different models of tourism, outlining how they contribute to or detract from island conservation efforts.

History of tourism on Australia's islands

Island tourism in Australia has its roots in low-volume, mostly *ad hoc* visitation to near-shore islands, often utilising infrastructure that was originally developed for farming or fishing enterprises, lighthouses or even military or penal outposts. Arguably one of the earliest island tourism enterprises was on Phillip Island (Victoria) with the opening of the first hotel, the Isle of Wight, in 1870 and, soon after that, the nearby Phillip Island Hotel. Rottnest Island off Fremantle in Western Australia, a former penal and strategic military asset, first offered tourism opportunities in 1902 and these have been continually developed to its present almost exclusive use. While most states have similar histories of evolving island tourism, arguably the most numerous and rapid growth in business ventures has occurred on Australia's tropical islands. Heron Island in the Great Barrier Reef region was a turtle cannery in the 1920s, but the green turtle (*Chelonia myda*) resource was quickly depleted and the island abandoned, before being revived as one of Queensland's first island resorts in the 1930s. Within the iconic Whitsunday Islands, 1929 saw the establishment of the first resort, on Lindeman Island. Others soon followed on Daydream, Long and Hayman Islands. The latter were bundled into tourism destinations by Reg Ansett in 1947, as one of the first major investments in Australia in island tourism. Norfolk Island (south-west Pacific) owes its current tourism industry to former military assets – the airfield was constructed during World War II and it allowed easy access for visitors after that period. This post-war period was a boon for island tourism, perhaps partly due to readily available ex-defence transport assets. The venerable Fairmile motor launches, for example, became virtually ready-built ferries that were extensively used to access island resorts off Queensland. Flying boats provided much easier and faster access to offshore areas and needed virtually no other infrastructure. They led to a surge in post-war tourism, for example to Lord Howe Island. Quicker, easier access to islands, coupled with growing availability and affordability of domestic airfares and the increasing affluence of Australians fuelled island tourism to a peak between the 1960s and 1980s. Added to this was the increased affordability of international airfares and subsequent rise in visitors to Australia. Indeed during this period, Australia might have had legitimate claim to being a world leader in island tourism. Great innovation was at work in the Great Barrier Reef region, for example. The establishment of underwater observatories such as that on Green and Hook Islands in the 1960s was considered 'The greatest tourist drawcard ever to be established in the Whitsunday area' (Whitsunday Regional Council 2016). Hamilton Island opened in stages commencing in 1982 and included a runway capable of landing passenger jets, creating seamless flight access from overseas destinations. In 1985 Hayman Island, also in the Whitsundays, received a A$300 million upgrade, a massive financial investment for the time, and became one of the world's most luxurious island resorts. In 1989, the airline pilots' dispute caused catastrophic, albeit short-term, damage to Australia's tourism industry. However, perhaps more significant was the increasing availability and affordability of overseas destinations for Australian travellers. Australian Bureau of Statistics (ABS 2016) data paint a clear picture of a trend commencing in the early 1990s for Australians to travel to overseas destinations: outbound Australian (resident short-term) departures to Fiji in 1991 numbered 90 100 but by 2015 were 348 900. Similar increases occurred in Australians visiting other international island destinations: Vanuatu, 18 500 to 46 000; Cook Islands, 4100 to 15 300; Samoa, 4000 to 28 100. However, most significant was the growing attraction of Australians to Indo-Asian island archipelagos such as Bali: South-east Asia combined went from 473 100 in 1991 to 2 835 000 in 2015, and the number of Australians visiting Indonesia skyrocketed from 174 700 in 1991 to 1 103 300 by 2015. Although data are not available for Australian island destinations alone, Australia's overall tourism trade deficit – a result of increased outbound traffic, in comparison to incoming visitors – stood at A$5 billion in 2009–10 compared to a A$3.6 billion surplus in 1999–2000 (TRA 2011). And the trend continues.

Current status of tourism on Australia's islands

The lure of cheap exotic island destinations abroad, and the contribution of the tourism trade deficit, have undoubtedly had a profound impact on island tourism in Australia, particularly in the traditional family resort holiday sector. For example, most resorts in the iconic Whitsunday Islands, and many others across the Great Barrier Reef region, have closed their doors over the last 10 years. Others such as Norfolk Island continue to struggle with ~20% fewer visitors than in previous peak periods. It is not difficult to imagine that Australia's relatively high labour and regulatory costs could prevent its family resort holiday sector from competing with island nations such as those in the south-west Pacific or South-east Asia. However, other iconic island destinations such as Rottnest (Western Australia), Bruny (Tasmania), Kangaroo (South Australia) and Magnetic Islands (Queensland) are recording stable or increased visitor numbers. These are similar in that they are large islands with diverse accommodation offerings and day visitation opportunities, and provide a mix of recreational and boutique cultural experiences. There are also many smaller islands across Australia which continue to successfully provide tourism offerings such as fishing and/or scuba diving/snorkelling and/or nature-based activities such as bird and marine turtle viewing; for example Green, Heron and Lady Elliot Islands in the Great Barrier Reef, Christmas and Sweers Island (Gulf of Carpentaria) and Wigram Island (Northern Territory). In addition, more exclusive resorts in places such as Hayman, Orpheus and Lizard Islands in the Great Barrier Reef, Satellite Island in Tasmania, Kangaroo Island in South Australia, Lord Howe Island in New South Wales, and others, provide higher-end products. These destinations are supplemented by a multitude of opportunities across Australia for visiting island protected area camping grounds.

Many island communities are looking towards tourism as a way to improve and sustain their local economies. It has been reported that Australia's islands form the foundation of an estimated A$13 billion/annum nature-based tourism industry. Islands like North Stradbroke (Queensland) and Christmas (Indian Ocean), with economies based largely around mining, have produced tourism strategies to identify market opportunities, infrastructure and skills requirements so they can grow that local industry (Icon Tourism 2010; Queensland Government 2015). Tourism Australia's (2016a) research showed that visitor demand is escalating for aquatic and coastal experiences and that islands *per se* were ranked ninth of 19 most appealing attractions. Given this popularity and the likely economic benefits, it seems a lack that neither Tourism Australia nor Tourism Research Australia appears to collect, collate or analyse data on visitation statistics and economic contributions of islands nation-wide. However, these island examples give an indication of the magnitude of their contribution. In 2012 the Great Barrier Reef alone contributed A$5.7 billion to the Australian economy and sustained 69 000 full-time equivalent jobs (Deloitte Access Economics 2013). In 2013–14, the tourism industry to Phillip Island contributed an estimated A$655 million to the island's economy (39.4% of gross regional product) and employed ~6400 people (43.3% of regional employment) (Tourism Victoria 2015). During 2015 Kangaroo Island had 126 000 visitors who spent A$104 million (South Australian Tourism Commission 2016) and Bruny Island attracted 111 000 interstate and international visitors (an increase of 30% from the previous year) and over 80 000 intrastate visitors with a total expenditure estimated to be nearly A$33 million (Destination South Tasmania 2016).

It is clear that over the past two decades there has been a reimagining of the Australian island tourism industry. This is worthy of analysis because the reimagining offers great opportunity, but also significant risks to the natural assets upon which island tourism depends.

Types of tourism and interplay with biodiversity conservation

For many, eco-tourism might appear to be a long-standing and familiar concept, indeed, one that is almost synonymous with island tourism. This is despite the term 'eco-tourism' being regularly used in the literature only since the late 1980s (Weaver and Lawton 2007). Nevertheless, it is familiar in the broader public consciousness although, as argued below, it is a concept with seemingly ever-broadening usage outside its original intent (Fennell 2005) to the point that many enterprises appear to be using the term more as 'green-washing'. This is of great concern as eco-tourism in general is assumed to be a positive influence, if not neutral, on the natural environment, despite the growing body of evidence to the contrary (discussed further below). This concern was first raised by the early 1990s with efforts to define the types of enterprise being labelled as eco-tourism (Buckley 1994). For island managers developing or reviewing eco-tourism strategies, whether their focus is on conservation, cultural, social or economic management issues, a better understanding of the potential role of different types of eco-tourism and their advantages and disadvantages should be a precursor to planning, development and implementations. Buckley's (1994) conceptualisation, while not island-specific, appears to remain useful and is shown here (Fig. 11.1).

Within this framework, Buckley (1994) proposed that the centre overlap area should be considered the most restrictive definition of eco-tourism ('maximum conservation benefit', explained below). We can suggest to eco-tourism proponents that the greater number of overlaps among types of eco-tourism, the greater the potential to achieve ecological sustainability and optimal conservation outcomes. However, using this framework as a guide to developing such enterprises requires a sound understanding of the characteristics of each type of eco-tourism (Table 11.1).

Problems that arise

While tourism can be the salvation for waning island economies, it is not without impact. Therefore, the real test is whether the industry is giving back to the natural resources upon which it depends. Many of Australia's most highly visited islands are managed under strategic plans aimed at growing visitor numbers by offering improved services, more unique experiences or greater exposure to natural values. The direct feedback loop

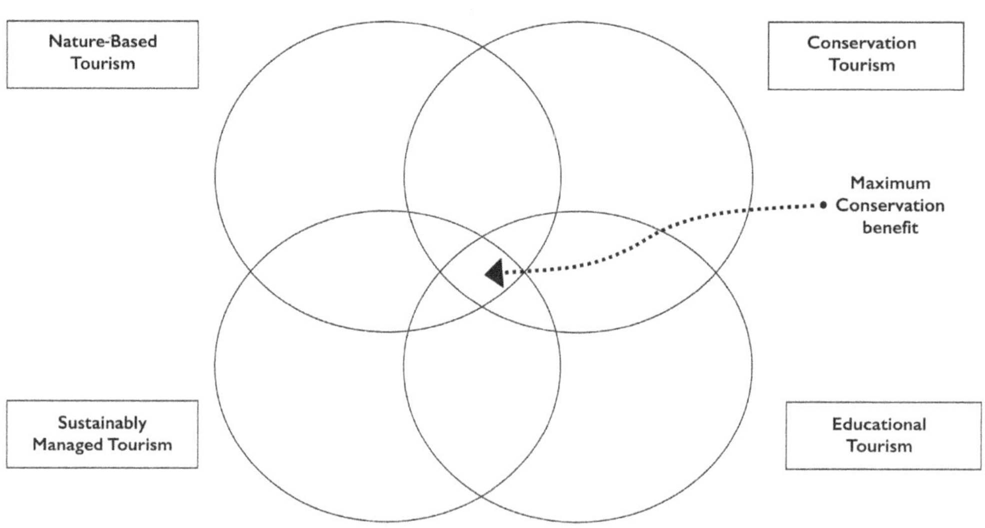

Fig. 11.1: A framework of different eco-tourism types. Source: after Buckley (1994).

Table 11.1. Characteristics of the types of eco-tourism

	Nature-based tourism	Sustainably managed tourism	Educational tourism	Conservation tourism
Intent	Use of natural assets as an attractive product to draw clients and derive a financial profit	Use of natural assets as an attractive product to draw clients with a longer-term view of profitability	Use of natural assets as educational resources to support and derive learning outcomes on a for-profit or not-for-profit basis	Involvement in managing natural assets is at least partly the product offering to clients on a for-profit or not-for-profit basis
Characteristics	User	User but with deliberate attempts to sustain business outcomes through minimisation of impacts	User and contributor, seeks to change behaviours and attitudes to drive sustainability	Contributor to improved or maintain a stable condition of natural assets
Overlap with conservation	Potentially negative effects if carrying capacity of the natural assets are exceeded	Goal is neutral impact. Seeks to remain within carrying capacity of the natural assets either by limiting client numbers or managing their impact	Potentially positive through creating attitudinal and behavioural change, providing broader sustainability is achieved within carrying capacity	Positive effects through improvement in condition of target natural assets

between visitor numbers, revenue generated and improved natural protection is often tenuous at best. For example, the phasing out of extractive industries like sand mining, to be replaced by nature-focused business, is evident in the reimagining of North Stradbroke Island whose vision is to become 'Australia's most desirable island community, striking a balance between sustainable economic growth and protection of the island's unique environment and cultural heritage' (Queensland Government 2015). The strategy for attaining this is to focus on sustainable tourism and education and training. While this sounds credible, the draft action plan, which is underpinned by over A$20 million of start-up funding, identifies few areas which directly contribute to nature conservation *per se*. Rather, it focuses on increasing visitor numbers to national parks by expanding visitor facilities and offering more nature- and adventure-based activities, including access to the previously restricted Peel Island.

Establishing or growing tourism on populated islands may not always be the panacea for small or ailing economies, and can lead to conflict if local communities have little measure of control or don't share equitably in the benefits. Island communities are diverse and typically a mix of Indigenous or generational landholders with a strong sense of place and ownership. Over time, newcomers take up residency to pursue different lifestyle choices, including off-island investors in businesses or real estate for private and rental purposes, consuming the most keenly sought island asset – space. Today many islands show the results of poorly managed and haphazard growth, inadequate infrastructure and increasing pressure on limited resources like water, energy and space. Delivering conservation in this socio-island environment involves embracing a community that may not consider conservation as its key priority. Many island communities can be overwhelmed by rapid growth that fails to meet local needs and aspirations. Often they want a healthy local economy – but not at the expense of their way of life, natural surroundings, local identity or their island turning into a homogenous tourist trap.

This situation contrasts with the strategies for Lord Howe Island, that maintain a strict policy of a maximum of 400 visitors on the island at any one time. Despite this restriction, available Australian

Bureau of Statistics data from 2000–08 (ABS 2011) shows a steady increase in median income of residents above the CPI. Lord Howe Island Board annual reports show a similar increase in total comprehensive income during the period 2011–16 (Lord Howe Island Board 2017). Thus, Lord Howe Island demonstrates that increasing visitor numbers is not the only avenue to economic growth.

In a study on local community perceptions of tourism on Bruny Island and Magnetic Island, Moyle *et al.* (2010) used social exchange and social representation theories to investigate island community perceptions of tourism development. They found that on Magnetic Island, the development of resorts, changes in the type of visitors and consequent changes to the social environment were reasons why locals did not interact with visitors. On Bruny Island, there was local dissatisfaction due to visitors overcrowding island infrastructure and pressure on services, roads etc. On both islands, the reason why locals did not interact with visitors was due more to the process of tourism development decision-making than to its consequent impacts. This finding highlighted the importance of community inclusion in sustainable island tourism policy and planning, both to facilitate the process of sustainable development and to educate the community about the likely consequences.

Future of conservation tourism

Australia's islands are critical for the conservation of biodiversity at all levels of definition (Lindenmayer *et al.* 2014; Woinarski *et al.* 2014): they serve as engine rooms for evolutionary diversification, as a refuge for species including many endemics and for breeding colonies, for coastal and other ecosystems and the landscapes which these components form. Islands are also arks for Indigenous and other social and cultural values that are intrinsically connected to a geographical place, and thus cannot be fostered and protected elsewhere. From a pragmatic view, islands will continue to be used for economic return, and tourism is by far the largest island-based industry in terms of number of islands used to derive that return. The question must be then, what models for future island tourism can provide a suitable economic return while having minimal impact on island values and, preferably, making a positive contribution to maintenance of those values?

There seems little doubt that tourism market forces – the choice of different market segments – will have a large influence on the future of island tourism in Australia, as it has in the past. However, there appear to be strong global trends that could be adapted to reimagine island tourism in Australia. One global megatrend in tourism has arguably produced the fastest-growing market sector, the experience-seeker. Tourism Australia (2016b) summarises this sector as people seeking:

- authentic social and personal experiences including opportunities to genuinely interact with local people at the destination;
- learning experiences outside their usual norm, i.e. different lifestyles and cultures;
- visits to 'authentic' places, not necessarily normal tourist destinations;
- challenges and participatory opportunities, including being able to see and do things that immerse the traveller in unique nature and cultural experiences.

It is not difficult to contend that this sector involves people who are well-travelled, early adopters, and more affluent and open-minded. Experience-seekers would be attracted to opportunities with a focus on low-volume, higher-value island tourism, and thus potentially have less impact. Further, due to this sector's desire to be challenged and participate at their destination, visitors would be willing to make positive contributions to environmental sustainability and nature conservation. We contend that this market sector would readily take up tourism offerings designed to overlap nature-based tourism, sustainably managed tourism, education tourism and conservation tourism (Fig. 11.2) and thus create the most ideal interplay between tourism and conservation.

Fig. 11.2: Middle Percy Island (GBR) 'A' Frame – a unique icon of cruising sailing history standing in face of mass tourism homogeneity.

We propose that development of such tourism offerings, for the experience-seeker sector, is the most logical and desirable future for tourism on Australia's islands. Fortunately, this process has already commenced, albeit slowly, but there are inspiring examples of how this could be furthered. Several case studies (Boxes 11.1 to 11.4) show examples of ways that might create the desired trajectory for island tourism in the future, and those best avoided.

Concluding comments

Island communities typically aspire to find ways to safeguard their natural places and open spaces, support locally owned businesses, encourage traditional vocations and retain a sense of their unique character and tradition. These customs and behaviours should be recognised as an essential component of any island conservation program.

Tourism operators who actively support the conservation of natural assets that underpin their business remain in the minority. While Australia's regulatory environment is no doubt demanding and costly (permitting, licensing, reporting and so forth), these are mostly transaction costs and rarely contribute to conservation of natural assets.

Robust and long-term planning is required so that growth meets local wishes, contributes to a sustainable economy, enhances the quality of community life and financially underpins conservation of the natural assets that make islands unique. Economic growth is not necessarily achieved only by growing visitor numbers.

There is great opportunity for island tourism to focus on points of unique difference, with the

Box 11.1: Rottnest Island

In 2010, Rottnest Island, Western Australia, celebrated 100 years of tourism. The small island reserve (1859 ha) has undergone many transformations since its discovery by a Dutch explorer in 1696, who named it 'Rats nest' due to the abundance of the well-known marsupial, the quokka (*Setonix brachyurus*). Located only 19 km west of Perth, visitor numbers have steadily grown (to over 600 000 in 2015) while recent estimates of the population of the quokka, a threatened species with dwindling numbers on the mainland, were still a viable 10 000–12 000 individuals. As a formally recognised protected marine and island reserve, the government-appointed Rottnest Island Authority is tasked with ensuring that the island's natural environment is conserved and enhanced while providing for the increasing visitor demand for the unique experiences of the island.

The island's rich cultural and social heritage has left its mark on the natural environment. Colonial exploits such as extensive vegetation clearing associated with farming, mining, a settlement, the development of World War II military defences and the introduction of exotic flora and fauna species have all had impacts. Catering for human habitation requires access to freshwater, power and management of wastewater and rubbish. There is an increasing demand for food, accommodation, facilities, transport and other infrastructure for residents and visitors.

The wetlands on Rottnest Island are of local, national and international significance. They provide habitat for quokkas and migratory birds and support unique microbiolite communities. However, groundwater monitoring has shown increasing salinity, raising concern about long-term aquifer health and the ability to meet increasing potable water demands and environmental water requirements in a drying climate.

The Rottnest Island Authority is required to meet its legal obligations while recovering its operating expenses. This has proven especially challenging as the Authority needs to provide all utilities on the island, operate to the highest environmental standard and protect the island's biodiversity, while competing for the provision of tourism services such as accommodation, food and recreational activities with much cheaper destinations such as Bali, easily accessible to locals and less than four hours flight time away.

In November 2014, the Authority released a Terrestrial Management Strategy that would provide an overarching management direction for ensuring that the condition and integrity of the flora, fauna, landforms, geology and hydrology are protected and enhanced where necessary. The strategy provides for a risk-based, adaptive and continually improving approach to meet or exceed best practices. It involves three specific management tools: the implementation and maintenance of an environmental management system (ISO14001) in Authority operations and by major contractors and lessees, the development and implementation of Conservation Action Planning, and achieving international eco-accreditation. Conservation action planning is an open-standard approach to conservation management developed through a joint venture of non-government organisations.

The application of EarthCheck accreditation led the Authority to benchmark its operations, thus facilitating the development of business cases for sustainability projects that led to reduced operating costs and smaller environmental footprints. These include the establishment of a volunteer-run native flora nursery for rehabilitation, developing a world-class walking trail to improve visitor access and reduce informal tracks, upgrade of the wastewater treatment plant (allowing use of recycled wastewater on the golf course and sports oval, thus significantly reducing water extraction from groundwater), establishing a glass crusher to reduce the volume of waste requiring transport to the mainland, and the commencement of a renewable energy project which will increase renewable energy penetration from 23% (wind generator) to 45% through the construction of a 600 KW solar farm with associated development of a demand management micro-grid.

The success of Rottnest Island Terrestrial Management Strategy requires agency-wide commitment to implementing and maintaining the environmental management system, addressing biodiversity and conservation management actions and continuing to achieve sustainability targets and objectives through the EarthCheck accreditation process. The island's administration must protect the unique ecology as its key asset. Such stewardship will ensure visitors can continue to experience the natural wonders this speck in the Indian Ocean offers current and future generations.

Box 11.2: K'gari-Fraser Island

K'gari-Fraser Island was listed as a World Heritage Area in 1992 after cessation of logging. It is the world's largest sand island, and is an outstanding example of ongoing biological, hydrological and geomorphological processes (Chapter 10). Its key landscapes include expansive beach habitats, sand cliffs and blowouts, perched freshwater lakes and tall rainforest growing on sand dunes. At the time of gazettal Fraser Island had more than 240 000 visitors per annum, a figure that has more than doubled to over 560 000 in 2015. Key recreational activities on Fraser Island include beach driving, camping and swimming in freshwater lakes, all based on key natural assets for which the island was inscribed on the World Heritage register. In a 2008 review, Thompson and Schlacher (2008) summarised the key impacts. 98% of the 122 km of ocean beaches are open to four-wheel driving, and camping zones cover 28.7 km (23%) of the dunes. Access tracks had destroyed over 20% of the dune environment within the camping zones, and there is evidence of accelerated erosion of beach habitats as a result of scalloping due to vehicle use. The condition of Fraser Island's freshwater lakes has also been of concern for some time, particularly those more oligotrophic in nature and thus more susceptible to changes in nutrient inputs (Hadwen *et al.* 2003). However, some impacts, presupposed to be a result of tourism, are not supported by evidence. The conservation significance of dingoes (*Canis familiaris*) on Fraser Island is unarguable, as it is a refuge from interbreeding with domestic dogs. This has often conflicted with the need to cull dingoes as a result of negative, and often dangerous, interactions with humans. However, a review by Allen *et al.* (2015) showed that despite the number of dingoes being culled (110 between 2001 and 2013 *v.* a population size of 76–171 during the whelping period in 2012), such culling would not have an adverse impact on the population. This was because most animals removed from the population were young males, not adults, and few females were removed.

Some aspects of tourism on Fraser Island have set benchmarks for more contemporary eco-tourism. The Kingfisher Bay resort has long been held in high regard for its environmental sustainability mandate, which includes sensitive design and building codes that seek to blend with the surrounding natural environment, minimisation of changes to the beach dune environment, integrated environmental educational activities, and energy and waste minimisation initiatives. Given the high level of visitation to the resort compared to its relatively small geographical footprint, it is likely that, despite being a built environment, it creates far less per person impact than more ephemeral activities such as beach camping.

It is clear that overall visitation levels to Fraser Island are above its inherent ecological carrying capacity, and degradation through excessive human use remains a threat to its World Heritage status.

approach of lower-volume, higher-value tourism enterprises in niche markets. The key challenge for island communities is retaining a high quality of life in the face of increasing pressure for growth, change and mass homogeneity.

Space is a premium asset on every island and 'nature' needs space to survive and flourish. The adoption of an island sustainability quotient, such as the visitor number limitations on Lord Howe Island, could be applied more widely. Such a quotient could transparently consider the resident and visitor carrying capacity of islands in terms of cost-effective provision of infrastructure relating to water, food, waste management and electricity generation etc., and protection of the natural assets of the island, ensuring adequate permanent reserves and no net loss of habitats, species and ecosystem services.

Box 11.3: Pennicott Wilderness Journeys, Tasmania

Pennicott Wilderness Journeys was established in Tasmania in 1999 by Robert Pennicott and Michaye Boulter and has grown from a small family business into a major eco-tourism enterprise that has won a plethora of national awards. Robert Pennicott has been inducted into the Hall of Fame as an Australian Tourism Legend. The business originally offered a scenic cruise around Bruny Island: by 2017 it offered multiple cruise options around Bruny Island, Tasman Island, the Iron Pot and a component of the Three Capes Track, serviced by a fleet of purpose-built high-speed boats and tour buses. The company attracts over 100 000 passengers annually, employs around 100 staff, has a retail shop and restaurant and plans to expand its operations to Wilsons Promontory, Victoria (Hudson 2016). The attraction of Pennicott Wilderness Journeys is its offer of exhilarating scenic cruises around three naturally rugged southern Tasmanian islands and the opportunity to view marine wildlife on roost sites, at haul-outs or frolicking freely. The tours take in spectacular dolerite columns, towering sea cliffs, deep-sea caves and rock arches, and regularly encounter seals, seabirds, dolphins and the seasonal passage of whales. In 2015 the multi-award-winning Bruny Island Cruise was voted by *Travel and Leisure* as one of the 100 greatest trips of the world (<www.travelandleisure.com>) and in October 2016 it won the Skal International Sustainable Tourism Award.

But what makes this island-based tourism enterprise unique is the owners' public declaration that it will donate 25% or more of the company's net profits back into the local community and environment (Stone 2012). This commitment is validated by a growing list of philanthropic donations, starting in 2007 when Pennicott established the Tasmanian Coast Conservation Fund through the environmental body Wildcare Inc. Pennicott then donated A$100 000 towards several coastal conservation projects, the most significant being a A$65 000 donation towards the eradication of feral cats from Tasman Island. Tasman Island (120 ha) holds Australia's largest breeding colony of fairy prions (*Pachyptila turtur*) and feral cats were estimated to kill ~50 000 seabirds per year. With a total project cost of A$250 000, Pennicott's donation expedited the work and the eradication project was declared a success in 2011 (PWS 2015). In 2011, Robert Pennicott launched the Pennicott Foundation to grow the company's philanthropic work further and by 2015 it supported over 300 projects, ranging from wildlife rehabilitation to beach clean-ups to school programs (www.pennicottjourneys.com.au/). In 2016, the Pennicott Foundation donated A$60 000 towards the eradication of black rats from Big Green Island (125 ha) and A$40 000 to the Bruny Island cat management program for Courts Island, and has committed additional funding to eradicating rabbits on the Breaksea Islands in Tasmania's south-west. In 2016, Pennicott reported that with his company still growing, each year it gives A$500 000–1 million to the Pennicott Foundation for environmental and humanitarian projects (Hudson 2016).

Pennicott Wilderness Journeys is 100% carbon offset and an Advanced Eco-tourism Certified operator, and is reported annually by EarthCheck as exceeding world's best practice standards for efficient fuel, energy and water use. This accreditation recognises high-quality nature-based tourism experiences with strong interpretation values and commitment to nature conservation and helping local communities. The business is relatively robust in its capacity to adapt to future climate change in Tasmania, having a diverse number of locations and asset placement that will combat future storm surges and increased bushfire risk. In fact, it may potentially benefit from increased visitor numbers with the predicted warmer winters in Tasmania (Tasmanian Climate Change Office 2013).

The commercial success of Pennicott Wilderness Journeys has had significant flow-on effects for Tasmanian tourism. It has especially increased the Chinese market and the popularity of Bruny Island, which in 2014 was voted one of Australia's best nature-based tourism attractions (Moyle *et al.* 2015). In April 2016, the Tasmanian government committed funds for an upgrade of the roads and jetty on Bruny Island and for an increased ferry capacity. It awarded the Pennicotts A$499 995 through the federal government's Regional Tourism Infrastructure and Innovation Fund towards the company's A$4.8 million investment in building a multi-purpose tourism centre at Adventure Bay (Tourism Tasmania 2016).

This case study demonstrates the enormous natural capital of three islands in Tasmania (Bruny, Tasman and the Iron Pot) and the significant benefits these can generate for local and regional economies and conservation more widely when showcased by conservation-driven, genuinely philanthropic tourism professionals.

Box 11.4: Wildmob

Wildmob was founded by dotcom entrepreneur Graeme Wood in 2008 to perform sensible, pragmatic conservation work at ground level using a volunteer-tourism model. This model allows paying volunteers to visit and enjoy spectacular natural areas, while contributing to the areas' conservation values through their work efforts. The initiative started with a project on Brampton Island in the Great Barrier Reef, where volunteers could contribute by undertaking restoration works including weed control within critically endangered littoral rainforest, and removing marine debris from marine turtle nesting beaches, reducing the risks of entanglement during a vulnerable period of their lifecycle. This initial project has expanded to undertaking management actions on around 50 islands in the Great Barrier Reef region, and now includes support for marine turtle and seabird surveys and research to inform management interventions. Wildmob's work has expanded to include Norfolk Island in the south-west Pacific, management of the endangered bridle-nail-tail wallaby in central Queensland and survey support in the mountain ash forests north of Melbourne.

Wildmob's activities include environmental education at both secondary school and university levels, supporting and/or providing meaningful experiential learning for young people considering or wishing to pursue careers related to nature conservation. This includes providing field ecological coursework in a range of environments including tropical and temperate marine ecology, coral reef ecology, and threatened species research and management.

In addition, Wildmob coordinates events that support its conservation initiatives, such as Birdweek and the Island Arks Symposium. These events provide some of the revenue required to support volunteers in the field, and to fund related conservation work in Wildmob's project areas. Importantly, the events also make a significant contribution to local economies, providing transparent financial value to natural assets that attract visitors. The Norfolk Island Birdweek in 2015 contributed over A$150 000 to the Norfolk Island economy in a single week.

Wildmob staff are involved in facilitation and coordination of conservation initiatives outside the scope of volunteer tourism, in their own fields of expertise. For example, Wildmob is developing an ecologically based fire management strategy for islands off the central Queensland coast, and has contributed to research needed to support translocation of the endangered Norfolk Island green parrot to an adjacent island.

In 2015–16, Wildmob volunteers numbered more than 400 and contributed 1900 person-days of conservation management work along the eastern seaboard of Australia and on Norfolk Island. Using Norfolk Island as an example, between mid-2014 and mid-2016, economic inputs to the island's economy was over A$500 000 because of Wildmob's volunteer, educational and nature-based events tourism. For such a small economy, that notable figure provides financial incentives for Norfolk's natural assets to be viewed as worthy of protection. That additional viewpoint influences the island community's desire to maintain the sustainability of conservation interventions, as natural assets are proven to have a real financial and cultural value.

References

Allen BL, Higginbottom K, Bracks JH, Davies N, Baxter GS (2015) Balancing dingo conservation with human safety on Fraser Island: the numerical and demographic effects of humane destruction of dingoes. *Australasian Journal of Environmental Management* **22**, 197–215. doi:10.1080/14486563.2014.999134

ABS (2011) *National Regional Profile: Lord Howe Island (Statistical Local Area).* Australian Bureau of Statistics, Canberra.

ABS (2016) *Overseas Arrivals and Departures, Australia, Dec. 2016.* Cat. No. 3401.0. Australian Bureau of Statistics, Canberra. <http://www.abs.gov.au/ausstats/abs@.nsf/featurearticlesbyCatalogue/961B6B53B87C130ACA2574030010BD05>

Buckley R (1994) A framework for tourism. *Annals of Tourism Research* **21**, 661–665. doi:10.1016/0160-7383(94)90126-0

Deloitte Access Economics (2013) *Economic Contribution of the Great Barrier Reef.* Great Barrier Reef Marine Park Authority, Townsville.

Destination South Tasmania (2016) *Bruny Island Destination Action Plan 2016–2018*. Destination South Tasmania and Department of State Growth, Hobart.

Fennell D (2005) *Ecotourism*. 2nd edn. Taylor and Francis, e-Library.

Hadwen WL, Arthington AH, Mosisch TD (2003) The impact of tourism on dune lakes on Fraser Island, Australia. *Lakes and Reservoirs: Research and Management* **8**, 15–26. doi:10.1046/j.1440-1770.2003.00205.x

Hudson S (2016) Tasmania's Pennicott Foundation is helping the world one project at a time. *Weekly Times*, 26 October. <http://www.weeklytimesnow.com.au/country-living/tasmanians>

Icon Tourism (2010) *Christmas Island Tourism Association: Marketing Strategy 2010–2015*. Icon Tourism, Perth.

Lindenmayer D, Dovers S, Morton S (eds) (2014) *Ten Commitments Revisited: Securing Australia's Future Environment*. 2nd edn. CSIRO Publishing, Melbourne.

Lord Howe Island Board (2017) *Annual Reports*. <http://www.lhib.nsw.gov.au/board/publications/reports>

Moyle BD, Croy WG, Weiler B (2010) Community perceptions of tourism: Bruny and Magnetic Islands. *Asia Pacific Journal of Tourism Research* **15**(3), 353–366.

Moyle B, McLennan C, Becken S, Battye R, Godfrey A (2015) *Nature-based Tourism Innovative Experiences*. Fact sheet. Griffith Institute for Tourism in partnership with Tourism Research, Australia.

PWS (2015) *Feral Cats Eradicated from Tasman Island*. Parks and Wildlife Service, Department of Primary Industries, Parks, Water and Environment, Hobart.

Queensland Government (2015) *Draft North Stradbroke Island Economic Transition Strategy*. Department of State Development, Brisbane.

South Australian Tourism Commission (2016) *Regional Tourism Reports 2015*. <http://tourism.sa.gov.au/research-and-reports/regional-tourism-profiles.aspx>

Stone G (2012) Traveller of the year: Robert Pennicott. *National Geographic*. <http://www.nationalgeographic.com/travel/travelers-of-the-year/robert-pennicott/>

Tasmanian Climate Change Office (2013) *Pennicott Wilderness Journeys. Adapting to a Changing Climate: Case Study*. <http://www.dpac.tas.gov.au/divisions/climatechange/adapting/climate_futures/climate_futures_for_tasmania_reports>

Thompson LMC, Schlacher TAJ (2008) Physical damage to coastal dunes and ecological impacts caused by vehicle tracks associated with beach camping on sandy shores: a case study from Fraser Island, Australia. *Journal of Coastal Conservation* **12**, 67–82. doi:10.1007/s11852-008-0032-9

Tourism Australia (2016a) *Coastal and Aquatic Experiences*. Fact sheet. <http://www.tourism.australia.com/statistics/consumer-demand-research.aspx>

Tourism Australia (2016b) *Global Experience Seekers*. <http://www.tourism.australia.com/documents/corporate/Global_Exp_Seekers.pdf accessed 18th November 2016>

Tourism Tasmania (2016) *New Investment in Bruny's Popularity*. <http://www.tourismtasmania.com.au/news/releases/2016/21-apr-investment>

Tourism Victoria (2015) *Value of Tourism to Phillip Island 2013–14*. Research fact sheet. Tourism Victoria, Melbourne.

TRA (2011) *What is Driving Australians' Travel Choices?* Tourism Research Australia, Canberra.

Weaver DB, Lawton LJ (2007) 20 years on: the state of contemporary ecotourism research. *Tourism Management* **28**, 1168–1179. doi:10.1016/j.tourman.2007.03.004

Whitsunday Regional Council (2016) *Hook Island Underwater Observatory*. <https://www.whitsunday.qld.gov.au/DocumentCenter/View/1177>

Woinarski J, Ball D, Burbidge A (2014) Islands. In *Ten Commitments Revisited: Securing Australia's Future Environment*. 2nd edn. (Eds D Lindenmayer, S Dovers and S Morton) pp. 117–128. CSIRO Publishing, Melbourne.

12

Partnerships for island conservation: it's all about people

Sally L. Bryant and Peter B. Copley

Summary

Australia's islands are incredibly rich natural repositories yet many face ongoing exploitation or degradation, or suffer from benign neglect. With tenacity and commitment, community-driven programs and non-government organisations are demonstrating what good partnerships, collective effort and sheer hard work can achieve for island conservation when few other resources are available. The conservation communities that catalyse around island environments not only contribute intellectual capital and millions of dollars of volunteered time, but are often the lynchpin which sustains conservation effort long-term. With varied structures and business models, environmental groups are becoming more sophisticated and expanding their level of expertise and involvement. This chapter reviews a range of community-driven programs on Australia's islands to illustrate the levels at which they operate and the magnitude of their contribution, and to reinforce that strong community partnerships are essential if we are to achieve lasting island conservation.

Introduction

Islands represent 10 of the world's 34 conservation hotspots and their conservation and sustainable use of natural resources are recognised internationally as critical to achieving the Strategic Plan for Biodiversity 2011–2020 and the Aichi Biodiversity Targets (Secretariat of the Convention on Biological Diversity 2014). Yet, despite their seminal role, island ecosystems are being increasingly degraded by invasive pests, pressures for development and the impacts of global threats, and their taxa have experienced rates of extinction that are unsurpassed in modern times (Baldacchino 2007; Secretariat of the Convention on Biological Diversity 2014). Even in Australia, 30% of extinctions since European settlement have been island-endemic taxa, and this figure is about to rise as three more island species are added to the list (Chapter 7). How then does Australia, a large island nation, address these threats and strengthen its efforts to meet its international obligations under the Convention on Biological Diversity?

Australia has around 8300 islands within its territorial waters (Geoscience Australia 2004). They extend from Heard, McDonald and Macquarie Islands in the sub-Antarctic, as far north as Christmas Island just 10° south of the equator and east to Norfolk Island in the South Pacific Ocean. The Great Barrier Reef alone comprises nearly 4000 reefs and islands scattered along the Queensland coast and is recognised universally as one of the seven wonders of the natural world. This huge geographic spread and climatic diversity presents serious challenges for island management but, compared to other parts of the world, Australia's island network has two significant advantages which aid its conservation. First, most of Australia's islands are unoccupied; in fact, excluding Tasmania, less than 1% of Australia's total population live permanently on islands (ABS 1999). Second, the majority of Australia's islands remain in reasonably good ecological condition or at best can still be repaired if there is support to do so. These two factors place Australia in an enviable position of being better able to retain and restore its islands without the added complexity of managing human needs or expectations. To capitalise on these two natural advantages while they still exist, we need to proactively secure strong partnerships to help undertake the work needed.

This chapter focuses on the enormous contribution the community makes to island conservation and its capacity to fill the void and to lead change. It provides a range of case studies of community-focused programs that are operating at every level from the grassroots to national and international partnerships on populated and unpopulated islands in every Australian state and territory. These examples highlight the challenges encountered and the achievements made.

Community-driven action groups

Island communities are diverse but most have a strong sense of ownership and desire to protect the natural environment which is often intrinsic to their way of life. Achieving conservation in this socio-island environment can be challenging and may involve embracing a community with diverse needs and aspirations, among which conservation may not be the top priority. Many local resident action groups have evolved to safeguard and advocate for the conservation of their island in the face of perceived threats, especially when new developments or expansion of infrastructure is proposed within existing natural settings (e.g. Bribie and Bruny Islands). These grassroots groups are often an incredible source of knowledge about local values and deliver an enormous range of services to the island and wider community.

Bruny Island Environmental Network

The Bruny Island Environmental Network Inc. (BIEN) was formed in 2009 as a community-based collective of landholders with a strong passion for the natural and cultural heritage of Bruny Island. Bruny Island is located 70 km south-east of Hobart, Tasmania and is 362 km^2 with a resident population of around 700 people, which swells significantly with shack owners and visitors during the summer. One of BIEN's initial tasks was to develop a guiding policy for nature-based tourism operators, as it was obvious this industry was set to grow and needed to be ecologically sustainable (Bruny Island Environment Network 2010). In 2013–14 Bruny Island was considered one of the best nature-based tourism attractions in Australia (Moyle *et al.* 2014) and in the year ending December 2015 it attracted 111 000 interstate and international visitors and over 80 000 intrastate visitors (Destination South Tasmania 2016). Over time, BIEN's tireless submissions on proposed developments, its provision of community advice, awareness-raising and direct action such as weed removal, shorebird protection and restoration work has made it a leading voice for environmental issues on the island. In partnership with the Kingborough Council and Bruny Island Community Association, BIEN helped develop the first phase of the Bruny Island cat management program, which in 2016 received Commonwealth funding as one of five Australian islands nominated to become cat-free.

BIEN also celebrates the cultural and artistic nature of island life, and every two years co-hosts the Bruny Island Bird Festival. This festival has grown into a four-day event showcasing the island as an international birding hotspot and stronghold for Tasmania's nationally threatened bird species. In 2016 the program was opened by Michael Leunig – Australia's national living treasure, cartoonist, poet and devout nature-lover – and attracted over 500 visitors who stayed and bought locally, generating an estimated A$200 000 for island businesses. BIEN has a membership of over 100 and its modest finances are generated from membership fees, small grants, donations and profits from successful events and publications.

Bribie Island Environmental Protection Association

The Bribie Island Environmental Protection Association (BIEPA) began 35 years ago as a self-funded group of residents committed to the conservation of Bribie Island, located ~65 km north of Brisbane in the Moreton Bay region. The island's unique vegetation communities and ecosystems are richly diverse and over 80% of the island is protected in national parks and conservation areas. Over time BIEPA has grown into a cohesive conservation organisation that manages educational events, mentors students, produces field guides and other publications and delivers a wide range of on-ground projects, especially dune and wetland restoration and species monitoring for the Queensland government (BIEPA 2017). It consults widely and, like many local groups, operates on small amounts of funding generated through memberships, small grants and donations. Over the years BIEPA has mounted formidable public campaigns opposing a range of proposed developments such as a pulp and paper mill, a marina at Buckley's Hole, and a massive A$2 billion desalination plant initially supported by both major political parties. Bribie Island is a low-lying sand island and particularly vulnerable to any impacts on its groundwater systems (Maynard *et al.* 2010), so potential developments involving water use need to be thoroughly scrutinised.

There are many island-based community-driven environmental organisations in Australia like BIEN and BIEPA that invest enormous amounts of volunteer time and intellectual capital into caring for the natural condition of their island homes. Their attributes exemplify what community-driven conservation can achieve when smart, professional, self-reliant, innovative and determined people drive local issues. Their service often extends at every level from on-ground works to representing submissions through development processes and sometimes legal hearings. They play a critical role in providing informed local, independent and professional advice on issues which potentially threaten their island's natural values and which often serve as a reference point for others.

Friends of islands partnerships

Over the past 20 years the exponential growth in 'friends of' groups has seen committed teams of volunteers provide a massive contribution towards protecting national parks and reserves around the country, including those on islands. In partnership with respective state government park authorities, 'friends of' groups now operate in nearly every Australian state and territory and, although their governance structures vary, the model appears successful. With proper training, volunteers can fulfil a range of significant roles in education, research and adaptive management including the collection of ecological data, often incorporating sophisticated operating procedures (Greenwood 2007). Once established, these groups quickly become self-motivated, need minimal supervision and often bring additional skills through their wide member base. Their success is inherent in a genuine and durable partnership containing key attributes such as shared passion and values, inclusive participation, willingness to engage and foster collaboration, mutually reinforcing activities towards achieving long-term goals, good governance, planning and dedicated coordination support. However, due to their partnership arrangements, many volunteer groups operate in an increasingly complex bureaucratic environment and compete for meagre resources in a highly competitive space. This is especially so for island-based groups, who face the added pressure of risk insurances due to remoteness, boat transport, biosecurity and timeframes dictated by the weather. Ritchie (2011) discusses the range of challenges faced by island-based groups in New Zealand, and these apply equally to 'friends of' groups in Australia. Ritchie says the key to group success is to demonstrate credibility and methodical project management through good planning and organisation. The best plans are those that are succinct, easily understood by all involved, and followed. A mountain of paperwork can kill enthusiasm and lead to burnout, whereas simple plans with easy language and clearly expressed timelines and processes have the greatest chance of success. These principles were borne out during the Pamona Island restoration program in New Zealand, where a review of success found that unless there is effective organisation, planning, commitment, partnerships and communication then member frustration can lead to a perceived lack of progress at every stage (Shaw *et al.* 2011).

Friends of Lord Howe Island

Arguably one of the most ambitious weed eradication programs in the world is underway on Lord Howe Island, located ~600 km east of Port Macquarie, New South Wales. With a timeframe of 30 years to achieve success, during its first 10 years of operation the program has removed over 2 million weed plants, eradicated six weed species and reduced the number of priority weeds swamping the island's ecosystem by 80% (Lord Howe Island Board 2015; Bower 2016). These achievements have been due in part to the contribution of dedicated locals and volunteers, who are vital for the program's ongoing success (Lord Howe Island Board 2015). So far, the eradication effort has taken 120 480 h of labour to which the Friends of Lord Howe Island have contributed 77 self-funded weeding trips and 25 000 volunteer hours, helped fund small projects and fostered community awareness of the problems weeds pose (Friends of Lord Howe Island 2015a, 2015b). Officially formed in 2001, the Friends of Lord Howe Island have focused on discrete community projects using traditional manual weeding methods, thus freeing professional staff to undertake the rope- and helicopter-work needed to tackle the more inaccessible areas on the island. Their weeding trips and other naturalist-style trips, often championed by local experts, help promote the island's unique wildlife and generate income for island businesses.

Tasmania's 'friends of' islands groups

In 1997 Wildcare Inc. was established as a joint community–Tasmanian government partnership and in the two decades since it was formed it has grown into the state's largest incorporated not-for-profit volunteer group with over 7000 members contributing over 500 000 h of voluntary project work each year (Wildcare 2016). Wildcare Inc. has 13 'friends of' groups focusing specifically on island-based initiatives. Their core tasks range from wildlife rescue, weed removal, habitat restoration, species monitoring and heritage maintenance to acting as caretakers on lighthouse islands or quarantine stations. For example, for the past 14 years the Friends of Bass Strait Islands have been removing African boxthorn (*Lycium ferocissimum*) from 10 outer islands in the Furneaux Region and, although voluntary, they replicate a professional weeding team. Groups of 10 or more volunteers spend up to a fortnight per trip, sometimes twice a year, chainsawing, hand-cutting, poisoning and burning weeds to protect seabird colonies, historical and Aboriginal sites and other important island values. In 2016 alone, Friends of Bass Strait Islands contributed over 3000 volunteer hours treating 145 ha of boxthorn with their costs of transport, equipment, accommodation and even food all underpinned by organisational grants and in-kind local contributions (Wild Times 2016). Each trip typically costs about A$10 000, so finding the funding needed is a constant challenge.

The Friends of Maatsuyker Island is another example of a volunteer group which makes a substantial contribution to maintaining the cultural and natural values of Maatsuyker Island in Tasmania's Wilderness World Heritage Area. Their work began in 2003, focusing on removal of a range of invasive weeds and has since expanded to monitoring and mapping of short-tailed shearwater (*Ardenna tenuirostris*) colonies and cataloguing historic sites and outbuildings on the island. It also includes lighthouse preservation and caretaker duties. Members adhere to strict biosecurity protocols and safety procedures during island visits. Friends of Maatsuyker Island regularly host and participate in community events to help showcase the island's unique natural and cultural history and the work they do (Fig. 12.1). Few of these island services would be achieved without the Friends partnership, and their commitment and capacity to assemble teams of experienced volunteers and generate their own finances is reflected by the island's improving natural condition.

Wildcare Inc. provides its registered groups with insurance cover, small grants for projects, training in the use of equipment and first aid, and

Fig. 12.1: Few of Maatsuyker Island's natural and historical values would be maintained without the volunteer Friends of Maatsuyker Island, Wildcare group, Tasmania. Photo: Marina Campbell.

facilitates larger sponsorship and donations through its partner programs. The Parks and Wildlife districts are responsible for identifying priorities and assisting groups to the best of their capacity, often despite declining budgets. Groups are responsible for managing their programs, for ensuring their teams are properly trained and equipped, for project governance and adequate funding to ensure project delivery.

Non-government organisations

One of the biggest challenges for any conservation entity is that of financial stability, especially for island-based work which is inherently expensive. Depending on their purpose, non-government organisations (NGOs) can adopt any number of financial structures ranging from not-for-profit to for-profit commercial business models. A preliminary study into six pest-free island sanctuaries in New Zealand found that many of their governing bodies were transitioning from being charitable trusts or incorporated not-for-profits into social enterprises with the primary purpose of proactively making money (Campbell-Hunt 2008). To do this they were taking on commercial business ventures, like tourism and education programs, which sat comfortably within their conservation objectives. However, in New Zealand, as in Australia, it is increasingly difficult to achieve full financial self-reliance, so establishing strong partnerships with a mutual benefit is the key to long-term success.

Environmental non-profit organisations in Australia involved with island conservation, such as the World Wide Fund for Nature – Australia (WWF), BirdLife Australia, the Invasive Species Council and global initiatives like Island Conservation, can seek public donations and philanthropy to help support their work as well as relying on memberships, grants and consultancies etc. These broad-reach community organisations fill an important niche by performing a myriad of valuable roles such as deploying on-ground support at priority hotspots, collecting, reviewing and interpreting large datasets, compiling strategic reviews, orchestrating large fundraising and awareness campaigns and lobbying governments for legislative change. For example, WWF was a strong partner with the Tasmanian and Commonwealth governments in the successful delivery of the Macquarie Island pest eradication program. Also, its online petition to halt industrial destruction of the Great Barrier Reef was delivered to the UNESCO World Heritage Committee with over half a million signatures. WWF's role in promoting and improving conservation issues across the entire country is well-known and has been long-standing.

In 2016, a new partnership forged between Australia's Invasive Species Council and Island Conservation is seeking to deliver a more strategic approach to tackling the number one threat to Australia's islands – invasive species. This alliance will focus attention on the successes of pest eradication programs globally and hopefully strengthen the need to prioritise island eradications on the national agenda. Already the alliance is gathering momentum and is a stakeholder in the Australian government's national environmental biosecurity forum dealing with quarantine issues affecting islands.

BirdLife Australia

BirdLife Australia contributes to island conservation in significant ways, especially in the provision of much-needed data on bird species distribution and status. For example, since 1998 over 7000 volunteers have contributed over 420 000 surveys comprising over 7.1 million bird records through its citizen science Australian Bird Atlas Project. This project forms a vital link between volunteers collecting data, scientists undertaking research and decision-makers managing the plight of Australia's birds, particularly shorebirds and marine species dependent on coasts and islands. This information is pivotal in tracking population trends, assigning status and underpinning key strategic documents, particularly the status of Australia's birds, islands and birds, and birds in a changing climate. The work of BirdLife Australia is also vital in helping inform the global standard for designating Key Biodiversity Areas – Nature's Hotspots, many of which are islands (BirdLife Australia 2017). None of this information would be available without BirdLife Australia's mobilisation of the expertise and enormous number of citizen scientists committed to bird conservation.

WildMob

The not-for-profit organisation WildMob has adopted a business enterprise model for its island conservation initiatives, which it delivers through on-ground work embedded into local communities. WildMob's vision is predicated on the 4Cs philosophy – Conservation, Community, Culture and Commerce – and its commercial enterprises are a way of injecting much-needed financial support into small, often struggling island communities.

Working in partnership with diverse stakeholders including governments, WildMob takes on the role of administering the project and supplying the workforce to deliver pest baiting, weeding, surveys, beach restoration or whatever is needed. WildMob offers people from all walks of life a chance to participate in a unique holiday adventure and do meaningful work in an amazing island location, and participants' payments help defray costs. When reviewed as an educational tool for university students, Griffith University (2012) found the WildMob model provided an excellent applied learning experience and improved the capacity of overseas students to contribute to their own communities when they returned home. In 2014–15, WildMob's achievements (WildMob 2015) translated to conservation work on 50 islands and along 500 km of the Great Barrier Reef, involving:

- 333 volunteers, 1843 field days and A$440 000 of volunteer time;
- 154 students taught in six outdoor classrooms;
- over A$500 000 spent in local communities.

NGO-owned islands

Two examples of privately owned islands managed specifically for conservation are Faure Island in Western Australia and Avoid Island in Queensland.

The Australian Wildlife Conservancy purchased the pastoral lease to Faure Island (4561 ha) in Shark Bay World Heritage Area in 1999. Since then it has worked closely with the Western Australian government and other organisations to eradicate pests, restore habitats and reintroduce a range of threatened species including boodie (*Bettongia lesueur*), banded hare-wallaby (*Lagostrophus fasciatus*), western barred bandicoot (*Perameles bougainville*) and the Shark Bay mouse (*Pseudomys fieldi*). Over the years its baseline surveys and ecological monitoring program have helped measure the successes of its multi-pest species removals and restoration efforts (Australian Wildlife Conservancy 2013), which are due to most of the staff being based in the field doing practical work. The Conservancy's work is underpinned by good conservation science and its findings are communicated and published widely, including in peer-reviewed journals. This not-for-profit entity has a robust governance structure, employs a sizeable team of ecologists and enlists volunteers to help deliver its work.

Avoid Island (83 ha), located ~125 km south-east of Mackay in Queensland, was purchased by the Queensland Trust for Nature in 2006 mainly to protect the breeding beaches of the threatened flatback turtle (*Natator depressus*). Staff and volunteers monitor the turtles during the breeding season (with around 7000 eggs laid annually) and remove weeds and debris from the beaches to maintain this habitat (Queensland Trust for Nature 2014). However, its work on the island has expanded and, through a partnership with the Great Barrier Reef Marine Park Authority, Mackay students are able to visit the island to learn about reef life and marine ecosystems through the Reef Guardian schools program. Avoid Island is an exemplar of how not-for-profit organisations, corporate partners and members of the public can come together to deliver lasting scientific and conservation outcomes. Avoid Island was recognised as one of Australia's most environmentally sustainable ecosystems at the 2015 Banksia Foundation Sustainability Awards.

Island philanthropy

A growing number of philanthropic trusts and business enterprises contribute to island conservation across Australia. Here we showcase a range of programs currently operating in Tasmania; all have a different objective and all rely on partnerships for their delivery. Since 1989 the Princess Melikoff Trust has financially supported the Tasmanian government's marine mammal conservation program, enabling staff to undertake a range of work in purpose-built vessels such as tracking cetaceans, rescuing marine mammals and monitoring seals on breeding and haul-out islands around the coast. Philanthropy from the Tasmanian-owned business Maria Island Walk supports habitat restoration on spectacular Maria Island National Park and contributes to ongoing maintenance of its heritage buildings. Philanthropy from another Tasmanian-owned business, Pennicott Wilderness Journeys (Hudson 2016), has been instrumental in supporting several significant island conservation initiatives including the successful Tasman Island cat eradication program, the eradication of rats on Big Green Island and the Bruny Island cat management program. Pennicott Wilderness Journeys has publicly committed to reinvest in the local environment and has pledged assistance towards restoration work on the

Fig. 12.2: The Hamish Saunders Memorial Trust field team after completing the first-ever biological survey of remote Inner Sister Island, Bass Strait in 2010. Photo: Mike Driessen.

Breaksea Islands in south-west Tasmania and other Bass Strait islands in the future.

The Hamish Saunders Memorial Trust Island Survey Program was established in 2003 as a partnership between a New Zealand family trust and the Tasmanian government, as the legacy of a volunteer who tragically died on Pedra Branca Island. Its mission is to nurture future leaders in marine and coastal conservation, and each year a multi-disciplinary survey is undertaken on a data-deficient Tasmanian island (Fig. 12.2). By 2017 the program had completed 11 comprehensive island expeditions including, for example, the Macquarie Harbour islets, Prime Seal, Inner Sister, Flinders, Schouten, Three Hummock, Hunter and Sloping Islands. Its findings have underpinned several initiatives such as the eradication of feral cats from Tasman Island, monitoring post-fire recovery on lungtalanana (Clarke Island), confirming the status of forty-spotted pardalote (*Pardalotus quadragintus*) on Flinders Island and providing baseline data for the Bruny Island cat management program (Natural and Cultural Heritage Division 2015). Not only has this international partnership expanded the biological inventory of Tasmania's islands, it has also provided emerging New Zealand ecologists with an opportunity to participate in multi-disciplinary surveys and better understand the fragility of island ecosystems in wild remote places.

Government frameworks for community delivery

There are many solid community/stakeholder/government partnerships supporting nature conservation on Australia's islands. Below we outline some of the most significant models.

Indigenous owners

Aboriginal and Torres Strait Islander peoples around the Australian coastline hold deep and abiding connections to islands associated with their peoples' sea countries, and these cultures and traditions are integral to the maintenance of Australia's islands (Chapter 8). Today this is recognised and supported nationally through several Australian government programs that assist Aboriginal and Torres Strait Islander people to manage their country, including:

- the Indigenous Protected Areas program;
- Indigenous Ranger programs, including Working on Country.

At the state/territory level, a range of statutory mechanisms and programs also enable Aboriginal and Torres Strait Islander peoples to assert and fulfil their continuing management interests and governance roles in island conservation:

- native title determinations;
- Indigenous Land Use Agreements;
- land acquisition programs, e.g. Indigenous Land Corporations;
- land transfer programs;
- collaborative/joint or shared management of protected areas;
- memoranda of understanding;
- Land and Sea Ranger funding programs.

The management activities undertaken by Aboriginal and Torres Strait Islander Ranger groups and communities on and around islands are critical in maintaining and restoring their diverse marine and terrestrial resources, and may involve:

- protection of cultural places, sites and stories;
- maintenance and passing on of cultural practices;
- intergenerational Indigenous knowledge transfer and community education;
- provision of, and specialist cultural interpretation for, tourism enterprises;
- surveillance/patrols of coastal and marine areas;
- conducting quarantine checks, regulatory compliance and enforcement activities;
- controlling pest plant and animal populations;
- applied fire management (at times referred to as 'right way' fire);
- monitoring sustainable harvest of traditional resources (e.g. food plants, some fisheries, other marine species, muttonbirds etc.);
- monitoring significant species and natural values including marine, terrestrial, migratory, rare or threatened.

High-visitation islands

It is often thought that on popular tourist islands where nature is the basis of the attraction, some of the profit generated from this industry should be directed back into maintaining the island values. However, the level of tourism support staying on-island can vary greatly, and in some cases there may be no direct benefit to conservation at all. This can leave a void that the government or community is expected to fill. Even on high-visitation islands where tourism is closely regulated through business enterprises, the community still steps up to strengthen the conservation initiatives. Here we highlight several examples of community-driven contributions on popular tourist islands.

Phillip Island, Victoria

10 000 ha Phillip Island, 90 min from Melbourne, is a major tourist drawcard that contributed A$1.3 billion in total tourism output in 2013–14 and supported the direct and indirect employment of 6400 people (Tourism Victoria 2015). Much of the island's tourism is nature-based and centred around the 1805 ha Phillip Island Nature Parks where a major attraction is the internationally renowned Penguin

Parade. The Nature Parks is a not-for-profit organisation overseen by a board of management appointed by the Victorian Environment Minister. With high visitor numbers to its ticketed visitor centres (over 1 million visitors in 2015–16), Nature Parks can fund its conservation work from entry fees without recurrent operational funding from the state government. These visitor numbers are forecast to grow at ~2.2% p.a. or by a total of ~55% by 2035 (Bass Coast Shire Council 2015).

Due to its attractions and high visitor numbers, the Nature Parks can value-add to its investments by attracting a range of sponsors – including a local Penguin Foundation that has raised over A$1.8 million since 2005 – and by supporting several partnerships including many volunteer groups who typify the breadth and diversity of services the community can provide. For example, a local business grows the indigenous plants used in island restoration programs, community members undertake hooded plover watching and koala protection, there are coastal action groups, a Landcare group, penguin study groups, historical and conservation societies, activities generated by the local school and several groups undertaking habitat restoration. The Friends of Churchill Island Society, for example, raise funds to protect nearby 57 ha Churchill Island and assist with guiding, animal and pest control, maintenance, researching and archiving.

Rottnest Island, Western Australia

Rottnest Island, 19 km off the coast of Perth, is an internationally recognised A-class tourism island with unique natural heritage and a threatened flagship species, the quokka (*Setonix brachyurus*). The island's reserve comprises 1859 ha of land and 3800 ha of marine reserve, all of which is managed by the statutory Rottnest Island Authority. The Authority's main income streams are on-island accommodation and admission fees, sizeable grants to its foundation and other federal project moneys. In 2015–16 the Authority reported a profit of A$2.42 million (RIA 2016). Approximately 610 000 people visited the island during 2015–16 and the figure is projected to rise to 800 000 p.a. by 2034 (RIA 2016).

While the Authority's responsibilities are impressively wide-ranging, including managing all the island's environmental programs, it also operates in-house volunteers who deliver a major component of the Rottnest Island Terrestrial Conservation Action Plan, for example providing 1800 volunteer hours per year in the plant nursery propagating stock for restoration. The Authority also facilitates approximately 20 external volunteer groups who contribute in total around 35 000 volunteer hours annually (equivalent to 17 full-time equivalent jobs per year) in helping protect island values and providing unfunded services that otherwise may not be delivered.

Kangaroo Island, South Australia

Kangaroo Island, 22 km off the coast of South Australia and three hours travel from Adelaide, is Australia's third-largest island, covering almost 4500 km^2, with a population of ~4500 people. Known for its abundant native wildlife, including its iconic Australian sea-lion *(Neophoca cinerea)* colony, nearly half of the island retains its native vegetation cover and about a third is included in the state's protected areas network. As such, it is a major tourism destination; 126 000 people visited the island in 2015 and spent A$104 million, with government projections to grow this further (South Australian Tourism Commission 2016).

The Kangaroo Island economy is based largely on farming and tourism centred on its brand values – its native vegetation, abundant wildlife and 'clean green' image (Kangaroo Island Natural Resources Management Board 2016). The community recognises the advantages of this image, and has committed to make sure that economic activity on the island is not only sustainable but demonstrates best practice and showcases islander determination and innovation. One local non-government group, Eco-Action, works solely to protect the environment for its intrinsic value and for lifestyle reasons (http://www.eco-action.com.au/index.htm).

Unlike the Phillip Island and Rottnest Island examples, except for a few small, locally owned enterprises, there is little direct, line-of-sight reinvestment of tourism incomes into environmental projects on Kangaroo Island, by either business or government. Rather, management and restoration of the island's natural resources is largely dependent upon:

- state government funding for parks and wildlife (~A$4.65 million) and for natural resources management (NRM) (~A$2.7 million);
- a regional NRM land levy on local ratepayers (currently A$176 400 p.a., increasing to A$385 000 in 2017–18) collected by local government on behalf of the NRM Board;
- Australian government National Landcare Program funding (currently A$1.37 million p.a.);
- community volunteer projects.

The governance of NRM on the island is through a local NRM Board with strong community representation and the state environment department. The island's NRM work is guided by a regional NRM plan: the latest version includes 12 local landscape action plans, all of which capture the communities' vision for the environment, community and economy of the island. The implementation of these plans requires a solid partnership between all stakeholders if they are to succeed.

Community-driven promotion of the natural assets of Kangaroo Island

Kangaroo Island has a significantly large art community that has developed a strong focus on promoting the diversity of the island's unique natural assets. From an awareness-raising and conservation advocacy perspective, a most impressive initiative by the Kangaroo Island artists is their contribution to the annual South Australia's Living Artists (SALA) Festival. Every year since 2010, Kangaroo Island artists have assembled a nature conservation-focused exhibition at the National Wine Centre in Adelaide, titled 'How do we love thee?' Each year the exhibition features a different ecological theme; in 2011, for example, it was the endangered South Australian glossy black-cockatoo (*Calyptorhynchus lathami halmaturinus*) and its dependence on the seeds of the drooping she-oak (*Allocasuarina verticillata*) for food, and in 2016 it showed the green carpenter bee (*Xylocopa aerates*) and its relationship with the flower stems of the Kangaroo Island grass tree (*Xanthorrhoea semiplana tateana*).

The wonder and natural beauty of Kangaroo Island are also actively promoted through the Kangaroo Island Nature's Paradise Facebook site, where locals and visitors share photographs and other information about the island's amazing environment.

Active community participation in habitat restoration

In addition to three 'friends of' parks groups and a Friends of Glossy Black Cockatoo group, there has been a long-term recovery program for Kangaroo Island's nationally threatened plants. While this program has received government support, its success has largely been due to the generous landholders who have supplied their land for restoration and to the thousands of hours of volunteer labour (Box 12.1).

Concluding comments

The case studies presented here describe but a few of the community-driven contributions towards island conservation in Australia. They contain key messages about the value of strong partnerships, their scientific capacity, good project management, longevity and how meagre community resources can translate into significant conservation benefits. No longer do community groups perform a single task – their diversification and capacity has expanded to fill the void left by unfunded programs, changing priorities and declining government budgets. Through persistence and self-reliance, community groups are more robust and better-equipped to tackle the planning and bureaucratic obstacles that so often emerge in challenging island environments, and their professional skills base means they can contribute at every level from project planning to on-ground works, often solving complex problems with innovation and creative solutions. The undeniable contribution of citizen science means robust and often massive datasets

Box 12.1: Case study: recovery project for Kangaroo Island's nationally threatened plants

Many of Kangaroo Island's flora and fauna species face ongoing and increasing pressures from habitat fragmentation, weed infestations, altered fire regimes and overbrowsing by herbivores. Although the island retains around 47% of its remnant vegetation, the eastern third of the island has less than 15% remaining which, consequently, is where most of the island's threatened plant species occur. The Kangaroo Island threatened plant project has covered 15 nationally threatened plant species, with the aim of recovering their populations by reinstating large patches of their habitat through diverse and intensive plantings next to important areas of remnant bushland.

Project work was based on comprehensive research into the ecology of local plant communities. To mimic the diversity and composition found in intact remnants, a minimum of 120 species were propagated and planted out at densities of 2000–3000 seedlings per hectare in each of the conservation plantings (Fig. 12.3).

Five species at risk of extinction were given attention: Kangaroo Island phebalium (*Leionema equestre*), MacGillivray spyridium (*Spyridium eriocephalum* var. *glabrisepalum*), small flower daisy-bush (*Olearia microdisca*), Kangaroo Island turpentine bush (*Beyeria subtecta*) and Kangaroo Island pomaderris (*Pomaderris halmaturina* subsp. *halmaturina*).

An array of planting methods was trialled, including direct seeding, direct planting and topsoil spreading with some of the local soil-seed bank. In addition, large-scale plantings were undertaken on sites where the former pastures and their fertile topsoils had been scalped and pushed aside, to reduce the densities of weedy species germinating and competing with the native species and to optimise the chances of native seedlings surviving.

Most planting sites have been on private land generously made available by the owners. On-ground planting activities in 2004–13 focused on a former farming property, Cygnet Park, in the Cygnet River Valley. The project engaged volunteers each year and from 2007 to 2013 a planting festival was held at Cygnet Park and other nearby properties, resulting in 400 000 seedlings from 145 plant species being planted over more than 200 ha. The 2011 festival saw 702 volunteers plant 120 000 seedlings across 53 ha at a rate of around 7000 seedlings per hour.

Fig. 12.3: At the 2011 Kangaroo Island Planting Festival over 700 volunteers planted 120 000 seedlings across 53 ha at a rate of around 7000 seedlings per hour. Photo: C. Wilson.

are available to decision-makers, that otherwise would be lacking. While this contribution, including its millions of dollars of volunteer labour, is becoming better recognised, it needs to be expanded and its power harnessed if we are to achieve more rapid and lasting conservation success.

Acknowledgements

Sincere thanks to Martine Kinloch, Heinrich Klein, Patrick Hodgens, Quentin and Dale Chester and Mark Bath on Kangaroo Island, and to Marg and Bob Graham and Dan Sprod from Bruny Island. We thank our referees Andrew Harwood and Michael Lockwood from the University of Tasmania and Gill Peacey from the South Australian government for their helpful comments, and Ellie Bock for guidance on Indigenous owners.

References

ABS (1999) *Australian Social Trends.* Australian Bureau of Statistics. <http://www.abs.gov.au/ausstats/abs@.nsf/2f762f95845417aeca25706c00834efa/65f1ae55ef772f18ca2570ec001117a4!OpenDocument>

Australian Wildlife Conservancy (2013) *Wildlife Matters: Summer 2013/2014.* Australian Wildlife Conservancy, Perth.

Baldacchino G (Ed.) (2007) *A World of Islands: An Island Studies Reader.* Island Studies Press, University of Prince Edward Island, Canada.

Bass Coast Shire Council (2015) *Phillip Island Tourism 2035: Developing a 20 Year Phillip Island Tourism Strategy.* Discussion paper. <http://basscoast.vic.gov.au>

BIEPA (2017) *BIEPA News.* February/March/April **24**, 1–16.

BirdLife Australia (2017) *Australian Birdlife* 6(1). BirdLife Australia, Melbourne.

Bower S (2016) Ten-year progress report on implementation of the Lord Howe Island Weed Eradication Program. Abstract. *Island Arks Symposium IV.* Norfolk Island. <http://islandarks.com.au/files/2016/02/Island-Arks-IV-Final-Program.pdf>

Bruny Island Environment Network (2010) *BIEN Policy on Ecotourism and Nature-based Tourism on Bruny Island.* < http://www.bien.org.au>

Campbell-Hunt D (2008) Community-driven ecological restoration in New Zealand. *Pacific News.* 30 July/12–15 August.

Destination South Tasmania (2016) *Bruny Island Destination Action Plan 2016–2018.* Destination South Tasmania and Department of State Growth, Hobart.

Friends of Lord Howe Island (2015a) Newsletter 40, Summer 2015.

Friends of Lord Howe Island (2015b) Newsletter 41, Winter 2015.

Geoscience Australia (2004) *Geodata Topo 100K.* Google Earth, 1:100,000 scale vector map data (DVD). Geoscience Australia, Canberra.

Greenwood JD (2007) Citizens, science and bird conservation. *Journal für Ornithologie* **148**, S77–S124. doi:10.1007/s10336-007-0239-9

Griffith University (2012) *Enhancing the International Student Experience: WildMob Wilderness Conservation Projects.* Final Report. Griffith University, Queensland.

Hudson S (2016) Tasmania's Pennicott Foundation is helping the world one project at a time. *Weekly Times*, 26 October, 1–4.

Kangaroo Island Natural Resources Management Board (2016) *Kangaroo Island Draft Natural Resources Management Plan 2017–2027.* <http://www.naturalresources.sa.gov.au/kangarooisland/get-involved/events/Plan-consultation-Kangaroo-Island-Natural-Resources-Management-Plan>

Lord Howe Island Board (2015) *Annual Report 2014–2015.* Lord Howe Island Board, Lord Howe Island.

Maynard S, James D, Davidson A (2010) *The Development of an Ecosystem Services Framework for South East Queensland.* Environmental Management. <http://link.springer.com/article/10.1007%2Fs00267-010-9428-z>

Moyle B, McLennan C, Becken S, Battye R, Godfrey A (2014) *Nature-based Tourism in Australia 2014.* Fact sheet. Griffith Institute for Tourism in partnership with Tourism Research Australia.

Natural and Cultural Heritage Division (2015). *South Bruny and its Offshore Islands, Natural Values Survey 2013.* Nature Conservation Report Series 15/1. Hamish Saunders Memorial Trust, New Zealand and Natural and Cultural Heritage Division, DPIPWE, Hobart.

Queensland Trust for Nature (2014) *Annual Review 2014.* Queensland Trust for Nature, Brisbane.

RIA (2016) *Annual Report 2015–2016.* Rottnest Island Authority, Fremantle. <http://ria.wa.gov.au/policy-and-reports/annual-reports>

Ritchie J (2011) Bagging them all in one go: personal reflections of a project manager about community-based multi-species animal pest eradication programmes in New Zealand. In *Island Invasives: Eradication and Management* (Eds CR Veitch, MN Clout and DR Towns) pp. 487–491. IUCN, Gland, Switzerland.

Secretariat of the Convention on Biological Diversity (2014) *Island Biodiversity: Island Bright Spots in Conservation and Sustainability.* World Trade Centre, Canada.

Shaw V, Whitehead JD, Shaw CT (2011) Creating an island sanctuary: a case study of a community-led conservation initiative. In *Island Invasives: Eradication and Management* (Eds CR Veitch, MN Clout and DR Towns) pp. 496–499. IUCN, Gland, Switzerland.

South Australian Tourism Commission (2016) *Regional Tourism Reports 2015.* <http://tourism.sa.gov.au/research-and-reports/regional-touri sm-profiles.aspx>

Tourism Victoria (2015) *Value of Tourism to Phillip Island 2013–14.* Research fact sheet. Tourism Victoria. <https://www.ecotourism.org.au/ass ets/...Hub-Ecotourism-Research/Philip-Island.pd>

Wild Times (2016) Weed control in the Furneaux Islands. *Wild Times,* Sept. **56**, 4.

Wildcare (2016) *Annual Report 2016.* Wildcare, Hobart.

WildMob (2015) *Annual Report.* <https://wildmob.org/wp content/uploads/2016/01/WildMob_AR 2015–2.pdf>

13

Value of islands for the marine environment

Mark Read, Graham Hemson and John Olds

Summary

By international standards, many Australian islands are in a relatively undisturbed condition. They support unique assemblages of flora and fauna including internationally significant populations of seabirds, shorebirds, marine turtles and pinnipeds. Most of these species are long-lived with complex life-histories that are inextricably linked to islands for part or all their life-history stages. Many of these species are migratory, spending only part of their lives on and around islands before returning to forage or breed elsewhere. The interconnectedness between islands and the species they support flows both ways, with many island habitats having changed over time because of the species that utilise them and the nutrients and seeds they transport to the islands. The islands and the species they support are managed through a complex framework of statutory and non-statutory tools, ranging from international agreements down to island- and species-level management plans. Identifying and minimising the impact of key threats on islands and to the species they support is critically important. To achieve this, managers are increasingly relying on decision-support and prioritisation tools to assist in allocating resources in an ever-competitive environment.

Introduction: the importance of islands to marine life

Australian islands support a number of nomadic and migratory species that visit islands to complete a critical element of their life-history. Examples include tiger sharks (*Galeocerdo cuvier*) migrating to Raine Island (Fig. 13.1) to feed on adult and hatchling green turtles (*Chelonia mydas*) (Fitzpatrick *et al.* 2012), migratory birds using the islands of north-west Western Australia as critical resting stops as part of a longer migration (DSEWPaC 2012a) and overwintering habitat for non-breeding migratory waders (Bamford and Moro 2011), to seabirds (Turner and Batianoff 2007), marine turtles (Limpus 2009; Waayers *et al.* 2015) and pinnipeds (Shirihai 2002) relying on islands as vital breeding sites. Islands support numerous discrete ecosystems which contribute to the maintenance of ecosystem functions as they provide defence against natural disasters and support nutrient cycling, and soil and sand formation (Mather and Bennett 1993). This chapter focuses on the value of Australia's islands to the marine environment. It examines why islands are so important to selected key species (birds, marine turtles and pinnipeds), what benefits islands provide for these species and what benefits these species provide for the islands. It

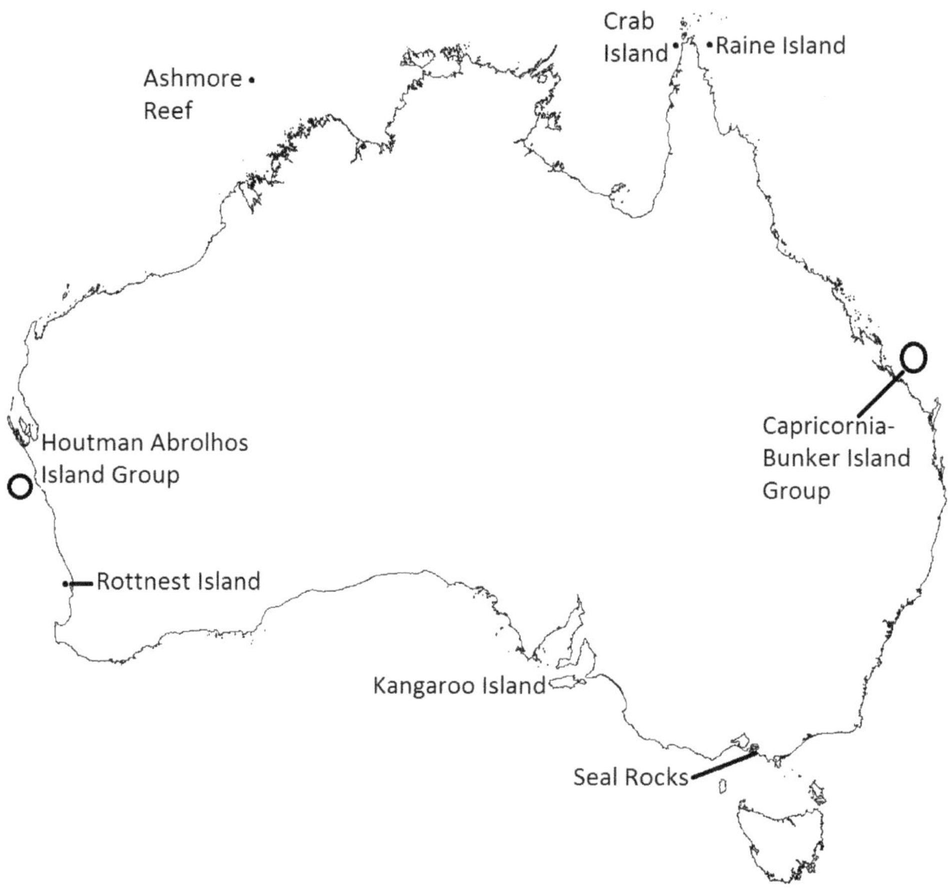

Fig. 13.1: Key locations described in this chapter.

also documents the interconnectedness of islands and identifies the critical elements to managing these unique ecosystems to support their natural values for current and future generations.

Birds

Seabirds, shorebirds, waterbirds and landbirds are some of the most conspicuous animal groups on Australian islands. Surveys for BirdLife Australia's Atlas on 219 (2.6%) of the estimated 8371 Australian islands have recorded 552 bird species, or 68% of the bird species recorded by the Atlas in Australian territories (O'Connor 2010). Some species are resident on islands all year round, some use islands temporarily for roosting or as a rest-stop as part of a migratory pathway (Battley and Rogers 2007), and others visit only to breed before spending the rest of the year at sea. Islands support high species diversity, with ~215 bird species from the islands and cays of the Great Barrier Reef (Stokes *et al.* 2004) and at least 72 species recorded from Macquarie Island (Shirihai 2002). Breeding populations of seabirds highlight the importance of islands to birds. Every year around 3.5 million seabirds, the majority of them penguins, arrive on Macquarie Island to breed and moult (DPIPWE 2015), and over 2 million birds from 35 different species breed on the 192 islands and cays of the Houtman Abrolhos Islands off Western Australia (Fig. 13.1) (DoF 2015). Most staggeringly, ~23 million short-tailed shearwaters (*Ardenna tenuirostris*) breed in ~285 colonies in south-eastern Australia, with an estimated 18 million nesting in burrows in at least 167 colonies in Tasmania (DPIPWE 2014a; Schumann *et al.* 2014). Historical accounts suggest that this population was once much larger (DPIPWE 2014a). Many resident bird species on Australian islands have been isolated for so long they are now recognised as distinct endemic species or subspecies. Tasmania has 12 bird species recognised as endemic (DPIPWE 2014b) and Lord Howe Island two (BirdLife International 2016). The sub-Antarctic islands of Macquarie, Heard and McDonald are among the most remote places on Earth, and not surprisingly they support endemic species of birds. The royal penguin (*Eudyptes schlegeli*) is endemic to Macquarie Island, with a population estimated at around 850 000 (DPIPWE 2015). Heard and McDonald Islands, which are located ~4000 km south-west of mainland Australia, support two endemic species, the Heard shag (*Leucocarbo atriceps nivalis*) and the black-faced sheathbill (*Chionis minor*) (DoE 2005).

Marine turtles

The terrestrial, intertidal and subtidal habitats associated with Australian islands provide critical habitat for marine turtles to complete several key phases of their life-history. Six of the world's seven species of marine turtle are found within Australian waters (green; loggerhead *Caretta caretta*; flatback *Natator depressus*; hawksbill *Eretmochelys imbricata*; olive ridley *Lepidochelys olivacea*; leatherback *Dermochelys coriacea*), and all will nest on islands (Limpus 2009). Breeding turtles exhibit natal homing, returning to the region where they hatched to breed. This trait has resulted in the evolution of discrete genetic stocks within each species that are defined by the presence of regional breeding populations that show low rates of interbreeding (Dutton *et al.* 2002). For some stocks, almost all the nesting effort is focused on a small number of islands (Box 13.1). For the eastern Australian stock of the flatback turtle, nearly all nesting occurs on only four continental islands in the southern Great Barrier Reef (Limpus 2009). For the northern Great Barrier Reef green turtle stock (the world's largest green turtle population), nearly 90% of the nesting occurs on Raine Island and nearby Moulter Cay (Fig. 13.1; Limpus 2009). Raine Island can host over 60 000 nesting females during the peak nesting season (Limpus *et al.* 2003) and over 27 000 green turtles have been recorded ashore during a single census of the nesting beach (Dunstan *et al.* 2015). Western Australia also supports significant nesting populations of green, flatback and hawksbill turtles in the ~3700 islands within its jurisdiction (Burbidge 2010; Waayers *et al.* 2015), and probably the third-largest loggerhead turtle population in the world (Limpus 2009). Several Western Australian rookeries occur on islands with significant oil and gas industry infrastructure, which presents some unique management challenges such as controlling light output which can disorient nesting females and emerging hatchlings (Pendoley and Kamrowski 2015). The intertidal and subtidal habitats surrounding islands also provide important habitat for foraging (Read and Limpus 2002; Bell 2013) and basking (Limpus 2009) marine turtles.

Pinnipeds

Pinnipeds are highly dependent on Australian islands. Ten species are recorded in Australian waters (DoE 2016a) and all use islands to haul-out (rest), moult, give birth, suckle and mate. Southern elephant seals (*Mirounga leonina*), the largest seal in the world (adult males reach 6.7 m and 3700 kg; Shirihai 2002), usually prefer low-energy sand or shingle beaches for hauling out and breeding (DoE 2016b), as their size precludes them from tackling more challenging terrain. Pinniped reproduction takes place at traditional breeding sites, with females coming ashore to give birth, mate and then continue to suckle their pup until weaning. Pupping is often synchronised, with 80% of southern elephant pups born in a space of three weeks (Shaughnessy 1999). For Australian fur-seals (*Arctocephalus pusillus*) at Seal Rocks, Victoria (Fig. 13.1), 90% of pups were born over a 26-day period (Warneke and Shaughnessy 1985). Pups are weaned at between three weeks for southern elephant seals (Slip *et al.* 1994) and six to seven months for Australian fur-seals (Warneke and Shaughnessy 1985). Pinnipeds also depend on islands when moulting. Most animals shed hairs year-round, but southern elephant seals shed in one event called 'catastrophic moulting'. Southern elephant seals cannot enter the water while moulting because they do not have proper insulation to keep them warm (Carrick *et al.*

Box 13.1: Crab Island

Crab Island (Fig. 13.1), a tiny sand island (280 ha) on the north-west tip of Cape York Peninsula, supports the world's largest aggregation of nesting flatback turtles (Fig. 13.2; Limpus 2009). Every winter up to 600 females per night come ashore to nest, laying around 55 eggs before returning to the water.

This concentration of potential prey means the adults and emergent hatchlings have to evade several predatory species to survive the journey across the beach and enter the water to disperse. The estuarine crocodile (*Crocodylus porosus*) targets both the nesting adults and hatchlings, with larger crocodiles attacking nesting turtles on the beach or at the water's edge. Sutherland and Sutherland (2003) recorded a predation rate of 1.17 nesting females per week during July 1997. Crocodiles also target the hatchlings when they emerge from the nest. Emergence usually happens at night, and crocodiles target the hatchlings either by sight or by using the integumentary sensory organs which line their jaws and legs (Leitch and Catania 2012). These sensory organs allow the crocodiles to detect the vibrations of the hatchlings as they crawl across the beach. Hatchlings that escape the crocodiles and other land-based predators such as ghost crabs (*Ocypode* sp.) and night herons (*Nycticorax caledonicus*) are then vulnerable to marine predators like silky sharks (*Carcharhinus falciformis*) and giant trevally (*Caranx ignobilis*). While predation on the beach and in the water contributes to the mortality of both adults and hatchlings, another threat to the rookery is from feral pigs (*Sus scrofa*) accessing the island and digging up the nests. Less than 1 km of water separates the island from the mainland at low tide, and on the adjacent mainland beaches feral pigs have been recorded impacting upon 40–90% of all flatback and olive ridley turtle nests (Limpus 2009; Whytlaw *et al.* 2013). Research has indicated that nest depredation is usually the result of individual mature boars displaying learned behaviour, rather than a general foraging behaviour exhibited by individuals (Whytlaw *et al.* 2013). To reduce the risks to the Crab Island rookery, effective control of the broader feral pig population and the boars exhibiting depredation behaviour has been required. Control of feral pigs can be best achieved from an integrated approach combining aerial and ground shooting and aerial baiting. Suppressing the feral pig population on the mainland adjacent to Crab Island is likely to reduce the chances for individuals to engage in exploratory behaviour and attempt to swim to Crab Island.

Fig. 13.2: Flatback turtle (*Natator depressus*) nesting on Crab Island, north-west Cape York Peninsula. Photo: Ian Bell.

1962). They spend about four weeks onshore before returning to the sea until the next breeding season (Slip *et al.* 1994).

Island interconnectedness

Interconnectedness describes the relationship that exists between islands, the marine environment and other locations, the flora and fauna that either inhabit the island or have some connection with the island, and the biophysical and ecological processes that contribute to the flow of energy in some way. Interconnectedness occurs at a variety of spatial scales, from small-scale such as pied imperial pigeons (*Ducula bicolor*) flying <10 km between inshore islands and the mainland to feed (Winter *et al.* 2016), to the international scale such as shorebirds that nest in Eurasia and migrate each year to feed at Ashmore Reef, Western Australia (Fig. 13.1; DSEWPaC 2012a). This interconnectedness operates at a variety of temporal scales, with some species undertaking daily movements. For example, daily flights between foraging and nesting grounds by wedge-tailed shearwaters (*Ardenna pacifica*) feeding chicks (Congdon *et al.* 2014) through to the five to 10-year remigration intervals for some female loggerhead and green turtles nesting on islands in the southern Great Barrier Reef (Limpus 2009). One of the critical elements of interconnectedness is the transfer of nutrients from marine ecosystems to islands. Seabirds, marine turtles and pinnipeds play a key role in bringing nutrients back to nutrient-poor island ecosystems (Bouchard and Bjorndal 2000; Schmidt *et al.* 2004; Wolfe *et al.* 2004). These nutrients come from faeces, regurgitated food, dead eggs, and young and adults killed by disease or misadventure. For example, black noddies (*Anous minutus*) deposit an estimated 45 t of guano on Heron Island per year, which represents the transfer of ~1.4 t of phosphorus per year from the surrounding waters to the island's soil (Allaway and Ashford 1984). Harrison (2006), studying the influence of seabird-derived nutrients on island ecosystems off south-western Australia, found that total nitrogen concentrations of soils and plants on most of the nine islands examined were significantly higher than those at three mainland sites. Analyses confirmed that the increase in nutrients was the result of guano inputs from seabirds. The plants on many islands, such as pisonia (*Pisonia grandis*), thrive on these nutrients (Walker 1991a; Harrison 2006). In addition, burrowing seabirds modify the soils of their nesting colonies by the transfer of nutrients and through the processes of bioturbation and bioerosion. On Rottnest Island (Fig. 13.1), wedge-tailed shearwater colonies support a low-diversity vegetation community dominated by short-lived succulents, with double the bare-soil exposure compared with a nearby heath community (Bancroft *et al.* 2005). Pinnipeds are also significant contributors to the transfer of ocean-derived nutrients to islands. A study of Australian sea-lions (*Neophoca cinerea*) on Kangaroo Island (Fig. 13.1) found that the resident population of ~1100 individuals contributed ~3800 kg nitrogen per year into the Seal Bay Conservation Park. If the sea-lion population were to decline, the nitrogen availability and coastal productivity of Kangaroo Island are likely to be reduced (Lavery *et al.* 2015).

Management of islands to support marine species

The management of islands to support marine species is achieved through an overlapping (and often complex) mosaic of statutory and non-statutory instruments. These may start at the international level through multi-lateral agreements like the Convention on the Conservation of Migratory Species of Wild Animals, to the national level through threat abatement plans (e.g. *Threat Abatement Plan for Predation by Feral Cats*, Commonwealth of Australia 2015), to the regional level (e.g. *Marine Bioregional Plan: North-west Marine Region*, DSEWPaC 2012b) down to the island-group level (e.g. *Barrow Group Nature Reserves Management Plan*, DPW 2015). Management plans usually provide a statement of policies, guidelines and a summary of on-ground actions that include the management of fire and pests and Indigenous and non-Indigenous heritage. Research and monitoring is an essential component of this management framework, as the information collected enables managers to make evidence-based decisions and determine the effectiveness of any

Box 13.2: The Capricornia Cays

The Capricornia and Bunker group and Lady Elliot Island consist of 22 reefs and 16 coral cays positioned at the far south of the Great Barrier Reef. The archipelago supports more than 50 species of birds, important nesting habitat for green and loggerhead turtles and more than 80% of Australia's pisonia, *Pisonia grandis* (Planning Services QPWS 2014; Walker 1991b). The Capricorn white-eye (*Zosterops lateralis chlorocephalus*) is found only on these 16 cays; however, the area is best known for the annual breeding of hundreds of thousands of wedge-tailed shearwaters and black noddies (Fig. 13.3), estimated to be half Australia's population of wedge-tailed shearwaters (Dyer *et al.* 2005).

The combination of size, stability, habitat, proximity to productive oceanic features, and remoteness from nest predators makes these cays excellent breeding sites. The Eastern Australian Current passes the edge of the Great Barrier Reef until it rounds the end of the Swains Reefs and cuts back towards the coast, creating a large sheltered backwater that favours the formation of a static mesoscale eddy (Kleypas and Burrage 1994). This Capricorn Eddy draws up nutrient-rich water from the deep, boosting planktonic productivity near the surface and driving a surge in trophic activity (Weeks *et al.* 2010). Small fish and squid build in numbers, drawing in pelagic predators such as tuna and mackerel that aggregate 'bait balls' of prey against the surface and within the range of seabirds (Goyert *et al.* 2014).

Food is, however, only half of the equation. The Capricornia Cays are the largest and geomorphologically most stable cays in the Great Barrier Reef (Heatwole and Walker 1989). North West Island, where most shearwaters and noddies nest, is the largest cay in the Great Barrier Reef and supports more than 100 ha of forested habitat. The soils of these cays illustrate this stability, with significant deposits of guano and organic matter adding much-needed fertility to the once barren coral sands. The fertility helps support abundant plant growth, which in turn provides nesting habitat for black noddies and helps to bind the sands, creating stable burrowing habitat for shearwaters.

As well as cycling nutrients from the ocean to the islands, seabirds are a key vector for plant dispersal

Fig. 13.3: Black noddies (*Anous minutus*) nesting in an octopus bush (*Argusia argentea*) on Heron Island. Photo: Andrew McDougall.

(Young *et al.* 2011). Pisonia seeds bear rows of sticky prickles; they readily adhere to incautious birds and are transported between islands (Burger 2005). Unfortunately, patterns of human visitation largely dictate the diversity of non-native plants on the islands (Heatwole and Walker 1989). While only a few islands, such as Erskine, have avoided human-mediated seed dispersal associated with the activities of guano miners, people harvesting green turtles to make soup, and tourists, most have benefited from management by the Queensland Parks and Wildlife Service (Burwell *et al.* 2010; Planning Services QPWS 2014).

To date goats (*Carpa hircus*), black rats (*Rattus rattus*), feral cats (*Felis catus*) and chickens (*Gallus gallus domesticus*) have all been eradicated, leaving only populations of domestic mice (*Mus domesticus*), sparrows (*Passer domesticus*), invasive ants (African big-headed ant *Pheidole megacephala*, pennant ant *Tetramorium bicarinatum*, black crazy ant *Paratrechina longicornis*) and several weeds on the islands. With a growing emphasis on biosecurity, including surveillance, further establishment of invasive species should be prevented, preserving the cays for future generations of turtles, trees, seabirds and people alike (Olds 2015).

management actions. For example, a recent study by Pendoley *et al.* (2014) delineated a migratory corridor along the north-west coast of Australia by tracking flatback turtles as they returned to their foraging grounds after nesting. They calculated a core corridor of 1150 km length and 30 800 km^2, and found that migratory tracks for humpback whales (*Megaptera novaeangliae*) overlapped with 96% of the core corridor, while the tracks of three other species overlapped by 5–10% (blue whales *Balaenoptera musculus*; olive ridley turtles; whale sharks *Rhincodon typus*). This study provides valuable information for proposing new locations requiring protection, as well as identifying high-priority network linkages between existing marine protected areas.

The management of pests (Box 13.2, Box 13.3) and fire are also critical for islands. It has been identified that the most pervasive threat to seabirds worldwide is invasive species (BirdLife International 2012): managers rely on early detection and targeted programs to eradicate pests before they get established, as well as focusing on the eradication of already established pest populations (Gregory *et al.* 2014). Many islands also support fire-sensitive vegetation that can be adversely affected by inappropriate fire regimes (Lyons *et al.* 2014) and managers use planned burns to maintain the resilience and complexity of island ecosystems (Specht and Specht 1999; Williams *et al.* 2001). With so many islands around Australia, it is increasingly important that managers prioritise management actions to maximise conservation outcomes. The process of spatial conservation prioritisation provides guidance on how available conservation effort and funds should be allocated (Moilanen *et al.* 2009; Wiens and Hobbs 2015; McCreless *et al.* 2016), and this type of decision-support framework is now commonly used by resource managers. Adaptive management and intervention is also part of business-as-usual, and conservation agencies are now more willing to deliberately intervene to address threats on islands. The Raine Island Recovery Project is such an example, where the Queensland government, Great Barrier Reef Marine Park Authority, BHP Billiton, Traditional Owners and the Great Barrier Reef Foundation are collaborating on a five-year, A$7.95 million project to protect and restore Raine Island's critical habitat to ensure the future of key marine species including green turtles and seabirds. This project includes direct intervention – building fences to minimise turtle deaths from cliff-fall, and re-profiling the nesting beach to improve nesting and hatching success by the turtles (DEHP 2016). There have been immediate positive outcomes from this intervention, with a dramatic decline in turtles dying from cliff-fall and a significant improvement in nesting success and a doubling in hatchling productivity (based on hatchling counts) from the re-profiled beach (Table 13.1; Dunstan *et al.* 2015).

Concluding comments

Many Australian islands and surrounding intertidal areas provide important nesting and foraging habitats to support internationally significant populations of marine turtles, seabirds, shorebirds and pinnipeds. The high population density

Box 13.3: Managing pisonia forests on the Capricornia Cays

Pisonia, *Pisonia grandis*, a large-leafed tree, is found almost exclusively on Indo-Pacific islands between the Tropics of Cancer and Capricorn. Outside Australia, pisonia is frequently cleared for plantations, development and guano mining. Australia's 190 ha of pisonia forests are internationally significant, as the majority are on island protected areas (Walker 1991b). They provide critical nesting habitat for seabirds including the large

Fig. 13.4: Plots on Tryon Island showing the results of replanting of pisonia, *Pisonia grandis*, following deforestation of this important tree by scale insects. (a) Before (in 2006). Photo: Joy Brushe. (b) After (in 2016). Photo: John Olds.

population of black noddies, and contribute to soil stabilisation and nutrient cycling. Since the early 1990s outbreaks of the scale insect *Pulvinaria urbicola* have been threatening pisonia throughout its global range. From 1993 to 2000, a scale outbreak destroyed over 90% of the pisonia forest at Tryon Island in the Capricornia Cays (Fig. 13.1; Kay *et al.* 2003). The forest at Coringa Island in the Coral Sea was also destroyed by an outbreak during the same period (Smith *et al.* 2004). Ants exacerbate scale insect outbreaks by protecting them from predators and parasitoids and moving them between host plants while reaping a reward in honeydew, the sugary exudate produced by the scale (Buckley 1987). Ants associated with known outbreaks on Australian islands are the introduced African big-headed ant (*Pheidole megacephala*) (Kay *et al.* 2003; Olds, unpub. data) and the pennant or Guinea ant (*Tetramorium bicarinatum*) (Smith *et al.* 2004). On mainland Australia *P. urbicola* is not considered a pest as it appears to exist in balance with several natural enemies such as parasitic wasps and its most effective predator – the Australian ladybird (*Cryptolaemus montrouzieri*). Scale outbreaks most likely occur because of a reduction in one or more of these natural enemies, a situation more likely to occur on smaller and less biodiverse islands, such as the cays where pisonia grows (Burwell *et al.* 2010). In 2001, an outbreak of *P. urbicola* in a pisonia forest on North East Herald Island, another Coral Sea cay, provided collaborating scientists and island managers with an opportunity to trial response options: *C. montrouzieri* and two species of parasitic wasp were released. They successfully controlled the outbreak (Smith *et al.* 2004). In May 2006, an outbreak was detected on Wilson Island in the Capricornia Cays and managers responded by baiting the African big-headed ant and releasing *C. montrouzieri*. By December 2006 the outbreak was contained and the forest was in recovery (Olds, unpub. data). In 2007, it became clear that Tryon Island's pisonia forest could not recover naturally. Subsequently, the Queensland Parks and Wildlife Service and volunteers began an ant-baiting and revegetation program to try and re-establish the pisonia forest there (Cruise *et al.* 2006). This program was successful with the replanted area now covering ~95% of the area previously occupied by pisonia, or ~40% of the island (Fig. 13.4). Weeds that were abundant in the early stages have been suppressed by the presence of the pisonia forest.

maintained by many of these island-dependent species means they play a vital role in the transfer of nutrients from the ocean to the island terrestrial ecosystems, which are often nutrient-poor. The nesting behaviour of marine turtles and the burrowing behaviour of some nesting seabirds, such as shearwaters and some species of penguin, also contribute to structuring and modifying the island ecosystems and vegetation communities through the processes of bioerosion and bioturbation. The obvious importance of Australian islands for supporting the marine environment and island-dependent species necessitates that their natural values are maintained (and where required, enhanced) and that threats are identified and where possible addressed. The most important message from this chapter is that the size of many islands means they are amenable to active management – intervention is a genuine option to achieve positive outcomes for island habitats and island-dependent species. Having transparent and defensible management frameworks developed using planning and prioritisation processes will ensure resources can be effectively allocated in an increasingly challenging budgetary environment.

Table 13.1. Changes in green turtle nesting success and hatchling counts in the re-profiled beach section compared with control sectors at Raine Island for the 2014–15 nesting season

	Re-profiled sector	Control sector A	Control sector B
Nesting success (%)	45.7 ± 3.1[a]	22.1 ± 3.0	32.2 ± 2.1
Hatchling counts[b]			
January 2015	1343 ± 196	1012 ± 177	529 ± 164
February 2015	2676 ± 284	1490 ± 100	1192 ± 215

[a] Mean ± standard error of the mean.
[b] Total number of hatchlings counted per 100 m pitfall trench, 6 p.m. to 6 a.m. the following morning.

References

Allaway WG, Ashford AE (1984) Nutrient input by seabirds to the forest on a coral island of the Great Barrier Reef. *Marine Ecology Progress Series* **19**, 297–298. doi:10.3354/meps019297

Bamford M, Moro D (2011) Barrow Island as an important bird area for migratory waders in the East Asian–Australasian flyway. *Stilt* **60**, 46–55.

Bancroft WJ, Roberts JD, Garkaklis MJ (2005) Burrowing seabirds drive decreased diversity and structural complexity, and increased productivity in insular-vegetation communities. *Australian Journal of Botany* **53**, 231–241. doi:10.1071/BT04079

Battley PF, Rogers DI (2007) Migration. In *Shorebirds of Australia* (Eds A Geering, L Agnew and S Harding) pp. 5–49. CSIRO Publishing, Melbourne.

Bell I (2013) Algivory in hawksbill turtles: *Eretmochelys imbricata* food selection within a foraging area on the Northern Great Barrier Reef. *Marine Ecology (Berlin)* **34**, 43–55. doi:10.1111/j.1439-0485.2012.00522.x

BirdLife International (2012) *Spotlight on Seabirds*. <http://www.birdlife.org/datazone>

BirdLife International (2016) *Endemic Bird Area: Lord Howe Island*. Fact sheet. <http://www.birdlife.org>

Bouchard SS, Bjorndal KA (2000) Sea turtles as biological transporters of nutrients and energy from marine to terrestrial ecosystems. *Ecology* **81**, 2305–2313. doi:10.1890/0012-9658(2000)081[2305:STABTO]2.0.CO;2

Buckley R (1987) Ant–plant–homopteran interactions. *Advances in Ecological Research* **16**, 53–85. doi:10.1016/S0065-2504(08)60087-2

Burbidge AA (2010) Australia's islands: biodiversity arks or rat-traps for future dodos? In *The State of Australia's Birds 2010: Islands and Birds* (Compilers J Kirkwood and J O'Connor) p. 2. Supplement to *Wingspan* **20**.

Burger AE (2005) Dispersal and germination of seeds of *Pisonia grandis*, an Indo-Pacific tropical tree associated with insular seabird colonies. *Journal of Tropical Ecology* **21**, 263–271. doi:10.1017/S0266467404002159

Burwell CJ, Nakamura N, Lambkin CL, Raven RJ, Wright K (2010) *Invertebrates of the Capricornia Cays: Human Impacts, Pest Species and Their Interactions with Fauna*. Department of Environment and Resource Management, Brisbane.

Carrick R, Csordas SE, Ingham SE, Keith K (1962) Studies on the southern elephant seal, *Mirounga leonina* (L.). III. The annual cycle in relation to age and sex. *Wildlife Research* **7**, 119–160. doi:10.1071/CWR9620119

Commonwealth of Australia (2015) *Threat Abatement Plan for Predation by Feral Cats*. Commonwealth of Australia, Canberra.

Congdon BC, McDuie F, Miller MGR, Weeks SJ, Steinberg C (2014) *Critical Seabird Foraging Locations and Trophic Relationships for the Great Barrier Reef*. Report to the National Environmental Research Program. Reef and Rainforest Research Centre, Cairns.

Cruise J, Olds JA, Melzer R (2006) *Tryon Island Restoration: Project Brief. Pest Arrest in Central Queensland*. Environmental Protection Agency, Queensland Parks and Wildlife Service, Rockhampton.

DEHP (2016) *Raine Island Green Turtle Recovery Project*. Department of Environment and Heritage Protection, Brisbane. <https://www.ehp.qld.gov.au/wildlife/animals-az/green-turtles-raine-island.html>

DoE (2005) *Heard Island and McDonald Island: Flying Birds*. Australian Antarctic Division, Department of the Environment, Hobart. <http://heardisland.antarctica.gov.au/nature/animals-of-himi/flying-birds>

DoE (2016a) *Seals and Sea Lions*. Department of the Environment, Canberra. <http://www.environment.gov.au/marine/marine-species/seals-and-sea-lions>

DoE (2016b) *Mirounga leonina in Species Profile and Threats Database*. Department of the Environment, Canberra. <http://www.environment.gov.au/sprat>

DoF (2015) *The Abrolhos Islands Information Guide*. Department of Fisheries, Perth.

DPIPWE (2014a) *Short-tailed Shearwater, Puffinus tenuirostris*. Department of Primary Industries, Parks, Water and Environment, Parks and Wildlife Service, Hobart. <http://www.parks.tas.gov.au/?base=5100>

DPIPWE (2014b) *Endemic Birds.* Department of Primary Industries, Parks, Water and Environment, Parks and Wildlife Service, Hobart. < http://dpipwe.tas.gov.au/wildlife-management/animals-of-tasmania/birds/endemic-birds>

DPIPWE (2015) *Animals and Plants: Macquarie Island.* Department of Primary Industries, Parks, Water and Environment, Parks and Wildlife Service, Hobart. <http://www.parks.tas.gov.au/index.aspx?base=623>

DPW (2015) *Barrow Group Nature Reserves Management Plan.* Department of Parks and Wildlife. Perth,.

DSEWPaC (2012a) *Species Group Report Card: Seabirds and Migratory Shorebirds. Supporting the Marine Bioregional Plan for the North-west Marine Region.* Department of Sustainability, Environment, Population and Communities.

DSEWPaC (2012b) *Marine Bioregional Plan for the North-west Marine Region.* Department of Sustainability, Environment, Population and Communities.

Dunstan D, Ariel E, Smithers S, Pickford J, Bell I, Read M (2015) Raine Island Turtle Recovery Project: conserving the world's largest green turtle nesting site in northern GBR, Australia. In *Proceedings of the Second Australian and Second Western Australian Marine Turtle Symposia.* Perth, 25–27 August 2014. (Compilers SD Whiting and AD Tucker) p. 22. Department of Parks and Wildlife, Perth.

Dutton P, Broderick D, Fitzsimmons N (2002) Defining management units: molecular genetics. In *Proceedings of the Western Pacific Sea Turtle Cooperative Research and Management Workshop* (Ed. I Kinan) pp. 93–101. Western Pacific Regional Management Council, Honolulu.

Dyer PK, O'Neill P, Hulsman K (2005) Breeding numbers and population trends of wedge-tailed shearwater (*Puffinus pacificus*) and black noddy (*Anous minutus*) in the Capricorn Cays, southern Great Barrier Reef. *Emu* **105**, 249–257. doi:10.1071/MU04011

Fitzpatrick R, Thums M, Bell I, Meekan MG, Stevens JD, Barnett A (2012) A comparison of the seasonal movements of tiger sharks and green turtles provides insight into their predator–prey relationship. *PLoS One* **7**, e51927 doi:10.1371/journal.pone.0051927

Goyert HF, Manne LL, Veit RR (2014) Facilitative interactions among the pelagic community of temperate migratory terns, tunas and dolphins. *Oikos* **123**, 1400–1408. doi:10.1111/oik.00814

Gregory SD, Henderson W, Smee E, Cassey P (2014) *Eradications of Vertebrate Pests in Australia: A Review and Guidelines for Future Best Practice.* PestSmart Toolkit publication. Invasive Animals Cooperative Research Centre, Canberra.

Harrison SA (2006) The influence of seabird-derived nutrients on island ecosystems in the oligotrophic marine waters of south-western Australia. MSc thesis. Edith Cowan University, Bunbury, Western Australia.

Heatwole H, Walker TA (1989) Dispersal of alien plants to coral cays. *Ecology* **70**, 787–790. doi:10.2307/1940230

Kay A, Olds J, Elder R, Bell K (2003) *The Impact and Distribution of the Soft Scale Pulvinaria urbicola in the Pisonia Grandis Forests of the Capricornia Cays National Parks, 1993–2002.* Report on the Scale Insect and Vegetation Monitoring Program on Tryon Island and the Scale Insect Surveys on Other Capricornia Cays. Internal Report. Queensland Parks and Wildlife Service.

Kleypas JA, Burrage DM (1994) Satellite observations of circulation in the southern Great Barrier Reef, Australia. *International Journal of Remote Sensing* **15**, 2051–2063. doi:10.1080/01431169408954227

Lavery TJ, Roudnew B, Mitchell JG (2015) Nitrogen transport from sea to land by a threatened and declining population of Australian sea lions (*Neophoca cinerea*) on Kangaroo Island, South Australia. *Australian Mammalogy* **37**, 92–96. doi:10.1071/AM12029

Leitch DB, Catania KC (2012) Structure, innervation and response properties of integumentary sensory organs in crocodilians. *Journal of Experi-*

mental Biology **215**, 4217–4230. doi:10.1242/jeb.076836

Limpus CJ (2009) *A Biological Review of Australian Marine Turtles*. Environmental Protection Agency, Brisbane.

Limpus CJ, Miller JD, Parmenter CJ, Limpus DJ (2003) The green turtle, *Chelonia mydas*, population of Raine Island and the Northern Great Barrier Reef: 1843–2001. *Memoirs of the Queensland Museum* **49**, 349–440.

Lyons MN, Keighery GJ, Gibson LA, Handasyde T (2014) Flora and vegetation communities of selected islands off the Kimberley coast of Western Australia. *Records of the Western Australian Museum* **81**, 205–244. doi:10.18195/issn.0313-122x.81.2014.205-244

Mather P, Bennett I (1993) *A Coral Reef Handbook: A Guide to the Geology, Flora and Fauna of the Great Barrier Reef*. Surrey Beatty & Sons, Sydney.

McCreless EE, Huff DD, Croll DA, Tershy BR, Spatz DR, Holmes HD, Butchart SHM, Wilcox C (2016) Past and estimated future impact of invasive alien mammals on insular threatened vertebrate populations. *Nature Communications* **7**, 12488. doi:10.1038/ncomms12488

Moilanen A, Wilson KA, Possingham H (2009) *Spatial Conservation Prioritization: Quantitative Methods and Computational Tools*. Oxford University Press, Oxford.

O'Connor J (2010) Monitoring birds on Australia's islands. In *The State of Australia's Birds 2010: Islands and Birds* (Compilers J Kirkwood and J O'Connor) pp. 4. Supplement to *Wingspan* **20**(4).

Olds J (2015) *A Biosecurity Framework for Queensland Parks and Wildlife Service-managed Islands*. Ecological Assessment Unit, Operational Support, Department of National Parks, Sport and Racing, Brisbane.

Pendoley K, Kamrowski RL (2015) Influence of horizon elevation on the sea-finding behaviour of hatchling flatback turtles exposed to artificial light glow. *Marine Ecology Progress Series* **529**, 279–288. doi:10.3354/meps11305

Pendoley KL, Schofield G, Whittock PA, Ierodiaconou D, Hays GC (2014) Protected species use of a coastal marine migratory corridor connecting marine protected areas. *Marine Biology* **161**, 1455–1466. doi:10.1007/s00227-014-2433-7

Planning Services QPWS (2014) *Management Plan: Capricornia Cays National Park and Adjoining State Waters*. Department of National Parks, Recreation, Sport and Racing, Brisbane.

Read MA, Limpus CJ (2002) The green turtle, *Chelonia mydas*, in Queensland: feeding ecology of immature turtles in Moreton Bay, southeastern Queensland. *Memoirs of the Queensland Museum* **48**, 207–214.

Schmidt S, Dennison WC, Moss GJ, Stewart GR (2004) Nitrogen ecophysiology of Heron Island, a subtropical coral cay of the Great Barrier Reef, Australia. *Functional Plant Biology* **31**, 517–528. doi:10.1071/FP04024

Schumann N, Dann P, Arnould JPY (2014) The significance of northern-central Bass Strait in south-eastern Australia as habitat for burrowing seabirds. *Emu* **114**, 234–240. doi:10.1071/MU13048

Shaughnessy PD (1999) *The Action Plan for Australian Seals*. Environment Australia, Canberra.

Shirihai H (2002) *The Complete Guide to the Antarctic Wildlife: Birds and Marine Mammals of the Antarctic Continent and the Southern Ocean*. Princeton University Press, New Jersey.

Slip DJ, Hindell MA, Burton HR (1994) Diving behavior of southern elephant seals from Macquarie Island: an overview. In *Elephant Seals: Population Ecology, Behavior, and Physiology* (Eds BJ Le Boeufey and RM Laws) pp. 253–270. Berkeley, University of California Press, Berkeley.

Smith D, Papacek D, Hallam M, Smith J (2004) Biological control of *Pulvinaria urbicola* (Cockrell) (Homoptera: Coccidae) in a *Pisonia grandis* forest on North East Herald Cay in the Coral Sea. *General and Applied Entomology* **33**, 61–68.

Specht RL, Specht A (1999) *Australian Plant Communities: Dynamics of Structure, Growth and Biodiversity*. Oxford University Press, Oxford.

Stokes T, Dobbs K, Mantel P, Pierce S (2004) *Fauna and Flora of the Great Barrier Reef World Heritage Area*. Great Barrier Reef Marine Park Authority, Townsville.

Sutherland RW, Sutherland EG (2003) Status of the flatback turtle (*Natator depressus*) rookery on Crab Island, Australia with notes on predation by crocodiles. *Chelonian Conservation and Biology* **4**, 612–619.

Turner M, Batianoff GN (2007) Vulnerability of island flora and fauna in the Great Barrier Reef to climate change. In *Climate Change and the Great Barrier Reef: A Vulnerability Assessment* (Eds JE Johnson and PA Marshall) pp. 621–666. Great Barrier Reef Marine Park Authority and the Australian Greenhouse Office.

Waayers D, Mau R, Mueller A, Smith J, Pet-Soede L (2015) A review of the spatial distribution of marine turtle nesting and foraging areas in Western Australia. In *Proceedings of the Second Australian and Second Western Australian Marine Turtle Symposia*. Perth, 25–27 August 2014. (Compilers SD Whiting and AD Tucker) pp. 83–86. Department of Parks and Wildlife, Perth.

Walker TA (1991a) Pisonia islands of the Great Barrier Reef. I. Distribution, abundance and dispersal by seabirds of *Pisonia grandis*. *Atoll Research Bulletin* **350**, 1–23. doi:10.5479/si.00775630.350-1.1

Walker T (1991b) Pisonia Islands of the Great Barrier Reef. III. Changes in the vascular flora of Lady Musgrave Island. *Atoll Research Bulletin* **350**, 31–41. doi:10.5479/si.00775630.350-3.31

Warneke RM, Shaughnessy PD (1985) *Arctocephalus pusillus*, the South African and Australian fur seal: taxonomy, evolution, biogeography and life history. In *Studies of Sea Mammals in South Latitudes* (Eds JK Ling and MM Bryden) pp. 53–77. South Australian Museum, Adelaide.

Weeks SJ, Bakun A, Steinberg CR, Brinkman R, Hoegh-Guldberg O (2010) The Capricorn Eddy: a prominent driver of the ecology and future of the southern Great Barrier Reef. *Coral Reefs* **29**, 975–985. doi:10.1007/s00338-010-0644-z

Whytlaw PA, Edwards W, Congdon BC (2013) Marine turtle nest depredation by feral pigs (*Sus scrofa*) on the Western Cape York Peninsula, Australia: implications for management. *Wildlife Research* **40**, 377–384.

Wiens JA, Hobbs RJ (2015) Integrating conservation and restoration in a changing world. *Bioscience* **65**, 302–312. doi:10.1093/biosci/biu235

Williams AAI, Karoly DI, Trapper N (2001) The sensitivity of Australian fire danger to climate change. *Climatic Change* **49**, 171–191. doi:10.1023/A:1010706116176

Winter J, Green D, Thorsborne M, Parsons M (2016) Population trends over 50 years of a pied imperial-pigeon breeding colony on North Brook Island, a tropical Great Barrier Reef island, Australia. *Emu* **116**, 14–21. doi:10.1071/MU15026

Wolfe KM, Mills HR, Garkaklis MJ, Bencini R (2004) Post-mating survival in a small marsupial is associated with nutrient inputs from seabirds. *Ecology* **85**, 1740–1746. doi:10.1890/03-3091

Young HS, McCauley DJ, Dirzo R (2011) Differential responses to guano fertilization among tropical tree species with varying functional traits. *American Journal of Botany* **98**, 207–214. doi:10.3732/ajb.1000159

14

Ecological restoration on New Zealand islands: a history of shifting scales and paradigms

David Towns, Keith Broome and Alan Saunders

Summary

Progressive advances in planning and technology have led to increasingly complex eradications of invasive mammals, which have now been cleared from over 100 islands around New Zealand. Here we review the context for these achievements and demonstrate how the eradications have enabled ecological restoration within the entire Mercury Islands archipelago. We discuss how resilience, which is the capacity for biological systems to respond to disturbance, varies by species, space and time. Furthermore, the temporal resilience (longevity) of restoration attempts may be influenced by social and cultural factors. These socio-ecological risks have been offset by widespread public involvement in island and mainland restoration projects. We have found that conservation actions within single islands, such as seabird population recovery, can have transboundary outcomes due to long-distance migration. The technology developed and lessons learned from New Zealand islands are being transferred offshore and will likely be applicable to many islands around Australia.

Introduction

Islands around New Zealand have long been recognised as special places for biodiversity, containing many species absent from the much larger North and South Islands. Many of these island species represent relict populations; a result of extensive destruction of mainland habitats through logging, land clearance for farming and the introduction of terrestrial mammals. This destruction so alarmed 19th-century environmentalists that they pressured colonial administrations to find solutions to the catastrophic declines of native birds. The government eventually established a network of islands as bird sanctuaries starting with Resolution in 1891, then Kapiti and Te Hauturu-o-Toi (Little Barrier) in 1897. On Resolution Island, the government caretaker, Richard Henry, released hundreds of flightless kiwi (*Apteryx* spp.) and kakapo (*Strigops habroptilus*) captured from nearby mainland Fiordland. Sadly, Henry's efforts ultimately failed when in ~1900 stoats (*Mustela erminea*) swam to the island and destroyed the translocated populations (Young 2004). Extinctions continued, adding dramatically to those following Polynesian settlement around the 13th century. Last-minute rescues did save some species. The last populations of South Island saddlebacks (*Philesturnus carunculatus*) were evacuated from Taukihepa (Big South Cape Island) in the early 1960s after ship rats (*Rattus rattus*) became established. Unfortunately, a bat (*Mystacina robusta*), bush wren (*Xenicus longipes*) and snipe (*Coenocorypha iredalei*) could not be saved from the rats on Taukihepa and by 1967 those three species were extinct (Tennyson and Martinson 2004).

In parallel with the disastrous spread of invasive mammals came growing recognition that the manageable size and isolation of islands presented unique opportunities to eradicate pests. The removal of island populations of feral cats (*Felis cattus*) from Takapourewa (1925), goats (*Capra hircus*) from Kapiti (1928) and Great King (1946) and pigs (*Sus scrofa*) from Aorangi (1936) (King 2005) demonstrated the effectiveness of removing larger mammals. Threatened species need not be evacuated to safe islands; eradicating pest mammals might enable recovery of threatened species *in situ*. The deliberate removal of three species of rats from a range of small islands in the early to mid-1980s was a significant step (Thomas and Taylor 2002; Towns and Broome 2003), not only because the chances of eradicating rodents were viewed as remote (Dingwall *et al.* 1978) but also because restoring native ecosystems was viewed as fanciful (King 1984). Over the last 40 years, developments in technology, best practice and skills have reached the point where all invasive mammals, including rodents, have been removed from more than 100 islands up to 11 000 ha across New Zealand (Towns *et al.* 2013). Research and ongoing monitoring provides compelling evidence of the impacts of rodents and other invasive mammals, and demonstrates that spectacular responses can follow their removal (see case study below). As institutional capability developed, some species recovery plans have aimed to establish populations through translocations to safe island refuges, or by eradicating from islands the threat to their survival. One example is the ancient reptilian tuatara (*Sphenodon punctatus*), which became confined to offshore islands. By the early 1990s about one-third (10) of these island populations were threatened by introduced rodents. Two populations were lost, but rats were removed from the remaining eight islands, and a further eight populations re-established on islands cleared of invasive mammals within the species' original range (updated from Cree 2014). Collectively, these actions changed the threat status of tuatara in the wild.

In this chapter, we review key components and conceptual changes behind the evolution of ecological restoration of island ecosystems in New Zealand. Accounts of technical advances, particularly against rodents, are reviewed elsewhere (Russell and Broome 2016) and not repeated here. We instead focus on paradigm shifts: the way thinking by island practitioners has become bigger and bolder, from single-species eradications to complex operations involving the eradication of multiple pests from larger and more remote islands. On islands cleared of invasive mammals, activities have shifted from those akin to selective zookeepers focused on individual species to considering the former and current ecosystem processes affecting ecosystem states. For example, restoring seabird islands involves a focus on reinstating guano-rich soils and their associated biotic communities. Our case study on the Mercury Islands illustrates how mobile native species within the archipelago can influence restoration goals originally designed for single locations. Approaches developed in New Zealand are being applied offshore and some of these insights will likely assist the growing number of island restoration projects in Australia.

The social dimension

At least 83% of eradications in New Zealand have been on uninhabited islands, most (72%) of which are public reserves vested in the Department of Conservation (DOC). Almost all successful eradication operations to date have been led by DOC, although other stakeholders and funders have also contributed. Māori *iwi* (tribes) have traditions as

Fig. 14.1: Tiritiri Matangi Island coastline with (a) tui (*Prosthemadera novaeseelandiae*) raiding a feeder provided for hihi and (b) takahe feeding on exotic pasture.

kaitiaki (guardians) and co-governance arrangements between DOC and *iwi* are increasingly used to ensure traditions, aspirations and cultural practices are appropriately reflected in management plans. Other stakeholders, including community groups, landowners, local authorities, non-government conservation groups and philanthropists are increasingly involved in island restoration initiatives around New Zealand. Social drivers have thus become an important component of island management. For example, of 19 island management projects listed on the Sanctuaries of New Zealand website (www.sanctuaries.org.nz), 11 were initiated by community groups.

One well-documented example of community-led restoration is on Tiritiri Matangi Island in the Hauraki Gulf (Fig. 14.1) where Supporters of Tiritiri Matangi implemented a restoration plan developed in partnership with DOC. Restoration involved reforestation of the previously farmed island by propagation and planting of 280 000 native trees over 10 years, assistance with funding the eradication of Pacific rats (*R. exulans*) by DOC in 1993, and the translocation of 11 species of birds, tuatara, two species of lizards and one large species of flightless invertebrate. Released birds include small, intensively managed populations of the threatened takahe (*Porphyrio hochstetteri*), kokako (*Callaeas wilsoni*) and hihi (*Notiomystis cincta*). The island now has over 25 000 visitors annually (Galbraith and Cooper 2013).

Advances and challenges in biosecurity

The first mammals introduced to New Zealand were Pacific rats and Polynesian dogs (kuri *Canis familiaris*), travelling with early Polynesian settlers in ~1280. Beginning with Cook's voyages in the late 18th century, offshore islands were deliberately stocked with rabbits (*Oryctolagus cuniculus*), goats and pigs to provide readily available food for island residents and future castaways. Additional species of rodents stowed away in cargo or made their own way ashore from moored, anchored or beached vessels. Closer to the mainland, islands were not only more frequently visited by people but many were also within the swimming ranges of some introduced mammalian pests.

Only 158 small islands (all less than 1000 ha) remained entirely pest mammal-free, although none of the invaded islands had the complete range of pest species found on the two main islands (Parkes and Murphy 2003). Early in the 20th century, isolation and the smaller suite of pests on islands led observers such as Guthrie-Smith to warn that the spread of pests such as rodents to islands could destroy vulnerable native species. Before the development of successful rodent eradication technologies, prevention was viewed as the only way of protecting islands from invasions. However, as the ability to eradicate pest mammals developed, biosecurity to prevent subsequent reinvasions became a crucial component of eradication projects in New Zealand.

For vulnerable islands of high natural value, biosecurity is likely to include quarantine measures targeting all potentially harmful organisms, including pathogens and seeds of potentially invasive plants. Through their isolation and distinct borders, islands also present opportunities to effectively apply rapid response protocols following pest incursions. Outreach programs promoting biosecurity alongside other island features may also be effective. However, such programs seek to change human behaviour in a context where compulsion is limited, incentives are few and human values vary. Unlike eradication projects which often provide demonstrable and dramatic outcomes, successful biosecurity produces no obvious change to the status quo but involves ongoing costs to manage nebulous risks. In contrast, a biosecurity failure produces tangible outcomes, which may lead to significant ongoing impacts on island ecosystems.

Ecological restoration of an archipelago and the roles of resilience: a case study

The Mercury Islands are an archipelago of seven islands and several smaller islets off the north-eastern North Island of New Zealand (36°35′S, 175°55′E). Biologically, Green (3 ha) and Middle (or Atiu, 13 ha) Islands are particularly significant because they were never occupied by invasive mammals. Both islands are inhabited by dense and diverse assemblages of reptiles, including tuatara and up to 10 species of skinks (Scincidae) and geckos (Diplodactylidae), and extraordinary densities of burrowing procellariform seabirds dominated by diving petrels and flesh-footed shearwaters, with burrow densities of over 10 000 per hectare (Thoresen 1967). Also present is a diversity of flightless invertebrates, including numerous species of minute land snails (Towns and Atkinson 2004) and an array of flightless crickets. The largest and most bizarre of these is the Mercury Islands tusked wēta (*Motuweta isolata*), distinctive for its large size (60 mm) and tusk-like appendages on the head of males (Fig. 14.2).

All larger islands in the group were at some time invaded by introduced mammals, including Pacific rats on Double (27 ha) and Red Mercury (225 ha) Islands, Pacific rats and rabbits on Korapuki (18 ha) and Stanley (100 ha) Islands, and Pacific rats plus ship rats, cats and goats (*Capra hircus*) on Great Mercury (1872 ha) Island. These species have been progressively removed, starting with goats from Great Mercury in ~1986, then Pacific rats and rabbits from Korapuki Island in 1986–87. The last invasive mammals were removed with the eradication of two species of rats and cats from Great Mercury in 2014 (Towns *et al.* 2016a). As these eradications progressed a new conceptual problem developed: is it possible to restore entire island ecosystems? The considerable biological challenges involved led Simberloff (1990) to call island restoration 'reconstructing the ambiguous'.

Here we outline why restoration is interwoven with cultural and social influences, which can greatly affect the longevity (resilience) of restoration projects. Resilience is a system's ability to absorb the effects of disturbance (Folke 2006). The concept was based on observations that more complex (biodiverse) ecosystems appeared to have greater stability and hence increased resilience or

Fig. 14.2: Seabird-burrowed milktree forest on Middle Island with (a) tusked weta, (b) Whitaker's skink and (c) robust skink (*Oligosoma alani*) as examples of threatened species now translocated to other islands.

capacity for renewal when subjected to disturbance (McCann 2000). Renewal after human disturbance is also the foundation for restoration ecology. With appropriate habitats made available, species from local source populations should naturally recolonise these restored ecosystems, which then, in theory, should become more resilient.

Biogeographic context

Until ~12 000 years ago the Mercury Islands were contiguous with the adjacent North Island mainland. As sea levels rose after the last glaciation, connection with the mainland was lost and ~8000 years ago the 'archipelago' was a single isolated landmass. Continued sea-level rise fragmented this Greater Mercury Island into the archipelago visible today (Fig. 14.3; Towns *et al.* 2016a).

These previous connections within the archipelago, and with the mainland, are reflected in the diversity of reptiles and flightless invertebrates. The biota became further fragmented with the arrival of people and commensal mammals. The small uninvaded islands provide reference data for those now cleared of introduced mammals, but some of the larger islands indicate the past existence of a range of communities wider than those at the reference sites (Towns *et al.* 2016a). Since 1986, natural recovery of the ecosystem following the removal of introduced mammals, together with restoration practices such as species translocations within the archipelago, have led to increased abundance – or reclaimed range – of 14 species of plants, six species of invertebrates, nine species of reptiles and 11 species of native birds (updated from Towns *et al.* 2013).

Biological resilience

Such examples of increased species abundance say little about their role in ecosystems and the capacity for passive recovery versus the need for active restoration. Resilient ecosystems should not need active restoration. To understand which components of the systems would need active restoration in the Mercury Islands, we must address three

Fig. 14.3: Great Mercury and the southern islands of the Mercury archipelago. Photo: R. Chappell.

questions: How do the terrestrial ecosystems function? How have invasive species affected these functions? What is the capacity of each system for natural recovery?

Ecosystem function

In response to the first question, the influence of seabirds on island systems through burrowing (soil disturbance) and nutrient subsidy of marine origin was viewed as fundamental when an archipelago-based focus was first proposed for the Mercury Islands (Towns *et al.* 1990). However, the extent to which seabirds drive these systems only became apparent from studies of islands in northern New Zealand with and without seabird predators (Fukami *et al.* 2006) and through global comparisons (Mulder *et al.* 2011). These studies demonstrated that seabirds loosen and desiccate soils through burrowing, add marine-derived nitrogen and phosphorus, reduce soil pH, reduce the density of seedlings, influence plant assemblage composition and can positively influence the abundance of at least eight orders of terrestrial invertebrates (Towns *et al.* 2009; Mulder *et al.* 2011). In addition, seabird burrows and the loosened soils around them provide shelters and sources of food for tuatara, at least three species of large nocturnal skinks and large invertebrates such as the Mercury Island tusked wēta and giant centipede (*Cormocephalus rubriceps*). Interestingly, other than the substrates and space for nesting, seabirds do not consume resources from within terrestrial island ecosystems.

Birds can act as transformative species on the islands. For example, in addition to the transformative effects of seabirds on ecosystems, other resident species can transform the composition and viability of communities. For example, a keystone (and hence transformative) species in these islands is an extremely mobile fruit pigeon, kērēru (*Hemiphaga novaeseelandiae*), which influences vegetation structure through the dispersal of seeds within and between the islands. Kērēru are the only birds able to ingest large fruit of tawapou (*Pouteria costata*), kohekohe (*Dyxoxylem spectable*), taraire (*Beils-*

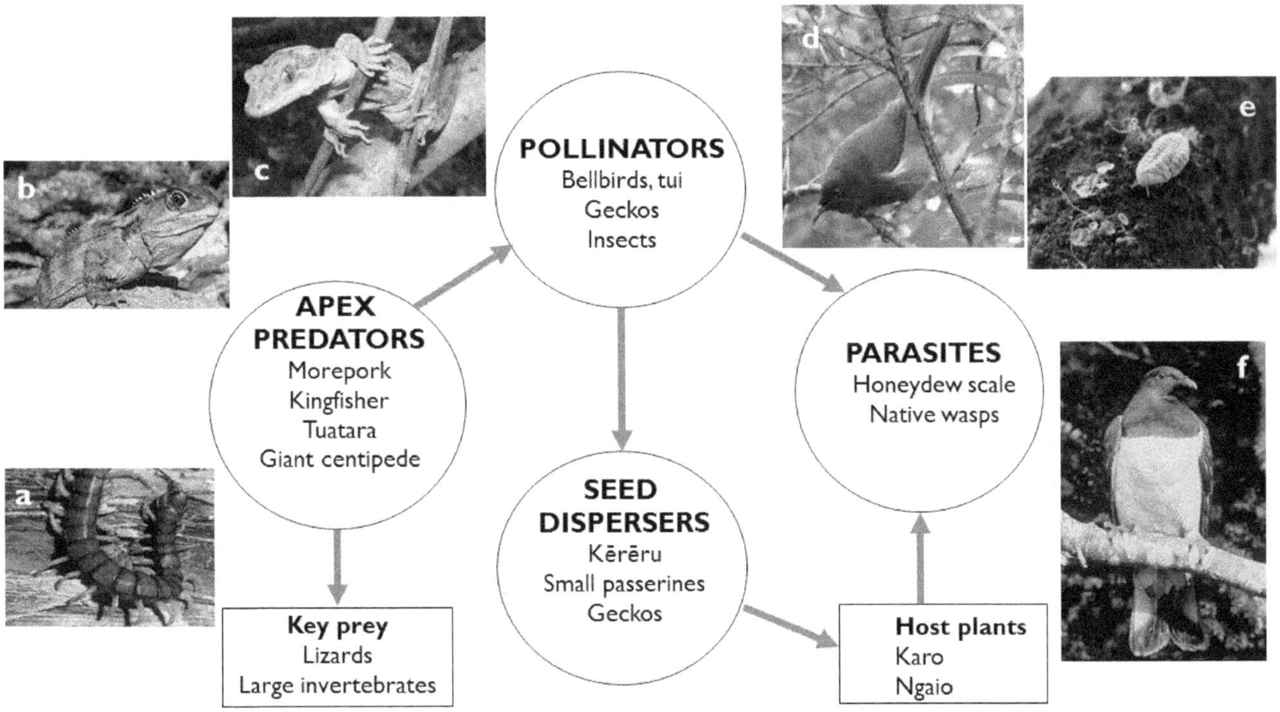

Fig. 14.4: Feedback loops and linkages between guilds of species that can influence ecosystem function in the Mercury Islands, with selected guild members including (a) giant centipede, (b) tuatara, (c) Duvaucel's gecko, (d) bellbird, (e) honeydew scale and (f) kērēru.

chmedia tarairi) and karaka (*Corynocarpus laevigatus*), which are canopy species now spreading within islands cleared of invasive mammals (Campbell and Atkinson 1999).

Other groups of species (guilds) have interactive effects within and between guild members. For example, the native pollinator guild includes three species of geckos and two of birds. Studies of two of the species of geckos (Evans *et al.* 2015) and the bellbird (*Anthornis melanura*; Beggs 2001) indicate that they are heavy users of honeydew, which in the Mercury Islands is produced by the scale insect *Coelostomidia zealandica*. Distribution of the scale insects is directly related to the abundance of two species of host plants (*Pseudopanax crassifolium* and *Myoporum laetum*) whose seeds are distributed by forest birds. There may thus be an indirect relationship between the distribution of plants that support scale insects and the abundance of pollinators such as birds and geckos. Other subtle relationships include the predator–prey link between geckos and apex predators such as the native owl (morepork *Ninox novaeseelandiae*) and tuatara, and the crucial role of large flightless invertebrates and other lizards as prey for this predator guild (Fig. 14.4). However, the resilience of such guilds can be challenged by invasive mammals.

Effects of invasive mammals

The five species of invasive mammals removed from the Mercury Islands vary in the form and intensity of their effects on native ecosystems and species (Table 14.1). Red Mercury and Double Islands were the only islands with a single species of pest mammal (Pacific rats). Although apparently the least destructive of commensal rat species, Pacific rats can reduce seabird density and diversity, suppress tuatara and diurnal lizard populations, extirpate selected species of large flightless invertebrates, geckos and nocturnal skinks, and modify forest composition through selective seed predation (Towns *et al.* 2006). With additional pest

Table 14.1. Invasive mammals removed from the Mercury Islands, and their effects in New Zealand island ecosystems

Species	Effects in New Zealand	Other island examples	References
Goat	Suppression of regeneration and complete removal of forest understorey; can induce close-cropped ground layer	Raoul, Macauley, Great King, Cuvier	King (2005), Bellingham *et al.* (2010)
Rabbit	Suppression of broadleaf forest regeneration through removal of forest understorey	Stanley, Korapuki, Ohinau	Hicks *et al.* (1975)
Pacific rat (kiore)	Seed predation and inflorescence browse on numerous species of forest plants; predation and localised extirpation of large flightless invertebrates; recruitment failure of tuatara; localised extirpation of nocturnal lizards; recruitment failure and extirpation of selected species of seabirds; implicated in historic extinctions of small terrestrial birds (e.g. wrens)	Te Hauturu-o-Toi, Tiritiri Matangi, Marotere Islands	King (2005), Tennyson and Martinson (2006), Towns *et al.* (2006, 2007), Campbell and Atkinson (2002)
Ship rat	Likely similar effects as Pacific rats but confounded by more recent arrival of ship rats into previously modified habitats; local extirpation of seabirds and heavy suppression of forest birds through predation at nests; implicated in historic extinctions of many species of forest birds	Islands around Great Barrier, Great Mercury	King (2005), Towns *et al.* (2006)
Feral cat	Predation of reptiles; predation and local extirpation of seabirds; local extirpation of ground-feeding birds and implicated in extinctions of some species	Raoul, Te Hauturu-o-Toi, Stephens, Herekopare	King (2005), Nogales *et al.* (2004)

mammals (such as rabbits) the intensity and breadth of habitat change and species loss increases, sometimes because of synergistic effects (Mulder *et al.* 2011). For example, feral cats often rely on populations of rodents for food when seabirds are absent, thereby supporting sustained destruction of nesting seabird colonies. This rodent–cat combination is implicated in the catastrophic declines of about half of the world's rarest petrels (Le Corre 2008).

By the mid-1980s, all Mercury Islands with pest mammals had heavily suppressed populations of seabirds, locally extirpated or few (<20) tuatara, only half the lizards of pest-free islands, and reduced diversity of flightless invertebrates including local extirpation of honeydew scale and large species of insects such as darkling beetles (*Mimopeus* spp.), tree wēta (*Hemidiena thoracica*) and tusked wēta. The effects varied by island, mammal species combination and anthropogenic modification (Towns *et al.* 1990), but were reflected in suppressed engineering effects of seabirds and disrupted predator–prey relationships. For example, on Korapuki Island, the apex predator guild (Fig. 14.4) was virtually destroyed with the loss of tuatara and its key beetle (*Mimopeus opaculus*) and wēta prey. The only remaining apex predators comprised a small population of kingfishers (*Todiramphus sanctus*) and rare giant centipedes.

Resilience through recolonisation

Studies of natural recovery in the Mercury Islands indicated that the capacity of these systems to adapt to disruption (i.e. their resilience) varies. For example, burrowing seabirds (Procellariformes) were traditionally regarded as strongly philopatric with limited capacity to recolonise new sites. However, more recent studies found some species to be effective colonists and, once established, social attraction can increase immigration rates, including those of unrelated species (Buxton *et al.* 2014). This attraction appears to be most effective within a radius of ~20 km of established breeding sites, which therefore encompasses all of the Mercury Islands (Buxton *et al.* 2014; Borrelle *et al.* 2015). Such predictions were supported by the rapid reappearance or expansion of breeding grey-faced petrels, diving petrels, Pycroft's petrels and fluttering shearwaters within two years of the removal of predators from Great Mercury Island (P. Corson, *pers. comm.*). Many other islands cleared of invasive mammals are outside these recruitment halos, so passive recolonisation is unlikely (Borrelle *et al.* 2015). There can thus be a spatial influence on resilience. Terrestrial birds including parakeets (kākāriki *Cyanoramphus novaezelandiae*), forest parrots (kākā *Nestor meridionalis*), bellbirds and kēreru reappeared and rapidly spread on Great Mercury within 12 months of the mammal eradications, demonstrating the mobility of some terrestrial species within the archipelago (P. Corson, *pers. comm.*). Most plants, flying invertebrates and spiders with ballooning behaviour are also effective colonists. On Korapuki Island, kingfishers have become one of the more commonly encountered forest birds. However, moreporks did not re-establish until 23 years after mammal removal, presumably when suitable prey was consistently available (Towns *et al.* 2016a), indicating temporal and behavioural influences on biological resilience.

Unlike the effective colonisers, numerous species appear incapable of crossing marine barriers to recolonise when extirpated (e.g. by non-native species). Examples include tuatara, lizards and flightless invertebrates such as wēta, darkling beetles, female scale insects, and land snails. Without management intervention, gaps in assemblages of poor dispersers would be permanent. Such species have often been the focus of conservation efforts because their vulnerability to habitat change and invasive species has so reduced their numbers that they are identified as nationally threatened. One example is the burrow-inhabiting nocturnal Whitaker's skink (*Oligosoma whitakeri*), a nationally vulnerable species (Hitchmough *et al.* 2016) reintroduced to Korapuki Island in 1988 (Towns *et al.* 2016b).

Restoration activities, such as those in the Mercury Islands, can have implications within and beyond the archipelago. For example, although Great Mercury Island is partly farmed, the remaining forest is still larger than the area of all other islands in the group combined. Species such as kēreru could reach numbers not seen for centuries and their seed dispersal abilities will likely influence forest structure on all islands in the archipelago. Without precedents for such effects across an archipelago, the outcomes for forest composition are unpredictable. Burgeoning seabird populations in the Mercury Islands will inevitably have transboundary effects elsewhere, increasing populations of non-breeding birds in areas such as eastern Australia, the north Pacific and the west coast of South America (Table 14.2).

Social determinants of resilience

The temporal resilience of island restoration projects is likely to be influenced by social and cultural factors. The first of these – the social environment – assumes that people will value the outcomes of restoration efforts. This in turn requires socially mandated invasive species eradications. Invasive mammals on the four smaller Mercury Islands were eradicated by DOC acting under the *Reserves Act 1977*, which requires that exotic species should where possible be removed from Nature Reserves (Towns *et al.* 2013). On privately owned Great Mercury Island the eradications were completed through a partnership between DOC and the owners at a cost of about NZ$1.4 million. In effect, the earliest eradications were legally mandated, but those on Great Mercury were in part socially motivated. One justification for the latter project was biosecurity – the reduced risk of invasions by rodents spreading from Great Mercury to other islands in the archipelago. Long-term protection of the investment in these islands relies on regular checks for new invasions, by staff on Great Mercury and DOC staff on the other islands. There have been two known post-eradication incursions, both of ship rats found in traps. One was on Korapuki Island in ~1987 and the second on Great Mercury Island in 2016 (Dilks and Towns 2002; P. Corson, *pers. comm.*). Given the distance to wild rat populations, both rats most likely reached the islands by boat. Landing on the Nature Reserves requires a permit; however, access is largely unrestricted to Great Mercury. The ongoing risk of future incursions in these insular ecosystems needs to be managed by communities valuing the islands' contribution to nature conservation. Long-

Table 14.2. Burrowing seabird species in the Mercury Islands and their areas of transboundary influence

Species	Island distribution	Forage area
Little penguin (*Eudyptula minor*)	All islands	Adults sedentary, juveniles may forage several hundred kilometres
Grey-faced petrel (*Pterodroma macroptera*)	Expanding colonies on Korapuki, Stanley and Great Mercury	After breeding, forage around the edge of the continental shelf; Tasman Sea as far as eastern Australia
Pycroft's petrel (*P. pycrofti*)	Korapuki, Double, Stanley and Rede Mercury; first sighted on Great Mercury in 2016	Only known from islands around New Zealand continental shelf with Red Mercury the largest colony; non-breeders migrate to central and north-eastern north Pacific
Flesh-footed shearwater (*Puffinus carneipes*)	Green and Middle (1000s); small populations (100s or less) on Korapuki, Double, Stanley and Red Mercury	Trans-equatorial migration to coasts of Korea, Japan and western North America
Sooty shearwater (*P. griseus*)	Small numbers reported on Korapuki	Trans-equatorial migrant to north Pacific between Japan and North America and as far north as Gulf of Alaska
Fluttering shearwater (*P. gavia*)	Abundant on Korapuki, first found breeding on Great Mercury Island in 2015	Trans-Tasman migrant to coast off eastern and southern Australia
Little shearwater (*P. assimilis*)	All islands except Great Mercury	Largely confined to waters around New Zealand
White-faced storm petrel (*Pelagodroma marina*)	Occasional sightings on Middle, breeding unconfirmed	Migrate to tropical Pacific off coast of Ecuador and Peru
Common diving petrel (*Pelecanoides urinatrix*)	Extremely abundant on Green and Middle, increasing on all other large islands	Appear to be confined to waters around New Zealand

Source: Data from Heather and Robertson (2015) and Borrelle *et al.* (2015).

term resilience thus relies on respect for landing restrictions, by the boating public. Standardised biosecurity protocols for border control (e.g. the quarantine procedures and rodent checks outlined earlier) need to continue. However, to date, there has been little conversation with locals about their understanding of why invasive mammals are a risk and what can be gained when the animals are removed. Instead, off-island biosecurity has relied on agency-based regulation (permitting) and awareness-raising campaigns such as the 'Treasure Islands' initiative (Bassett *et al.* 2016).

Great Mercury Island may be particularly useful for raising awareness and gaining the social (community) licence to manage biosecurity across this archipelago. On Great Mercury Island, seabird-modified island habitats, and many of the species found in them, could in time be experienced by the public as an alternative to the more fragile Nature Reserves. In the meantime, people with access to boats are being asked by island managers to value a resource that few can experience. There is increasing local interest in the Mercury islands. For example, Mercury Bay Area School assists with surveys for invasive plants on Great Mercury Island (M. Fay, *pers. comm.*). Habitat Tuateawa, a community conservation group, is attempting to control pests on the Coromandel mainland to levels where burrowing seabird populations could be established using the Mercury Islands as a source (A. Saunders, *pers. obs.*).

Analyses of the most appropriate targeted messages, who they should be aimed at and how to efficiently spread them through communities on the adjacent mainland would be a useful investment in the long-term sustainability of what is becoming a multi-million-dollar biodiversity conservation program.

Cultural impacts on resilience

A second influence on biological resilience involves cultural acceptance and impact. Particularly strong emotional attachments to the Mercury islands are held by local *iwi*. Ngati Hei gifted Middle and Green Islands to the Crown as tuatara reserves and view Great Mercury (Ahuahu) as the first landing place (Hawaiiki) of their Polynesian ancestors (J. Davis, *pers. comm.*). Recent archaeological research on the island (McIvor and Ladefoged 2016) indicates that Great Mercury may be one of New Zealand's earliest settled sites (S. Holdaway, *pers. comm.*), which underpins the significance of the islands to Ngati Hei. In the smaller Mercury Islands, indications of periodic use by Māori include fortifications, dwelling sites, rock terraces and the presence of Pacific rats. Oral history and soil profiles indicate that all the islands were periodically burned which, coupled with various species of introduced mammals, has influenced the subsequent patterns of regeneration (Atkinson 2004). Polynesian dogs on Great Mercury (S. Holdaway, *pers. comm.*), likely devastated populations of seabirds and tuatara (Mulder *et al.* 2011). A complex interplay between island size and cultural history may thus be responsible for some idiosyncratic species distributions. For example, the tree species *Pisonia brunoniana* is abundant on some islands within the Mercury archipelago but absent from others. Similarly, Duvaucel's gecko *Hoplodactylus duvaucelii* is on the smaller Mercury Islands but inexplicably absent from Red Mercury.

DOC regularly consulted with local *iwi* and has directly involved interested participants in research and management in the Mercury Islands. In response, Ngati Hei requested assistance from DOC to remove rabbits, Pacific rats and mice (*Mus musculus*) from nearby Ohinau Island (45 ha), a task completed in 2005 (Towns *et al.* 2016a). These actions protected a rare population of flesh-footed shearwaters and provided a new site for tusked wēta. They can also rekindle cultural practices and the songs and legends associated with the island, reaffirm *ahikā roa* or rights of guardianship through use, and demonstrate working relationships between DOC and *iwi* (J. Davis, *pers. comm.*).

Discussion

Island-based conservation in New Zealand has shifted in scale, complexity and emphasis. Improved technologies have increased the scope of pest eradications from relatively small islands away from public view to the complexities of removing eight species of invasive mammals from Rangitoto-Motutapu Islands (combined area 3881 ha) in full view of 1.5 million Auckland residents. The latter project was not the initiative of government agencies, but was developed through a community-led initiative that gained political (including prime ministerial) support. Rangitoto-Motutapu Islands are important because they receive more than 100 000 visitors annually, with the attendant challenges for biosecurity. They represent a new dimension in community involvement, with pending development of a co-management agreement between DOC and a collective of *iwi*. Pest removals and island ecosystem restoration at increasing scale and complexity have also stimulated work on new frontiers: islands vulnerable to natural reinvasions, islands with large resident populations, and insular areas on the mainland (Fig. 14.5).

Islands susceptible to reinvasions

So far, the total offshore island area of New Zealand without invasive mammals has increased from 0.5% to 10% (Russell *et al.* 2015). Many more species and terrestrial ecosystems would benefit if multiple pest eradications were possible on inshore islands accessible to mainland pests, and on the larger inhabited islands.

The origins of pest incursions can now be tracked through DNA fingerprinting; this technology is being used to determine whether invasive mammals subsequently caught on islands are survivors of an eradication or new arrivals. Invasion

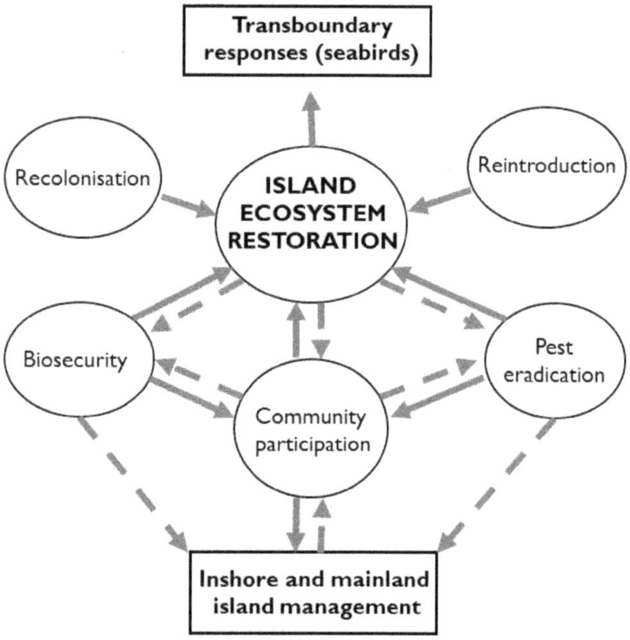

Fig. 14.5: Schematic representation of key components in island restoration (circles) and unpredicted spinoffs (rectangles), dependent linkages (solid lines) and indirect feedback (broken lines).

risk assessments can thus be determined, along with potential trade-offs between the costs of managing incursions and the potential benefits of the eradications for threatened species. For example, DNA fingerprinting has identified the most efficient trapping regimes for managing incursions of stoats from the South Island mainland to Secretary Island (Veale *et al.* 2013).

Islands with residents

If kakapo are to re-establish on islands within their original range, invasive mammals will need to be managed on Stewart Island, the third largest of New Zealand islands. Besides increased scale, management will involve adapting and applying established approaches and techniques around more than 300 residents' homes and backyards in a coordinated, effective and respectful way. The new field of conservation socio-ecology should be used to understand the aspirations of island residents and other stakeholders, provide ways to respect the special relationships island residents have with their environment, and understand wider social, cultural, economic and political issues affecting people's engagement and support.

Restoration of 'mainland islands'

Islands offer a restricted range of habitats. Hundreds of species under stress on the North and South Islands cannot be managed on offshore islands. The ecological responses to pest eradications on islands has encouraged attempts to affordably control pests to densities at which 'island-type' responses might be achieved at mainland sites, even if eradication is not feasible. Experimental management projects in the early 1990s at Mapara in the central North Island (Innes *et al.* 1999), and in the Eglinton Valley in Fiordland on the South Island, indicated the potential for intensive and sustained control of multiple pests. These attempts were followed by six mainland restoration projects initiated by DOC (Saunders 2000). These projects are still active, which attests to their economic viability and contribution to species recovery and ecological restoration goals. More importantly, following the example set on Tiritiri Matangi Island, mainland restoration projects have stimulated at least 600 voluntary community-based projects by private landowners, Māori *iwi* and conservation groups, often working in partnership with DOC, on islands and at mainland sites (Peters *et al.* 2015). For example, at least 30 community groups trap stoats and other predators as part of a national campaign to reduce threats to kiwi. The community-led Karioi-Maunga ki to moana (mountain to the sea) project on the west coast of the North Island aims to reduce predator numbers and restore colonies of seabirds at a mainland site (Fig. 14.6).

New forms of partnerships are emerging, often involving institutional donors, corporate sponsors and philanthropists, with the common goal of increased abundance and diversity of native species. Of national importance are ongoing settlements of historic land disputes between the Crown and Māori *iwi* under the Treaty of Waitangi, signed in 1840, which includes re-affirmation of guardianship (*kaitiakitanga*) over sites with cultural and ecological significance. Te Hauturu-o-Toi and Rangitoto-Motutapu

Fig. 14.6: Karioi Rangers deploying stoat traps as part of the Karioi-Maunga ki to moana project, Raglan, New Zealand. Photo: Kristel van Houte.

are two examples where there may be opportunities for achieving innovative social, economic and environmental objectives in island restoration.

Concluding comments: implications for Australian islands

Our experiences cover 30 years of island management, during which there have been surprising shifts in perceptions and paradigms in New Zealand. Some of the lessons learned may be transferrable to Australia. First, gradually scaling up the size and complexity of eradication campaigns demonstrated how rodents, then entire suites of introduced mammals, could be eradicated from medium-sized islands – frequently in one comprehensive operation. Second, the scale and subtlety of some ecological responses on islands following pest removal was unexpected, and provided lessons about the extent of ecological damage that invasive species can induce. Third, restoration actions at one site can have consequences elsewhere. The Mercury Islands case study indicated how the recolonisation of islands by seabirds can have transcontinental effects across some species' range. Fourth, the importance of community engagement and participation cannot be overstated. Public support for biosecurity is an essential part of protecting investments in island restoration and, given the chance, community organisations can develop new initiatives at island and mainland sites. This public willingness to engage overturns an early 20th-century belief that invasive species are here to stay and native species must cope or fail. Instead, the huge efforts undertaken by community groups provide evidence of widely held views that native species should persist, a commitment already enshrined in the Treaty of Waitangi but long overlooked. Finally, process and technological developments may also be transferrable. DOC's Island Eradication Advisory Group has already provided technical advice and peer review to eradication projects in several countries. New Zealand specialists have provided advice in the planning, implementation and evaluation of eradication and control operations in Australia, several Pacific states, the US, Mexico, Ecuador, Chile and Argentina. Online eradication resource kits prepared by the University of Auckland-based Pacific Invasives Initiative (http://pacificinvasivesinitiative.org) provide access to best practice and management project frameworks.

Lessons learned about the impacts of invasive mammals have been rapidly transferred from offshore islands to mainland sites. With assistance from community groups, populations of threatened species such as kokako have responded to intensive predator control (Innes *et al.* 1999). The numerous pest control and restoration projects led by community groups reflect a growing view that ordinary New Zealanders can halt declines of native biodiversity (Peters *et al.* 2015). The Predator Free 2050 program launched by the Prime Minister of New Zealand in July 2016 (Joyce *et al.* 2016) reflects advances in pest management, ecological restoration and political support for wide-scale pest control – and, if possible, eradication. Technical, social and economic challenges mean that eradicating the species targeted by Predator Free 2050 (such as possums *Trichosurus vulpecula*, ship rats and stoats) may at present be unachievable (Parkes *et al.* 2017). However, 40 years ago, the same belief applied to the removal of rats from islands.

Provided public expectations are carefully managed, and there is a coordinated and progressive approach to addressing the technical and social challenges involved, history suggests that further surprises will develop from New Zealanders' island restoration ethos.

Acknowledgements

All images were supplied by one of us (DRT) except the following: Fig. 14.1, Department of Conservation *Te Papa Atawhai*, Crown Copyright and published with permission; Fig. 14.2a, E. Humphries; Fig. 14.3, R. Chappell; Fig. 14.4a, D. Veitch; Fig. 14.4b, J. Gardiner; Fig. 14.4f, J. Kendrick; Fig. 14.6, K. van Houtte.

References

Atkinson IAE (2004) Successional processes induced by fires on the northern offshore islands of New Zealand. *New Zealand Journal of Ecology* **28**, 181–193.

Bassett IE, Cook J, Buchanan F, Russell JC (2016) Treasure Islands: biosecurity in the Hauraki Gulf Marine Park. *New Zealand Journal of Ecology* **40**, 250–266. doi:10.20417/nzjecol.40.28

Beggs J (2001) The ecological consequences of social wasps (*Vespula* spp.) invading an ecosystem that has an abundant carbohydrate resource. *Biological Conservation* **99**, 17–28. doi:10.1016/S0006-3207(00)00185-3

Bellingham PJ, Wiser SK, Wright AE, Cameron EK, Forester LJ (2010) Disperser communities and legacies of goat grazing determine forest succession on the remote Three Kings Islands, New Zealand. *Biological Conservation* **143**, 926–938. doi:10.1016/j.biocon.2010.01.001

Borrelle SB, Buxton R, Jones HP, Towns DR (2015) A GIS-based decision-making approach for prioritizing seabird management following predator eradications. *Restoration Ecology* **23**, 580–587. doi:10.1111/rec.12229

Buxton RT, Jones C, Moller H, Towns DR (2014) Drivers of seabird population recovery on New Zealand islands after predator eradication. *Conservation Biology* **28**, 333–344. doi:10.1111/cobi.12228

Campbell DJ, Atkinson IAE (1999) Effects of kiore (*Rattus exulans* Peale) on recruitment of indigenous coastal trees on northern offshore islands of New Zealand. *Journal of the Royal Society of New Zealand* **29**, 265–290. doi:10.1080/03014223.1999.9517597

Campbell DJ, Atkinson IAE (2002) Depression of tree recruitment by the Pacific rat (*Rattus exulans* Peale) on New Zealand's northern offshore islands. *Biological Conservation* **107**, 19–35. doi:10.1016/S0006-3207(02)00039-3

Cree A (2014) *Tuatara: Biology and Conservation of a Venerable Survivor.* Canterbury University Press, Christchurch.

Dilks P, Towns DR (2002) *Developing Tools to Detect and Respond to Rodent Invasions of Islands: Workshop Report and Recommendations.* Department of Conservation Science Internal Series 59, Wellington.

Dingwall PR, Atkinson IAE, Hay C (Eds) (1978) *The Ecology and Control of Rodents in New Zealand Nature Reserves.* Information Series No. 4. Department of Lands and Survey, Wellington.

Evans AE, Towns DR, Beggs JR (2015) The relative importance of sugar resources to endemic gecko populations in an isolated island ecosystem. *New Zealand Journal of Ecology* **39**, 262–272.

Folke C (2006) Resilience: the emergence of a perspective for socio-ecological systems analysis. *Global Environmental Change* **16**, 253–267. doi:10.1016/j.gloenvcha.2006.04.002

Fukami T, Wardle DA, Bellingham PJ, Mulder CPH, Towns DR, Yeates GW, Bonner KI, Grant-Hoffman MN, Williamson WM (2006) Above- and below-ground impacts of introduced predators in seabird-dominated island ecosystems. *Ecology Letters* **9**, 1299–1307. doi:10.1111/j.1461-0248.2006.00983.x

Galbraith M, Cooper H (2013) Tiritiri Matangi: an overview of 25 years of ecological restoration. *New Zealand Journal of Ecology* **37**, 258–260.

Heather B, Robertson H (2015) *The Field Guide to the Birds of New Zealand.* Penguin Random House, New Zealand.

Hicks GRF, McColl HP, Meads MJ, Hardy GS, Roser RJ (1975) An ecological reconnaissance of Korapuki Island, Mercury Islands. *Notornis* **22**, 195–220.

Hitchmough RA, Barr B, Lettinck M, Monks J, Reardon J, Tocher M, van Winkel D, Rolfe J (2016) *Conservation Status of New Zealand Reptiles 2015*. Department of Conservation, Wellington.

Innes J, Hay R, Flux I, Bradfield P, Speed H, Jansen P (1999) Successful recovery of North Island kokako *Callaeus cinerea wilsoni* populations by adaptive management. *Biological Conservation* **87**, 201–214. doi:10.1016/S0006-3207(98)00053-6

Joyce S, Guy N, Barry M (2016) *Predator-free NZ 2050 to be a Massive Team Effort*. Government press release. <http://www.beehive.govt.nz/release/predator-free-nz>

King C (1984) *Immigrant Killers*. Oxford University Press, Auckland.

King C (2005) *The Handbook of New Zealand Mammals*. Oxford University Press, Melbourne.

Le Corre M (2008) Rats, cats and seabirds. *Nature* **451**, 134–135. doi:10.1038/451134a

McCann KS (2000) The diversity-stability debate. *Nature* **405**, 228–233. doi:10.1038/35012234

McIvor IH, Ladefoged TN (2016) A multi-scalar analysis of Māori land use on Ahuahu (Great Mercury Island), New Zealand. *Archaeology in Oceania* **51**, 45–61. doi:10.1002/arco.5080

Mulder CPH, Anderson WB, Towns DR, Bellingham PJ (Eds) (2011) *Seabird Islands: Ecology, Invasion and Restoration*. Oxford University Press, New York.

Nogales M, Martín A, Tershy BR, Donlan JC, Veitch D, Puerta N, Wood B, Alonso J (2004) A review of feral cat eradication on islands. *Conservation Biology* **18**, 310–319. doi:10.1111/j.1523-1739.2004.00442.x

Parkes J, Murphy E (2003) Management of introduced mammals in New Zealand. *New Zealand Journal of Zoology* **30**, 335–359. doi:10.1080/03014223.2003.9518346

Parkes J, Nugent G, Forsyth DM, Byrom AE, Pech RP, Warburton B, Choquenot D (2017) Past, present and two potential futures for managing New Zealand's mammalian pests. *New Zealand Journal of Ecology* **41**, 151–161. doi:10.20417/nzjecol.41.1

Peters M, Hamilton D, Eames C (2015) Action on the ground: a review of community environmental groups' activities and partnerships in New Zealand. *New Zealand Journal of Ecology* **39**, 179–189.

Russell JC, Broome KG (2016) Fifty years of rodent eradications in New Zealand: another decade of advances. *New Zealand Journal of Ecology* **40**, 197–204. doi:10.20417/nzjecol.40.22

Russell JC, Innes JG, Brown PH, Byrom AE (2015) Predator-free New Zealand: conservation country. *Bioscience* **65**, 520–525. doi:10.1093/biosci/biv012

Saunders A (2000) *A Review of Department of Conservation Mainland Restoration Projects and Recommendations for Further Action*. Department of Conservation, Wellington.

Simberloff D (1990) Reconstructing the ambiguous: can island ecosystems be restored? In *Ecological Restoration of New Zealand Islands* (Eds DR Towns, CH Daugherty and IAE Atkinson) pp. 37–51. Department of Conservation, Wellington.

Tennyson AJD, Martinson P (2006) *Extinct Birds of New Zealand*. Te Papa Press, Wellington.

Thomas BW, Taylor RH (2002) A history of ground-based rodent eradication techniques developed in New Zealand, 1959-1993. In *Turning the Tide: The Eradication of Invasive Species* (Eds CR Veitch and MN Clout) pp. 301–310. IUCN SSC Invasive Species Specialist Group, Gland, Switzerland.

Thoresen AC (1967) Ecological observations on Stanley and Green Islands Mercury Group. *Notornis* **14**, 182–199.

Towns DR, Atkinson IAE (2004) *Restoration Plan for Korapuki Island (Mercury Islands) 2004–2024*. Department of Conservation, Wellington.

Towns DR, Broome KG (2003) From small Maria to massive Campbell: forty years of rat eradications from New Zealand islands. *New Zealand Journal of Zoology* **30**, 377–398. doi:10.1080/03014223.2003.9518348

Towns DR, Atkinson IAE, Daugherty CH (1990) The potential for ecological restoration in the

Mercury Islands. In *Ecological Restoration of New Zealand Islands* (Eds DR Towns, CH Daugherty and IAE Atkinson) pp. 91–108. Department of Conservation, Wellington.

Towns DR, Atkinson IAE, Daugherty CH (2006) Have the harmful effects of introduced rats on islands been exaggerated? *Biological Invasions* **8**, 863–891. doi:10.1007/s10530-005-0421-z

Towns DR, Parrish GR, Tyrrell CL, Ussher GT, Cree A, Newman DG, Whitaker AH, Westbrooke I (2007) Responses of tuatara (*Sphenodon punctatus*) to removal of introduced Pacific rats from islands. *Conservation Biology* **21**, 1021–1031. doi:10.1111/j1523-1739.2007.00742.x

Towns DR, Wardle DA, Mulder CPH, Yeates GW, Fitzgerald BM, Parrish GR, Bellingham PJ, Bonner KI (2009) Predation of seabirds by invasive rats: multiple indirect consequences for invertebrate communities. *Oikos* **118**, 420–430. doi:10.1111/j.1600-0706.2008.17186.x

Towns DR, West CK, Broome KG (2013) Purposes, outcomes and challenges of eradicating invasive mammals from New Zealand islands: an historical perspective. *Wildlife Research* **40**, 94–107. doi:10.1071/WR12064

Towns DR, Borrelle S, Thoresen J, Buxton R, Evans A (2016a) Mercury Islands and their role in understanding seabird island restoration. *New Zealand Journal of Ecology* **40**, 235–249. doi:10.20417/nzjecol.40.27

Towns DR, Miller KA, Nelson NJ, Chapple DG (2016b) Can translocations to islands reduce extinction risk for reptiles? Case studies from New Zealand. *Biological Conservation* **204**, 120–127. doi:10.1016/j.biocon.2016.04.024

Veale AJ, Edge K-A, McMurtrie P, Fewster RM, Clout MN, Gleeson D (2013) Using molecular techniques to quantify reinvasion, survival and *in situ* breeding rates during control operation. *Molecular Ecology* **22**, 5071–5083. doi:10.1111/mec.12453

Young D (2004) *Our Islands, Our Selves.* University of Otago Press, Dunedin.

15

Managing islands in the context of climate change

Stephen T. Garnett and April E. Reside

Summary

Climate change seems one of the grimmest of threats – slow, inexorable and inevitable. For island species, the problem seems even worse as island taxa cannot readily escape to more suitable climates. There are several sources of comfort. First, most island taxa have experienced much climate change before, including hotter times with higher sea levels. Those that could not survive have been selected against already. Second, as now, climate extremes are likely to be ameliorated by the surrounding ocean. Third, island biodiversity managers are among the most innovative. Many taxa will adapt by themselves, and should be left to do so. 'No action' may be the best action unless monitoring indicates a problem. If species are reacting adversely, the first step should be alleviating the pressure from other threats so that taxa have greater opportunity for natural selection to favour individuals best adapted to altered climates. Only then are more intensive interventions justified. Most will be *in situ* but a few taxa populations may need to be moved to more suitable climates, either on other islands or in captivity. In summary, climate change needs to be seen as another threat to be managed, but losses are not inevitable.

Introduction

That climate change is happening at a global level is unequivocal (Glikson 2016). There is empirical evidence that the changing climate is having a profound effect on biodiversity (Burrows *et al.* 2011; Poloczanska *et al.* 2013) and consequently on conservation management (Jones *et al.* 2016). There has already been one extinction attributable to sea-level rise driven by climate change (Watson 2016; Waller *et al.* 2017) – the Bramble Cay melomys (*Melomys rubicola*) – even though the extinction could have been prevented (Woinarski *et al.* 2016, 2017). This illustrates that climate change is a real and present danger to island biodiversity, alongside invasive species and habitat alienation (McCreless *et al.* 2016; Harter *et al.* 2015). It is also a pervasive long-term driver of change: all current action needs to be placed in the context of a rapidly changing climate. As a result, all conservation management needs to build in provision for anticipating climate change and reacting in time to ameliorate the threats that the changes will bring. Some of this management needs to be proactive because responding after an event, such as the waves that drowned the Bramble Cay melomys, may be too late. In other cases, delaying action will result in higher costs and fewer benefits (Hannah *et al.* 2007). Given that we are increasingly aware of the threats posed by climate change and their probabilities, a failure to take timely anticipatory management is inexcusable.

Action to counter the effects of climate change is also possible. Unlike finite manageable threats such as invasive species for which permanent solutions are often possible and proven, managing for climate change can seem overwhelming and so far beyond the control of island managers that it is tempting to focus instead on other threats. But climate change need not be disempowering. Rather, it needs to be incorporated into management in the same way as other threats. It may, however, require managers to adopt innovative approaches to conservation that would conventionally be unacceptable. Many conservation managers are still unwilling to accept innovative approaches (Hagerman *et al.* 2010) but, as chapters in this book attest, radical approaches have often been embraced by people managing islands (Alderman and Hobday 2016). As a result, they are often better equipped in terms of skills and 'can-do' attitudes than many managers of protected areas and threatened species in places where threats have not yet required the desperate conservation interventions too often necessary on islands.

This chapter reviews the interaction between climates and biodiversity, the impacts that can be anticipated and potential solutions. Then, using birds as a lens, we review the actions required to retain Australian island species that models suggest are the most vulnerable to climate change over the next 50 years. We conclude with a summary of approaches to management and climate change planning principles.

Past climate change and island biodiversity

The first point about managing islands in the face of climate change is to appreciate that island biodiversity is a product of past climate change – indeed, many islands exist only because of relatively recent sea-level rise. Despite often having small population sizes with low genetic capacity to cope with change and low functional redundancy or available area of habitat, many species must have persisted through a range of climates during their evolutionary history. For instance, following the last glacial maximum ~20 000 years BP, when the sea level was ~125 m lower than today and continental islands were connected to mainland Australia, sea levels sometimes rose at rates far faster than predicted for the near future: between 14 300 BP and 14 600 BP it rose ~5 cm a year, totalling ~16 m over the 300 years (Hanebuth *et al.* 2000, 2009). This means that coastal-dependent species have proven sufficiently adaptable in the past to survive rapid sea-level rise under the conditions at the time. Many island taxa also persisted on islands that were much smaller than they are today. During the Eemian, just 120 000 years ago – well within the evolutionary history of most taxa – sea levels were 10 m higher than they are currently (Rijsdijk *et al.* 2014). For continental islands, extinctions could have been reversed by recolonisation during subsequent glacial lowstands, but around Australia sea levels were over a metre higher than current levels for at least a thousand years, around 6000 BP (Lewis *et al.* 2013). During that time all islands would have been smaller than they are now. This fluctuation in sea levels is reflected in the higher than expected number of endemics on islands that were once much larger (Weigelt *et al.* 2016).

Other climatic trends that have affected the evolutionary history of island species are reductions in rainfall during the Pleistocene, when large volumes of the world's freshwater were frozen, the world experienced both lower and higher temperatures, major climatic drivers like the monsoon and the El Niño Southern Oscillation changed (Reeves *et al.* 2013) and patterns of winds and currents differed because major marine connections between oceans were closed (e.g. Torres and Bass Straits). However, unlike sea-level rise, our understanding of these patterns remains incomplete (Kohfeld *et al.* 2013), and we have little knowledge of their impact on the climate of individual islands. What we do know is that all island taxa have persisted through a range of climates during their evolutionary history.

Box 15.1: Helping Australian island birds cope with climate change

A recent analysis of the effect of climate change on Australian birds found that current climatic conditions of 101 Australian terrestrial and inland water bird taxa would no longer overlap their current range by 2085, that 16 marine birds have breeding sites for which the surrounding seas are predicted to be at least 10% less productive than today, and 55 terrestrial taxa are likely to be exposed to more frequent or intense fires (Garnett *et al.* 2013; Garnett and Franklin 2014). Among the places where birds were most exposed to climate change were Tiwi, King and Kangaroo Islands while marine birds nesting on Lord Howe and Norfolk Islands, the Great Barrier Reef and the Houtman Abrolhos were thought likely to face substantial declines in local marine productivity. Many seabirds were considered potentially highly sensitive to climate change, based on a set of ecological and morphological metrics, while small island taxa were the most likely to be both exposed and sensitive to climate change, followed by marine taxa, most of which use islands for breeding.

An adaptation plan was drawn up for each of the 101 most vulnerable bird taxa, of which 34 occurred on islands, 26 exclusively, classifying potential actions into those that could usefully be initiated immediately and those that may be necessary in the future. These actions were then divided into those that constituted single, one-off projects and those that were likely to be ongoing, such as monitoring. Finally, all projects were costed over a 50-year period at current costs (Table 15.1).

All taxa require monitoring from now onwards and a baseline survey was required, to enable monitoring to begin. Without monitoring, interventions cannot be justified and may not be needed. One intervention – for the orange-bellied parrot (*Neophema chrysogaster*) – was likely to need the purchase of habitat likely to support salt marsh as the sea level rises. The major costs, however, were for the eight taxa likely to require retention in captivity indefinitely (about A$420 million, largely for construction of appropriate facilities) and ongoing research to follow seabirds (about A$1.2 million) and determine whether and where greater regulation of fishing is required.

Table 15.1. Management actions and their estimated costs (A$'000) recommended for Australian island bird taxa predicted to be the most vulnerable to climate change up to 2070

Timing	Extent	No. of taxa	Habitat enhancement	Intensive management	Preservation	Research	Monitoring
Immediate	Defined	15	1600 (3)	1000 (1)	–	584 (12)	10 (1)
	Ongoing	34	–	107 (5)	220 (1)	-	2716 (34)
Future	Defined	18	–	2700 (3)	–	14 756 (26)	–
	Ongoing	20	371 (8)	4499 (17)	–	–	–
Total		34	1971 (8)	8306 (18)	220 (1)	15 340 (22)	2726 (34)

Immediate actions are those that should start now, future actions should start when there is evidence of climate change impacts. Defined actions are one-off; ongoing actions continue indefinitely. Figures in parentheses are number of taxa, with some taxa in multiple categories. Source: Garnett *et al.* (2013).

Indeed, for birds in Wallacea and the West Indies, Dalsgaard *et al.* (2014) found that the current fauna reflects current conditions with no echo of past climates, and suggested that climatic variation during the Pleistocene may have been too mild to induce extinction among endemic species. This means that, barring additional pressures, many species have an inherent capacity to cope with a wide range of climates, including those soon to be visited upon modern humans. Climate change will be a problem for some taxa (Box 15.1) but largely because the impacts are exacerbated by other changes wrought by humans.

Modelling future island climates

Modelling future changes in climate is done at a global level. Regionalisation of climate models is improving for mainland areas but is by no means certain (Fowler *et al.* 2007), and robust atmospheric climate data are available only for major landmasses such as continental Australia and some larger associated continental islands (Franklin *et al.* 2014). Nevertheless, a range of climate impacts that will affect island taxa can be predicted with modelling (CSIRO and Bureau of Meteorology 2015) for Australian regions and are likely to apply to nearby continental islands. These include temperature rises of 0.5–1.4°C by 2030 and up to 5°C by 2090, more frequent atmospheric and oceanic heat waves, heavier rainfall (when it occurs), more severe bushfire weather in the south and less frequent but higher-intensity cyclones in the north. The greatest uncertainty is around rainfall, with lower winter and spring rainfall predicted in the south and east, summer and autumn rainfall trends uncertain, and wet season rainfall uncertain in northern Australia.

The climate of Australia's six oceanic island groups – Norfolk, Lord Howe, Macquarie, Heard, Christmas and Cocos (Keeling) – is tied closely to surrounding marine environments. Given that the sea is likely to increase in temperature more slowly than land, the climate of these islands may be buffered from temperature extremes, as appears to be the case with biodiversity hotspots in New Caledonia (Pouteau and Birnbaum 2016). Furthermore, for Australian islands, projections of climate models suggest that changes in marine productivity in the southern hemisphere are likely to be lower than in the north (Cabré *et al.* 2015). Patterns of, and projected change in, marine productivity in Australian waters are complex (Steinacher *et al.* 2010) and constantly being refined. For example, recent and better models on eddy dynamics suggest that the Tasman Sea, previously expected to become less productive, may become more so (Matear *et al.* 2013).

Sea level will certainly rise. As evidenced by the Bramble Cay melomys, higher seas are already causing extinctions on low-lying atolls. However, such endemism is rare on atolls, which have always been relatively transient as sea level has fluctuated. By contrast, most Australian islands extend well above most projections of sea-level rise for the rest of the 21st century. Elsewhere, the impact of sea-level rise on biodiversity will be greatly exacerbated by the response of island peoples to occupy lands further inshore (Wetzel *et al.* 2012). However, most Australian islands with substantial biodiversity have small human populations, with little likelihood of major additional impacts on biodiversity should settlements be displaced.

Solutions

Potential options for conserving biodiversity under climate change have been reviewed by Mawdsley *et al.* (2009) and Shoo *et al.* (2013), and incorporating climate change into systematic conservation planning has been reviewed by Jones *et al.* (2016). Options include doing nothing in the hope that species will respond to climate change without assistance, maintaining and enhancing existing habitat, intensive species management and preserving populations *ex situ*. Underpinning all these strategies is the need to monitor so that we can understand how species and ecosystems are responding to the changes in climate, so that strategies can be adapted as required and intervention hastened.

Do nothing

Doing nothing to help species cope with climate change can sometimes be the best strategy because they may be able to cope with change on their own. However, monitoring needs to be in place, as well as contingency plans for when monitoring indicates intervention is necessary. As noted, most island species have had to cope with climate change in the past so may have retained an inherent capacity to cope in the future. Of course, the difference is that humans have greatly altered the environment of many island species, reducing ecosystem resilience and exposing vulnerabilities not previously evident (Brook 2008).

Maintain and enhance habitat

The least specific and easiest intervention is to increase ecosystem resilience, which generally means maintaining habitat in a favourable state, and reducing threats. Where habitat loss is the most pressing threat, the simplest and most passive form of this adaptation strategy is to expand the protected area estate. Retaining areas of natural habitat in protected areas is known to be an effective strategy for conserving species under climate change (Thomas *et al.* 2012; Thomas and Gillingham 2015). However, expanding the protected area estate without at least commensurate increases in management funds can be worse than doing nothing. More intensive is active habitat management to improve the quality for targeted species. This can include restoring degraded or lost habitat, for example replanting or controlling weeds to allow for natural regeneration. For Australian environments, this can include more active fire management, where burning has been poorly controlled, or invasive species control. For some islands, strengthening of biosecurity can significantly reduce the risks of invasive species establishment. These are standard procedures for conservation managers whether on islands or elsewhere.

Similarly, proactive identification of refugia followed by their protection is a management action that pre-empts climate change while benefiting current conservation objectives. For example, Keppel *et al.* (2015) applied a three-step process in Tasmania that involved:

- defining the scope, scale and resolution of refugia;
- identifying and quantifying their characteristics;
- prioritising them for conservation.

The most important refugia for plants on the largest of Australian islands were in cool, wet and topographically complex sites. For smaller islands, refugia are likely to be at correspondingly reduced scales with a heavy reliance on microclimatic conditions (Ashcroft *et al.* 2009, 2012). They will be at least partly defined by past land use with few alternatives, such as the last substantial remnants of Norfolk Island flora in the island's national park. Others will long ago have been identified as refugia through traditional knowledge, such as spring-fed rainforests on the Tiwi Islands (Tacon 2010; Hoverman and Ayre 2012). Such sites are refugia under current conditions, and will continue to be so as the climate changes.

Finally, habitat enhancement can include the creation of new habitats. Novel ecosystems are emerging as a direct result of climate change (Chapin and Starfield 1997) and potentially have an important role to play in climate change adaptation (Hobbs *et al.* 2009). Deliberate creation of novel ecosystems in the face of climate change can enhance both biodiversity conservation and the provision of ecosystem services (Lunt *et al.* 2013).

One management strategy that is commonly cited for mainland habitats but has little relevance for islands (except perhaps for the largest island landmasses) is the maintenance and extension of landscape connectivity to facilitate species moving to track suitable climate. In fact, in many cases islands maintain their qualities because they lack connectivity. Their isolation is both their major vulnerability and their source of distinctiveness.

Intensive species management

The next level up in intensity of management is to adopt approaches that aid target species rather than the habitats in which they live. This includes many of the standard management practices applied to islands and practices that might equally be grouped with habitat enhancement, such as predator control. The difference is one of degree: some invasive species control programs assume benefits to a range of species in the ecosystem, while others closely target invasive species known to have a deleterious effect on target native species.

This category of interventions also includes some approaches such as assisted colonisation, which has been used in conservation since the 19th century. However, this approach is now being used to avoid species going extinct because they are unable to disperse sufficiently to keep up with changing climates. While the introduction of novel species to an island ecosystem is often bad for species established there naturally, shifts in climate will mean that islands are sometimes the only place where the environment is likely to remain suitable. For example, the current distribution of the rufous scrub-bird (*Atrichornis rufescens*) is in north-east New South Wales and south-east Queensland, where the climate is thought likely to become unsuitable for this species over the next 50 years, but parts of Tasmania may develop a suitable climate (Garnett *et al.* 2014). A key element of any such intervention will not only be the standard risk assessments that are undertaken for translocations but policy changes that allow movements of taxa between states (Burbidge *et al.* 2011) and surveys of social acceptability (Garnett *et al.* 2017).

More subtle is the potential use of genetic enhancement to improve the capacity of species to adapt to climate change. The increasing sophistication of genetic analysis is allowing identification of genes that enable species to cope with increased temperatures (Grover *et al.* 2013) or drought (Deikman *et al.* 2012). It may also allow the introduction of such genes into vulnerable native populations in the same way as they are being engineered into commercial crops. Less targeted genetic enhancement is already being proposed for some small populations, such as those of the helmeted honeyeater (*Lichenostomus melanops cassidix*) (Harrisson *et al.* 2016), because levels of inbreeding are so high. New CRISPR-9 and gene drive technologies are allowing far more targeted interventions for gene fragments of increasing length. They create opportunities not only to introduce favourable genes to stressed populations but also to enhance invasive species control – although not without controversy (Webber *et al.* 2015; Corlett 2017).

Preserving populations

Ultimately some species may become so scarce and the probabilities of survival in the wild so low that their only chance will be to remove some individuals out of the wild population until conditions have ameliorated. *Ex situ* conservation can occur in two ways – through maintaining and breeding adult organisms and through the preservation of tissue or seeds.

For threatened plants, the contribution of *ex situ* collections of live plants to conservation is so well established (Knapp *et al.* 2014) that, by 2020, the Global Strategy for Plant Conservation (Convention on Biological Diversity 2011) aspires to have 'at least 75 per cent of threatened plant species in *ex situ* collections, preferably in the country of origin, and at least 20 per cent available for recovery and restoration programmes'. In relation to animals, the strategy is less accepted. While careful breeding using stud books to maximise retention of genetic variability can work for short periods, long-term consequences can be substantial because the process of captivity inevitably leads to selection of individuals that thrive under such circumstances, rather than under wild conditions (Tschirren *et al.* 2009). This potentially compromises return to the wild, should circumstances ever allow (Araki *et al.* 2007).

Plants also lead the way in cryopreservation, both for tissue (Popova *et al.* 2015) and seed storage (Fu *et al.* 2015). The role of this form of *ex situ* conservation for animals is still being debated but the steady advances in techniques suggest that such approaches may soon become reliable tools for genetic conservation (Mandawala *et al.* 2016).

Supporting tools

There are data and tools available to help guide selection and prioritisation of management actions for species conservation under climate change. For example, models can help predict the location of future suitable climate for species, and these are available online (http://climas.hpc.jcu.edu.au/ for Australian birds, Garnett *et al.* 2014) and for other vertebrates (Reside *et al.* 2013). Tools are available to use these climate predictions for species, to prioritise areas to be the focus of conservation action (Ball *et al.* 2009; Moilanen *et al.* 2014) and to prioritise the actions themselves (Pouzols and Moilanen 2013), bringing in priorities for species currently and in the future. Importantly, these tools can help quantify uncertainty surrounding the climate predictions, such as unknown severity of climate change, species' ability to disperse to new habitats and the population responses. Furthermore, the tools can assist with examining the trade-offs between different actions, or acting in different areas or at different times, and help to identify synergies between proposed actions. More data are clearly needed, as many Australian islands have not been extensively surveyed or monitored for plants or invertebrates (Chapter 3). However, the existing data and tools provide a good starting point that may reflect the broader changes across the island ecosystems.

Australian adaptations

Using birds as a basis (as an adaptation plan has been developed for all Australian birds including seabirds; Garnett *et al.* 2014), two approaches illustrate the range of options for managing climate change on islands. One is to describe the actions recommended to help Australia's most vulnerable taxa (Box 15.1). The other is to describe what might be done for a set of exemplar islands.

Case study islands

Houtman Abrolhos

This set of low-lying islands off the Western Australian coast supports large populations of seabirds, including the only population of the Australian lesser noddy (*Anous tenuirostris melanops*). There are also distinctive populations of landbirds including a subspecies of painted buttonquail (*Turnix varius scintillans*) that is among the taxa most likely to go extinct in the next 20 years (H. Geyle, *pers. comm.*) because of habitat degradation by introduced wallabies. While sea-level rise is unlikely to be a problem over the next 50 years (Zhu *et al.* 1993), shifts in the Leeuwin Current may affect both the availability of food in surrounding waters (Surman and Nicholson 2009) and the climatic suitability of the islands for the distinctive landbirds. Apart from more frequent monitoring, options for adaptation include supplementary feeding of the buttonquail, eradicating introduced wallabies and the establishment of new colonies of the lesser noddy on islands further south.

Norfolk Island

The climate of Norfolk Island is expected to be hotter, potentially with longer droughts (Director of National Parks 2011). Drought may reduce food availability for the distinctive suite of landbirds that persist in the small national park that is all that remains of the forest that once covered the island. Among the birds most threatened by climate change are the local subspecies of boobook (*Ninox novaeseelandiae undulata*), which was once reduced to a single female (Garnett *et al.* 2011). Hot dry weather could also increase the risk of fire, especially from a grove of eucalypts planted beside the native forest, which could destroy owl nesting habitat. Adaptation options include assisted colonisation of Lord Howe Island, where a closely related subspecies of the owl was extirpated in the 1930s, and replacement of the eucalypt plantation with local rainforest species. The island group also supports several seabird species which forage in seas expected to decline substantially in productivity (Steinacher *et al.* 2010). Options include tracking of foraging individuals to identify feeding sites, which may vary in location both within and between years, to identify areas that may require conservation action such as regulating fishing or provision of supplementary feeding.

Kangaroo Island

Models predict that Kangaroo Island will become drier, resulting in a climate unlike that which currently occurs there. This may make it difficult for a suite of endemic bird subspecies to persist. There is the option of establishing a costly captive population in perpetuity, which would be difficult for the endangered glossy black-cockatoo (*Calyptorhynchus lathami halmaturinus*), of which there are only a few hundred left (Crowley *et al.* 1996). Rather, for this and other island subspecies, genetic supplementation may be appropriate, importing genes from mainland populations that live in climates more closely resembling that predicted to occur on Kangaroo Island. Alternatively, cockatoos could be re-established on the mainland, where they used to occur in the 19th century. Reintroduced populations could be supported by irrigating and fertilising food trees to increase food availability, and refugia for the cockatoos could be identified through modelling and provided with special protection.

Tiwi Islands

The Tiwi Islands have a wetter climate than most of the adjacent mainland and contain distinctive fauna that includes endemic subspecies of masked owl (*Tyto novaehollandiae melvillensis*) and yellow-tinted honeyeater (*Ptilotula flavescens melvillensis*). Models predict a drying of the climate, so that the best future climate space for the Tiwi Islands taxa is Cape York Peninsula. However, long before monitoring reveals any climate-related decline, there is time to create detailed models from which refugia can be identified. These could then be the focus of management by the Tiwi natural resource management group, minimising threats and maximising the birds' productivity. The other alternatives are genetic supplementation, as on Kangaroo Island, or captive breeding.

Tasmania

Tasmania will be subject to less extreme warming than other parts of the country; therefore, one of the issues it will face is whether to receive biodiversity refugees from places further north that are likely to become too hot and dry. Such decisions are more political and ethical than technical, requiring a social licence if they are to be successful: initial research suggests that most Tasmanians are willing to entertain the idea (Garnett *et al.* 2017), even if they blanch closer to the event. However, there are technical questions on the potential for overstocking of refugia, some of which have already been identified for plants (Keppel *et al.* 2015). For instance, Tasmania may contain the last areas with habitat suitable for both species of scrub-bird – the noisy scrub-bird (*Atrichornis clamosus*) from south-western Australia and the rufous scrub-bird (*A. rufescens*) inland from Brisbane. One of the first stages of planning must therefore be some form of prioritisation. The other distinctive climate adaptation strategy for a Tasmanian bird is for the orange-bellied parrot (*Neophema chrysogaster*), a species that feeds largely in coastal salt marshes during the non-breeding season, including in northern and western Tasmania. Such areas will become inundated as the sea level rises. One potential adaptation response is to purchase land that is unlikely to be inundated when the seas rise and can act as a land bank for the parrots in the future.

King Island

Once covered in forest, the remnants of native vegetation support two bird taxa from the top five most likely to go extinct in the next 20 years (H. Gelye, *pers. comm.*) – the King Island subspecies of brown thornbill (*Acanthiza pusilla archibaldi*) and King Island scrubtit (*Acanthornis magnus greenianus*). As with the orange-bellied parrot, both need active management regardless of climate change. In fact, they are typical of many island species – in need of active conservation management to counter a whole suite of non-climate-related threats so that when climate change does start to become a more prominent driver of ecological processes, the King Island taxa have as much genetic diversity as possible on which natural selection can act.

Concluding comments

Climate change is one of many threats that conservation managers of islands need to incorporate into all their plans. The evolutionary history of the climates through which species have survived gives hope that they may cope with climate change again. Many, however, will need help. This can range from more passive habitat protection and improvement, though active interventions of the sort familiar to most island managers such as invasive animal and plant eradication or control. For a small number of taxa, *ex situ* conservation will be the most appropriate strategy, but that is likely to be some time coming. The most important message from this chapter, therefore, is that climate change is one of the more manageable environmental challenges that managers are likely to encounter in their working lives.

Fortunately, for Australian islands the dangers of climate change are far less than elsewhere because their human population are small, and probably diminishing as part of an ongoing process of urbanisation. Indeed, the danger in some areas is not that there are too many people but that there are too few with the skills and interest to undertake the necessary conservation management. Probably the single most important thing managers can do is to monitor what they have now so they have timely evidence of change as it occurs, which can act as triggers to put long-established contingency plans into play when needed.

Acknowledgements

We would like to thank Claire Johnston for editorial assistance and acknowledge the contributions of Don Franklin and others in the original climate change adaptation project.

References

Alderman R, Hobday AJ (2016) Developing a climate adaptation strategy for vulnerable seabirds based on prioritisation of intervention options. *Deep-Sea Research* **140**, 290–297. doi:10.1016/j.dsr2.2016.07.003

Araki H, Cooper B, Blouin MS (2007) Genetic effects of captive breeding cause a rapid, cumulative fitness decline in the wild. *Science* **318**, 100–103. doi:10.1126/science.1145621

Ashcroft MB, Chisholm LA, French KO (2009) Climate change at the landscape scale: predicting fine-grained spatial heterogeneity in warming and potential refugia for vegetation. *Global Change Biology* **15**, 656–667. doi:10.1111/j.1365-2486.2008.01762.x

Ashcroft MB, Gollan JR, Warton DI, Ramp D (2012) A novel approach to quantify and locate potential microrefugia using topoclimate, climate stability, and isolation from the matrix. *Global Change Biology* **18**, 1866–1879. doi:10.1111/j.1365-2486.2012.02661.x

Ball I, Possingham HP, Watts ME (2009) Marxan and relatives: software for spatial conservation prioritization. In *Spatial Conservation Prioritization: Quantitative Methods and Computational Tools* (Eds A Moilanen, KA Wilson and HP Possingham) pp. 185–195. Oxford University Press, Oxford.

Brook BW (2008) Synergies between climate change, extinctions and invasive vertebrates. *Wildlife Research* **35**, 249–252. doi:10.1071/WR07116

Burbidge AA, Byrne B, Coates C, Garnett ST, Hars S, Hayward MW, Martin TG, McDonald-Madden E, Mitchell NJ, Nally S, Setterfield SA (2011) Is Australia ready for assisted colonization? Policy changes required to facilitate translocations under climate change. *Pacific Conservation Biology* **17**, 259–269. doi:10.1071/PC110259

Burrows MT, Schoeman DS, Buckley LB, Moore P, Poloczanska ES, Brander KM, Brown C, Bruno JF, Duarte CM, Halpern BS, Holding J (2011) The pace of shifting climate in marine and terrestrial ecosystems. *Science* **334**, 652–655. doi:10.1126/science.1210288

Cabré A, Marinov I, Leung S (2015) Consistent global responses of marine ecosystems to future climate change across the IPCC AR5 earth system models. *Climate Dynamics* **45**, 1253–1280. doi:10.1007/s00382-014-2374-3

Chapin FS, Starfield AM (1997) Time lags and novel ecosystems in response to transient climatic change in arctic Alaska. *Climatic Change* **35**, 449–461. doi:10.1023/A:1005337705025

Convention on Biological Diversity (2011) *Updated Global Strategy for Plant Conservation 2011–2020.* <https://www.cbd.int/gspc/strategy.shtml>

Corlett RT (2016) A bigger toolbox: biotechnology in biodiversity conservation. *Trends in Biotechnology* **35**, 55–65. doi:10.1016/j.tibtech.2016.06.009

Crowley GM, Garnett ST, Pedler LP (1996) *Assessment of the Role of Captive Breeding and Translocation in the Recovery of the South Australian Subspecies of the Glossy Black-cockatoo Calyptorhynchus lathami halmaturinus.* Report No. 5. Birds Australia, Melbourne.

CSIRO and Bureau of Meteorology (2015) *Climate Change in Australia: Information for Australia's Natural Resource Management Regions.* CSIRO and Bureau of Meteorology, Canberra.

Dalsgaard B, Carstensen DW, Fjeldså J, Maruyama PK, Rahbek C, Sandel B, Sonne J, Svenning J-C, Wang Z, Sutherland WJ (2014) Determinants of bird species richness, endemism, and island network roles in Wallacea and the West Indies: is geography sufficient or does current and historical climate matter? *Ecology and Evolution* **4**, 4019–4031. doi:10.1002/ece3.1276

Deikman J, Petracek M, Heard JE (2012) Drought tolerance through biotechnology: improving translation from the laboratory to farmers' fields. *Current Opinion in Biotechnology* **23**, 243–250. doi:10.1016/j.copbio.2011.11.003

DNP (2011) *Norfolk Island National Park and Botanic Garden Climate Change Strategy 2011–2016*. Director of National Parks, Department of Sustainability, Environment, Water, Population and Communities, Canberra.

Fowler HJ, Blenkinsop S, Tebaldi C (2007) Linking climate change modelling to impacts studies: recent advances in downscaling techniques for hydrological modelling. *International Journal of Climatology* **27**, 1547–1578. doi:10.1002/joc.1556

Franklin DC, Ehmke G, VanDerWal J, Garnett ST (2014) The exposure of Australian birds to climate change. In *Climate Change Adaptation Plan for Australian Birds* (Eds ST Garnett and DC Franklin) pp. 7–26. CSIRO Publishing, Melbourne.

Fu YB, Ahmed Z, Diederichsen A (2015) Towards a better monitoring of seed ageing under *ex situ* seed conservation. *Conservation Physiology* **3**, cov026.

Garnett ST, Franklin DC (2014) *Climate Change Adaptation Plan for Australian Birds*. CSIRO Publishing, Melbourne.

Garnett ST, Olsen P, Butchart SHM, Hoffmann AA (2011) Did hybridization save the Norfolk Island boobook owl *Ninox novaeseelandiae undulata*? *Oryx* **45**, 500–504. doi:10.1017/S0030605311000871

Garnett ST, Franklin DC, Ehmke G, VanDerWal JJ, Hodgson L, Pavey C, Reside AE, Welbergen JA, Butchart SHM, Perkins GC, Williams SE (2013) *Climate Change Adaptation Strategies for Australian Birds*. National Climate Change Adaptation Research Facility, Gold Coast.

Garnett ST, Pavey CR, Ehmke G, VanDerWal J, Hodgson L, Franklin DC (2014) Adaptation outlines for species that are both highly sensitive and highly exposed. In *Climate Change Adaptation Plan for Australian Birds* (Eds ST Garnett and DC Franklin) pp. 79–258. CSIRO Publishing, Melbourne.

Garnett ST, Zander KK, Hagerman S, Satterfield T, Meyerhoff J (2017) Social preferences for adaptation measures to conserve Australian birds threatened by climate change. *Oryx* doi:10.1017/S0030605316001058

Glikson A (2016) Cenozoic mean greenhouse gases and temperature changes with reference to the Anthropocene. *Global Change Biology* **22**, 3843–3858. doi:10.1111/gcb.13342

Grover A, Mittal D, Negi M, Lavania D (2013) Generating high temperature tolerant transgenic plants: achievements and challenges. *Plant Science* **205**, 38–47. doi:10.1016/j.plantsci.2013.01.005

Hagerman S, Dowlatabadi H, Satterfield T, McDaniels T (2010) Expert views on biodiversity conservation in an era of climate change. *Global Environmental Change* **20**, 192–207. doi:10.1016/j.gloenvcha.2009.10.005

Hanebuth T, Stattegger K, Grootes PM (2000) Rapid flooding of the Sunda Shelf: a late-glacial sea-level record. *Science* **288**, 1033–1035. doi:10.1126/science.288.5468.1033

Hanebuth TJJ, Stattegger K, Bojanowski A (2009) Termination of the Last Glacial Maximum sea-level lowstand: the Sunda Shelf data revisited. *Global and Planetary Change* **66**, 76–84. doi:10.1016/j.gloplacha.2008.03.011

Hannah L, Midgley G, Andelman S, Araújo M, Hughes G, Martinez-Meyer E, Pearson R, Williams P (2007) Protected area needs in a changing climate. *Frontiers in Ecology and the Environment* **5**, 131–138. doi:10.1890/1540-9295(2007)5[131:PANIAC]2.0.CO;2

Harrisson KA, Pavlova A, Gonçalves da Silva A, Rose R, Bull JK, Lancaster ML, Murray N, Quin B, Menkhorst P, Magrath MJ, Sunnucks P (2016) Scope for genetic rescue of an endangered subspecies through re-establishing natural gene flow with another subspecies. *Molecular Ecology* **25**, 1242–1258. doi:10.1111/mec.13547

Harter DE, Irl SD, Seo B, Steinbauer MJ, Gillespie R, Triantis KA, Fernández-Palacios JM, Beierkuhnlein C (2015) Impacts of global climate change on the floras of oceanic islands: projections, implications and current knowledge. *Perspectives in Plant Ecology, Evolution and Systematics* **17**, 160–183. doi:10.1016/j.ppees.2015.01.003

Hobbs RJ, Higgs E, Harris JA (2009) Novel ecosystems: implications for conservation and restoration. *Trends in Ecology and Evolution* **24**, 599–605. doi:10.1016/j.tree.2009.05.012

Hoverman S, Ayre M (2012) Methods and approaches to support Indigenous water planning: an example from the Tiwi Islands, Northern Territory, Australia. *Journal of Hydrology* **474**, 47–56. doi:10.1016/j.jhydrol.2012.03.005

Jones KR, Watson JEM, Possingham HP, Klein CJ (2016) Incorporating climate change into spatial conservation prioritisation: a review. *Biological Conservation* **194**, 121–130. doi:10.1016/j.biocon.2015.12.008

Keppel G, Monkany K, Wardell-Johnson G, Phillips B, Welbergen J, Reside A (2015) The capacity of refugia for conservation planning under climate change. *Frontiers in Ecology and the Environment* **13**, 106–112. doi:10.1890/140055

Knapp Z, Boardman L, Brown B, West J (2014) What are we conserving? Living collections contributing to target 8 of the global strategy for plant conservation. *Australasian Plant Conservation: Journal of the Australian Network for Plant Conservation* **22**, 11–13.

Kohfeld KE, Graham RM, De Boer AM, Sime LC, Wolff EW, Le Quéré C, Bopp L (2013) Southern Hemisphere westerly wind changes during the Last Glacial Maximum: paleo-data synthesis. *Quaternary Science Reviews* **68**, 76–95. doi:10.1016/j.quascirev.2013.01.017

Lewis SE, Sloss CR, Murray-Wallace CV, Woodroffe CD, Smithers SG (2013) Post-glacial sea-level changes around the Australian margin: a review. *Quaternary Science Reviews* **74**, 115–138. doi:10.1016/j.quascirev.2012.09.006

Lunt ID, Byrne M, Hellmann JJ, Mitchell NJ, Garnett ST, Hayward MW, Martin TG, McDonald-Madden E, Williams SE, Zander KK (2013) Using assisted colonisation to conserve biodiversity and restore ecosystem function under climate change. *Biological Conservation* **157**, 172–177. doi:10.1016/j.biocon.2012.08.034

Mandawala AA, Harvey SC, Roy TK, Fowler KE (2016) Cryopreservation of animal oocytes and embryos: current progress and future prospects. *Theriogenology* **86**, 1637–1644. doi:10.1016/j.theriogenology.2016.07.018

Matear RJ, Chamberlain MA, Sun C, Feng M (2013) Climate change projection of the Tasman Sea from an eddy-resolving ocean model. *Journal of Geophysical Research: Oceans* **118**, 2961–2976.

Mawdsley JR, O'Malley R, Ojima DS (2009) A review of climate-change adaptation strategies for wildlife management and biodiversity conservation. *Conservation Biology* **23**, 1080–1089. doi:10.1111/j.1523-1739.2009.01264.x

McCreless EE, Huff DD, Croll DA, Tershy BR, Spatz DR, Holmes ND, Butchart SHM, Wilcox C (2016) Past and estimated future impact of invasive alien mammals on insular threatened vertebrate populations. *Nature Communications* **7**, 12488. doi:10.1038/ncomms12488

Moilanen A, Pouzols FM, Meller L, Veach V, Arponen A, Leppanen J, Kujala H (2014) *Zonation Spatial Conservation Planning Methods and Software: User Manual*. Version 4. C-BIG Conservation Biology Informatic Group, University of Helsinki, Finland. <www.cbig.it.helsinki.fi/software, www.cbig.it.helsinki.fi/software>

Poloczanska ES, Brown CJ, Sydeman WJ, Kiessling W, Schoeman DS, Moore PJ, Brander K, Bruno JF, Buckley LB, Burrows MT, Duarte CM (2013) Global imprint of climate change on marine life. *Nature Climate Change* **3**, 919–925. doi:10.1038/nclimate1958

Popova E, Shukla M, Kim HH, Saxena PK (2015) Plant cryopreservation for biotechnology and breeding. In *Advances in Plant Breeding Strategies: Breeding, Biotechnology and Molecular Tools* (Eds JM Al-Khayri, SH Jain and DV Johnson)

pp. 63–93. Springer International Publishing, Switzerland.

Pouteau R, Birnbaum P (2016) Island biodiversity hotspots are getting hotter: vulnerability of tree species to climate change in New Caledonia. *Biological Conservation* **201**, 111–119. doi:10.1016/j.biocon.2016.06.031

Pouzols FM, Moilanen A (2013) RobOff: software for analysis of alternative land-use options and conservation actions. *Methods in Ecology and Evolution* **4**, 426–432. doi:10.1111/2041-210X.12040

Reeves JM, Bostock HC, Ayliffe LK, Barrows TT, De Deckker P, Devriendt LS, Dunbar GB, Drysdale RN, Fitzsimmons KE, Gagan MK, Griffiths ML (2013) Palaeoenvironmental change in tropical Australasia over the last 30,000 years: a synthesis by the OZ-INTIMATE group. *Quaternary Science Reviews* **74**, 97–114. doi:10.1016/j.quascirev.2012.11.027

Reside AE, VanDerWal J, Phillips BL, Shoo LP, Rosauer D, Anderson BJ, Welbergen JA, Moritz C, Ferrier S, Harwood TD, Williams KJ, Mackey BG, Hugh S, Williams SE (2013) *Climate Change Refugia for Terrestrial Biodiversity: Defining Areas that Promote Species Persistence and Ecosystem Resilience in the Face of Global Climate Change.* National Climate Change Adaptation Research Facility, Gold Coast.

Rijsdijk KF, Hengl T, Norder SJ, Otto R, Emerson BC, Ávila SP, López H, Loon EE, Tjørve E, Fernández-Palacios JM (2014) Quantifying surface-area changes of volcanic islands driven by Pleistocene sea-level cycles: biogeographical implications for the Macaronesian archipelagos. *Journal of Biogeography* **41**, 1242–1254. doi:10.1111/jbi.12336

Shoo LP, Hoffmann AA, Garnett ST, Pressey RL, Williams YM, Taylor M, Falconi L, Yates CJ, Scott JK, Alagador D, Williams SE (2013) Making decisions to conserve biodiversity under climate change. *Climatic Change* **119**, 239–246. doi:10.1007/s10584-013-0699-2

Steinacher M, Joos F, Frolicher TL, Bopp L, Cadule P, Cocco V, Doney SC, Gehlen M, Lindsay K, Moore JK, Schneider B (2010) Projected 21st century decrease in marine productivity: a multimodel analysis. *Biogeosciences* **7**, 979–1005. doi:10.5194/bg-7-979-2010

Surman CA, Nicholson LW (2009) The good, the bad and the ugly: ENSO driven oceanographic variability and its influence on seabird diet and reproductive performance at the Houtman Abrolhos, eastern Indian Ocean. *Marine Ornithology* **37**, 129–138.

Tacon P (2010) Identifying ancient sacred landscapes in Australia: from physical to social. In *Contemporary Archaeology in Theory: The New Pragmatism* (Eds RW Preucel and SA Mrozowski) pp. 77–91. John Wiley and Sons, New York.

Thomas CD, Gillingham PK (2015) The performance of protected areas for biodiversity under climate change. *Biological Journal of the Linnean Society* **115**, 718–730. doi:10.1111/bij.12510

Thomas CD, Gillingham PK, Bradbury RB, Roy DB, Anderson BJ, Baxter JM, Bourn NAD, Crick HQP, Findon RA, Fox R, Hodgson JA, Holt AR, Morecroft MD, O'Hanlon NJ, Oliver TH, Pearce-Higgins JW, Procter DA, Thomas JA, Walker KJ, Walmsley CA, Wilson RJ, Hill JK (2012) Protected areas facilitate species' range expansions. *Proceedings of the National Academy of Sciences of the United States of America* **109**, 14063–14068. doi:10.1073/pnas.1210251109

Tschirren B, Rutstein AN, Postma E, Mariette M, Griffith SC (2009) Short- and long-term consequences of early developmental conditions: a case study on wild and domesticated zebra finches. *Journal of Evolutionary Biology* **22**, 387–395. doi:10.1111/j.1420-9101.2008.01656.x

Waller NL, Gynther IC, Freeman AB, Lavery TH, Leung LKP (2017) The Bramble Cay melomys *Melomys rubicola* (Rodentia: Muridae): a first mammalian extinction caused by human-induced climate change? *Wildlife Research* **44**, 9–21.

Watson J (2016) Bring climate change back from the future. *Nature* **534**, 437. doi:10.1038/534437a

Webber BL, Raghu S, Edwards OR (2015) Opinioils CRISPR-based gene drive a biocontrol silver bullet or global conservation threat? *Proceedings of the National Academy of Sciences of the United States of America* **112**, 10565–10567. doi:10.1073/pnas.1514258112

Weigelt P, Steinbauer MJ, Cabral JS, Kreft H (2016) Late Quaternary climate change shapes island biodiversity. *Nature* **532**, 99–102. doi:10.1038/nature17443

Wetzel FT, Kissling WD, Beissmann H, Penn DJ (2012) Future climate change driven sea-level rise: secondary consequences from human displacement for island biodiversity. *Global Change Biology* **18**, 2707–2719. doi:10.1111/j.1365-2486.2012.02736.x

Woinarski JCZ, Lindenmayer DB, Garnett ST, Legge SM (2016) Conservation practice: a very preventable mammal extinction. *Nature* **535**, 493. doi:10.1038/535493e

Woinarski JCZ, Garnett ST, Legge SM, Lindenmayer DB (2017) The contribution of policy, law, management, research, and advocacy failings to the recent extinctions of three Australian vertebrate species. *Conservation Biology* **31**, 13–23.

Zhu ZR, Wyrwoll KH, Collins LB, Chen JH, Wasserburg GJ, Eisenhauer A (1993) High-precision U-series dating of Last Interglacial events by mass spectrometry: Houtman Abrolhos Islands, Western Australia. *Earth and Planetary Science Letters* **118**, 281–293. doi:10.1016/0012-821X(93)90173-7

16

Australian islands: current thinking, emerging themes and a way forward to manage these offshore assets

Derek Ball, Dorian Moro and Sally L. Bryant

Summary

Australians need to reconsider their approach to conserve their island assets. Australia needs an island renaissance: a shared vision and a realisation that the current trajectory of loss of natural heritage on islands (and elsewhere) needs to change. As a nation, we have an opportunity to utilise offshore islands to make significant and highly cost-effective gains for the conservation of our natural heritage. While conservation successes are being achieved on some islands, it is clear these are not sufficient to stem the national loss of island biodiversity. Here, we elaborate on why and how we might do this within the context of Australia's islands. We suggest there is a need for sound public policy, and appropriate governance structures to guide a renaissance in island conservation. We identify six emerging themes that stem from current research, discussions and awareness from island practitioners around Australia. We conclude by suggesting future directions for the implementation of a public policy and collective national-level thinking to progress island conservation and management for Australia. We propose that the future of island management in Australia should incorporate four interdependant foundations that become the remit of a central, multi-faceted national organisation with a capacity to work across sectors.

Introduction

Australia's attempt to conserve its natural heritage is at crisis point. State of the Environment reporting (SoE 2016) and other learned narratives (Lindenmayer *et al.* 2014) continue to paint an unambiguous trend of ongoing species extinction, loss of ecosystem function and indeed loss of ecosystems, and gross degradation at landscape and seascape scales. Further, there is ample evidence of the failure of institutions to reverse these trends, despite impressive financial investments and a massive bank of knowledge and experience to draw upon. This storyline is particularly concerning for Australia's offshore islands, remote assets which contain some of Australia's most iconic fauna, flora and ecosystems and in many instances may be the only hope for saving values that are facing extinction on the mainland. Australia needs a new shared vision and a realisation that the current trajectory of loss of natural heritage on islands needs to change. To achieve this vision we must re-examine, and if need be discard, long-held beliefs. We need to explore new ideas and concepts, and question old paradigms. Most importantly, we need to build a holistic view of the opportunities that islands present and urge Australia as a nation to utilise its islands to make significant and highly cost-effective gains for conservation of its natural heritage. The process of developing an island strategy to achieve these gains offers the exciting prospect of reinventing the way in which conservation management interventions are designed and delivered for natural assets across Australia. Australia's islands are, in this respect, 'low-hanging fruit'. They represent a tiny fraction of the total Australian landmass and support a small part of the Australian population. Many, and perhaps most, have few if any competing land uses either presently or likely in the future. Additionally, they provide refuge for a massive amount of Australia's biological diversity and are economically valuable sources of income for local people and Australia as a nation. Thus Australia's islands are areas that offer high-return, low-cost and minimal-risk settings for developing new ways of conserving Australia's natural heritage.

In this chapter, we identify six themes, which have emerged from previous works on islands around Australia and nearby New Zealand:

1 the need to identify, better understand and monitor the values and connectedness of Australia's islands at regional and national scales;
2 many islands can only be effectively managed by incorporating socio-economic considerations within a risk framework;
3 the need to review national and international programs in the context of islands;
4 the need for explicit prioritisation of conservation interventions on islands;
5 the need to adapt island management within the context of climate change;
6 the need for holistic management in island restoration.

The chapter concludes with some innovative thoughts to progress island conservation and management for Australia.

Emerging theme 1: the need to identify, better understand and monitor the values and connectedness of Australia's islands at regional and national scales

Attempts to summarise the value of Australia's islands inevitably results in a list of impressive but nonetheless disparate numbers describing different features. Australia's islands collectively form a massive part of Australia's natural heritage. They comprise 40% of its coastline despite being only 1.26% of its total landmass (Geosciences Australia 2004), provide refuge or potential refuge for over 35% of Australia's threatened species (SPRAT 2017) and overall contain an incredibly diverse set of landscapes and component ecosystems. Derived economic benefits are significant, with upwards of A$13 billion generated annually by nature-based tourism alone (Chapter 11). The possibilities for conservation of Australia's natural heritage are equally impressive. Nowhere else but on islands can non-native species such as pest animals be eradicated and prevented from returning, cost-effectively. So too can unwanted species be prevented from getting to islands, through effective biosecurity measures. These features offer incredible opportunities for conserving components of Australia's biological diversity where it is not yet possible to do so on the mainland. Translocation of insurance populations of threatened species to islands, for example, offers cost-effective triage measures. Despite these inarguable national- and international- level values, islands do not feature as distinct landscapes within national or State of the Environment reporting. Thus, there are currently no transparent, effective, repeatable means by which to monitor islands or the natural values they protect. What is lacking is a set of national environmental accounts for Australia's islands that track their health and condition.

Chapter 2 in many ways summarises the work that commenced in 2011 in the lead-up to a symposium dedicated to islands (Island Arks, www.islandarks.com.au), during which interim and initial state and territory reports were presented. This material is an invaluable benchmark resource that can be reviewed, updated and included in Australia's next State of the Environment report, due in 2020.

Despite the highly significant economic value of islands, there is little by way of explicit process to measure, monitor and manage their financial sustainability even though island-based tourism contributes well over 10% of all tourism revenue in Australia (2014–15 Tourism Australia Annual reporting showed a total of A$107 billion revenue in that year). Neither Tourism Australia nor Tourism Research Australia provides island-specific research or reporting to underpin decision-making and planning within individual island, or island-based, tourism enterprises. Surely this will change with Tourism Australia's most recent consumer research that identifies aquatic and coastal experiences as the number one international visitor attraction, and 'island experiences' as ranking ninth of 19 overall attractions (Tourism Australia 2016).

Similarly, the Australian Bureau of Statistics provides little social or economic data that underpin planning for island communities. We argue that the fundamental principles of island management should include the imperative to undertake conservation interventions holistically; that is, in full cognisance of social, economic and cultural imperatives. In planning to do so, at a scale sufficient to significantly improve outcomes for island conservation (and island tourism) at a state and national level, socio-economic data become essential components of the holistic management vision. There seems an important collaborative role here for both tourism peak bodies and the Bureau of Statistics.

Removal of non-native species from islands and, conversely, the translocation of threatened species to islands to form insurance populations away from invasive threatening species has been implemented with success across Australia (Chapter 7). With new and improving technologies, these interventions are becoming more commonplace and cost-effective over time. However, they have been largely localised and, while they have made valuable contributions to island and threatened species conservation, they could be improved by prioritising translocations at a national-level context rather than a state-based context. Until we do so, we continue to underutilise one of the most important contributions that islands can make to the conservation of Australia's natural heritage: their inherent capacity to provide refuges for threatened species and ecosystems away from other impacts.

There are ecological links between terrestrial island ecosystems, their littoral zones and adjacent intertidal and marine ecosystems (Chapter 13). These are exemplified by species that use all or some of the different ecosystems to complete their lifecycles, such as seabirds, marine turtles and pinnipeds. Perhaps what is missing is information about less tangible ecosystem processes in these systems, such as nutrient and energy flows, and the ramifications of future changes to those connections. In addition, it seems that the scale of connectivity among islands, and the adjacent marine ecosystems, remains poorly understood. A better understanding of this dimension would help to progress complementary island and marine ecosystem management. For example, must an archipelago be managed spatially to encompass its offshore catchment of ocean? Should fisheries management consider the need to sustain seabird or pinniped breeding populations? Are there key islands along latitudinal gradients that require complementary management efforts because of a dependence on marine or terrestrial processes that interconnect them? Knowledge of these connections may support managers not just to design boundary maps for landscape realms, but to consider islands within defined ecosystem process zones such as marine protected areas, and vice versa. Jurisdictional issues are a culprit to some degree: islands typically are a state or territory responsibility but many surrounding offshore waters are under Commonwealth jurisdiction (including the majority of Australia's marine protected area network). There is progress to be made in ensuring a more ecologically functional approach to achieving overall conservation goals across terrestrial and ocean realms.

Emerging theme 2: many islands can only be effectively managed by incorporating socio-economic considerations within a risk framework

Socio-economic linkages need to be better understood and assessed during future island conservation interventions. Many Australian islands are remote, uninhabited, not held under any tenure other than that of the Crown, rarely if ever visited, or are already protected from development or habitation by various statutory instruments. These areas offer remarkable opportunities for natural ecosystems and landscapes to continue evolving and functioning with minimal human impact. Where impacts do occur, management interventions can be designed and implemented with virtually no considerations other than those normally associated with conservation ecology and its practice. However, there is a clear need to manage not only islands unencumbered by community needs or desires, but also islands where people regularly visit or live, and this will present additional challenges. In New Zealand, for example, conservation initiatives are rapidly transitioning to inhabited islands and successfully integrating interventions with holistic ecosystem repair (Chapter 14). Impacts from developments or from environmental pressures such as pest species or climate change require a more holistic and community-oriented approach ('Social determinants of resilience', Chapter 14).

Managing people's diverse beliefs about environmental impacts becomes a challenge when the levels of understanding and expectations of risk are varied. For example, gaining local community support for black rat (*Rattus rattus*) control or eradication, particularly when it involves the use of toxicants (rodenticides), is an important yet often difficult process. Ogden and Gilbert (2009) identified three reasons why gaining local support for rat eradications is typically challenging:

- a lack of appreciation of the ecological damage resulting from rats, and therefore a low priority placed on their elimination;
- suspicion by community members that conservationists want access to private lands and island-wide biosecurity;
- numerous regulatory barriers that delay decisions and actions, and ultimately result in disinterest in such projects.

There are many lessons from New Zealand and elsewhere that demonstrate the critical importance of community involvement and acceptance if island eradications or other conservation initiatives are to be successful and sustained long-term.

We therefore advocate an emerging theme for island conservation where perceived or real environmental problems on islands – by developments and/or anthropogenic factors – should be assessed within a risk framework involving genuine community consultation rather than beliefs based on dogma. This paradigm aligns with the framework of environmental impact assessments and, importantly, involves community members and stakeholders.

Some islands are subject to tourism and industrial development that has been viewed by society as inconsistent with uses such as nature conservation, local heritage values, and social and economic values. It is important to consider in advance how conflict among users might be resolved or reduced to ensure that the highest and best use is afforded to islands while ensuring that their intrinsic conservation, social and economic values are maintained. There are several commonly held beliefs that need to be challenged. For example, 'eco-tourism' or 'nature-based tourism' as general concepts are assumed to be positive influences, if not benign, on the natural environment, despite the growing body of evidence to the contrary (Buckley 1994; see also Chapter 11).

Many island authorities aspire to grow visitor numbers through nature-based tourism but have few mechanisms in place to ensure that profits help underpin the conservation asset, or have sustainability quotients to ensure visitor numbers do not exceed the island's carrying capacity. When both are ignored, eco-tourism can lead to haphazard growth in the number of people living on an island, and subsequent local community dissatisfaction. There is ample evidence, for example, that World Heritage-listed K'gari-Fraser Island, the world's largest sand island, is threatened by eco-enterprises (Chapter 11). No doubt this is not an uncomplicated evolution despite standards for management of World Heritage-listed properties being well-known (Stolton and Dudley 2012). Despite the standards, numerous related problems such as wavering and often inconsistent political will and direction, lack of management foresight and lack of coordination between sectors (e.g. conservation *v.* tourism) shine through as root causes behind a chronic problem of environmental degradation from overuse by tourism enterprises.

In comparison, the public might well consider the development of heavy industry on an island as an anathema, particularly if the island is of international conservation significance. Nevertheless, the Barrow Island experience has clearly shown that conservation and heavy industry can successfully co-exist (Chapter 9). The influences on Barrow Island seem to be robust project planning within an environmental risk framework, transparent and consistent will by both the proponent and government to protect the environment but also realise the economic needs of the venture, sustained workforce engagement including the development and maintenance of an ongoing environmental culture, and effective identification, mitigation and monitoring of environmental risks. Barrow Island has no doubt been a confronting experience for those who believe that protected areas are places for nature only, through the exclusion or rigid control of humans. However, Barrow Island does not stand in isolation in this regard.

The development and management of Indigenous Protected Areas (IPAs), within which Indigenous Australians continue to live and harvest natural resources, has been a successful approach to return custodianship of Indigenous lands. Creation of IPAs has resulted in culturally appropriate and thus ongoing safeguarding and care of island-based biocultural resources, and to an increasing degree the wider coastal areas and seascapes (Chapter 8). By 2017 there were 25 dedicated IPAs with islets, islands and/or sea country in immediate proximity to offshore islands.

The growing involvement of citizens who live and work on islands must be viewed as integral to island management. Community-based conservation groups and non-government organisations continue to grow in number, using a plethora of business and governance models. Many of the groups are well and truly part of mainstream conservation, if not the dominating force, and their success is inherently due to their genuine and durable partnerships. These partnerships reflect key attributes such as shared passion and values, inclusive participation, willingness to engage and foster collaboration, mutually reinforcing activities towards achieving long-term goals, good governance, robust project planning and dedicated coordination support (Chapter 12). As people with vested interests in their community group and organisation, as well as mutual interests with land managers to protect island assets, opposing viewpoints on how best to plan and deliver conservation are inevitable. Histories of conservation interventions have shown that island communities will not necessarily embrace changes unless given compelling reasons to do so (Walter and Hamilton 2014). Provided the community is engaged and supportive, we advocate that community engagement should shift the focus away from the impact and instead address environmental concerns within a risk assessment. Impacts are identified and workshops with stakeholders are facilitated to address the consequence of an impact by a development, and the likelihood of this occurring. Where a risk is assessed as too high, then developments may either not proceed or mitigations may be identified by stakeholders to reduce the risk to 'as low as reasonably practicable' (ALARP). ALARP criteria must be transparent and developed in partnership with the community; in so doing, people become engaged and involved, and participate in discussions to reduce perceived risks while understanding the likelihood of an impact.

Success can arise from open and transparent debate, the development of well-understood and agreed agendas and guiding policy and, above all, sound governance structures. The 4Cs model, which we consider has significant merit, emphasises the need to develop and promote *community*, *cultural* and *commercial* benefits as integral and essential components of *conservation* interventions (Chapter 4). However, this approach will itself provide challenges to the conservation practitioner. Developing programs of collective impact across these four areas of endeavour will require new skill sets within teams working on island conservation interventions. These teams will need a truly common agenda as a shared vision for change; they must have a shared view of what success looks like; team members must engage in mutually reinforcing activities, each taking ownership and charge of what they do best; they will need to be in regular communication so that each recognises and appreciates their common motivation; and they must have a backbone support organisation that facilitates and helps coordinate activities.

Finally, with the growing focus on the holistic management of islands from an ecological perspective, there is a growing realisation that social, cultural and economic elements must be included in the process to ensure transparent implementation of proposed interventions, and long-term sustainability of the outcomes. A notable example is the sophisticated biosecurity arrangements put in place on some islands to reduce the risk of reinvasions of pest species following their eradication, perhaps best demonstrated by the Treasure Islands approach in the Hauraki Gulf, New Zealand (Bassett *et al.* 2016). Potential weak links in these biosecurity pathways are human behaviours, thus island users need compelling reasons to comply with guidelines designed to prevent reinvasions. Importantly, we must cease to consider conservation interventions as independent of other human endeavours. Social needs and aspirations, cultural prerogatives and economic realities must be integral to the planning and implementation of long-term conservation goals for Australia's island values.

Emerging theme 3: the need to review national and international programs in the context of islands

Australia's islands have huge potential to contribute to improved conservation of Australia's natural heritage. We argue that the disparities between state/territory jurisdiction over islands and the Australian government over adjacent Commonwealth-controlled oceans and islands are the cause of inadequacies and inefficiencies in island conservation outcomes. These disparities result in a lack of national strategic focus in the use of islands to achieve maximum gains for biodiversity conservation, and a lack of complementary management of islands and surrounding marine ecosystems as required by their many ecological linkages.

From a pragmatic viewpoint, rewriting laws governing the Federation of Australia is not an option. However, there are highly successful approaches that have fostered better collaboration over other Australian landscapes in the past. For example, islands are probably the only iconic Australian landscape that has not been favoured with a cooperative research paradigm – tropical rainforests, desert ecosystems, Antarctic ecosystems, coral reefs, savannahs, coastal zones, estuaries and waterways have all been the focus of cooperative research centre programs. We propose that the values of islands to Australia, from the perspectives of biological conservation, tourism, industry and implicitly their social and economic values, merit a strategic and cooperative approach between governments and the private sector. Specifically, we invite such a cooperative program with the following agenda of works:

- establish and maintain a program for biodiversity inventory of Australia's islands;
- establish and implement a program to systematically and transparently monitor key biological, social and economic values of

islands, including tourism activity, using nationally agreed criteria;

- prioritise Australia's islands according to conservation and potential use. Where appropriate, systematically identify innovative tourism and other business opportunities for islands in Australia that embrace the principles of environmental sustainability, while also identifying islands that should not be subject to any human disturbance or use;
- champion and expedite the addition of islands with outstanding biodiversity values to the existing set of Matters of National Environmental Significance under the *Environment Protection and Biodiversity Conservation Act 1999*;
- develop and implement a systematic, comprehensive and strategic approach to the conservation of Australia's islands, including integrated terrestrial and marine management planning for all islands of significant conservation value.

Emerging theme 4: the need for explicit prioritisation of conservation interventions on islands

The fact that Australia's financial investment in threatened species and ecosystem recovery over the last two decades has failed to achieve any national conservation objectives is surely a clear signal for the need for change. Systematic conservation planning has evolved to be a robust science: starting with relatively humble efforts to resolve conflicts in multi-use forest planning exercises (Pressey 1998), to development of sophisticated software to assist in broader decision-making (Pressey *et al.* 2009), to the point where we have access to models for cross-landscape (realm) planning and explicit tools such as the Island Decision Support System (Chapter 5). The real challenge seems not a lack of tools with which to maximise return on investment, but the lack of willingness to use them. Reluctantly, we use the examples of Australia's most recent extinctions, all of which occurred on islands and within national parks, to illustrate that what was lacking was a willingness to prioritise action: the Christmas Island pipistrelle (*Pipistrellus murrayi*) in 2009, the Christmas Island forest skink (*Emoia nativitatis*) in 2014 and the Bramble Cay melomys (*Melomys rubicola*) between 2009 and 2014 (Gynther *et al.* 2016; Woinarski *et al.* 2017). Sadly, the plight of these species was known well before their final demise but, despite repeated requests, funding and subsequent action never materialised. It is concerning to note the management agency responsible for Christmas Island National Park received revenue well in excess of A$40 million in both 2008/09 and 2013/14 (https://www.environment.gov.au/topics/national-parks/parks-australia/publications) and the agency responsible for Bramble Cay more than A$225 million on average over the 2012/13 and 2013/14 years (https://www.npsr.qld.gov.au/about/corporatedocs/#annual_reports). If prevention of extinctions is not a priority in protected areas where the goal of nature conservation is paramount, we question the future of many Australian islands that are not protected areas.

Given the lack of national-level strategies to act on island priority issues, there seems little alternative but to join other commentators who have called for a major policy review into the way Australia manages threatened species and our natural heritage in general (Wentworth Group of Concerned Scientists 2014). We discuss this below, particularly in the context of Australia's islands being ideal settings for a demonstration of what public policy frameworks might work best in the future.

Emerging theme 5: the need to adapt island management within the context of climate change

Almost one-third of the world's countries, including Australia, are islands with peoples, cultures and ecosystems now threatened by natural disasters compounded by the real and serious threat of

Box 16.1: The Great Barrier Reef imperative

The Great Barrier Reef World Heritage Area (GBRWHA) is made up of ~3000 individual coral reefs, 900 islands, inter-reef communities, seagrass beds, mangrove and other ecosystems and is regarded as one of the seven wonders of the natural world. None of these systems occur in isolation: there is a plethora of ecological linkages sustaining their interconnectedness and function, many still poorly understood. The GBRWHA is under dire threat. Over time, catchment modification has resulted in substantial loads of sediment, inorganic nutrients and herbicide runoff into the reef lagoon. Whether to allow a new coal mine development, with associated emissions and dredging, is currently being decided. The reef environment continues to suffer impacts from elevated numbers of crown-of-thorns starfish (*Acanthaster planci*) and the stark loss in coral cover has never been more apparent. Climate change is manifesting in increasing water temperatures, ocean acidification, sea-level rises and other changes. The northern part of the GBRWHA has suffered two catastrophic climate change-induced bleaching events in as many years (2016, 2017) and its future looks bleak. Even though coral can migrate and re-establish if suitable habitat exists, there is little evidence to suggest the southern part of the system will be immune from major changes. The cliché 'death by a thousand cuts' is apt, as the causes and consequences of this scenario manifest in a major ecosystem collapse.

Publicly, the growing alarm and urgency of the plight of the GBRWHA is becoming more and more evident. The prevailing reaction globally is mourning the pending loss of a unique natural ecosystem – a World Heritage treasure. Across Australia, this view is shared by many including ecologists, naturalists, conservationists and those who for centuries have learned, marvelled and enjoyed themselves in the wonders of the reef. This concern is growing among fishers and all those who rely on the GBRWHA for their livelihood and income, especially tourism operators, torn between two narratives: the need to advocate for large and immediate efforts to mitigate threats, yet emphasising what remains so as not to deter visitation. This dilemma is especially poignant as Tourism Australia has embarked on a major documentary project with renowned English broadcaster and naturalist, Sir David Attenborough, to help promote the GBRWHA as a marine wonderland.

The GBRWHA is a metaphor for island conservation and the urgent need for change. Innovation, more effective processes and collective effort can address these problems if we act quickly. Australians are already embracing renewables at an accelerating rate, businesses large and small are pledging to reduce their climate change footprint, banking institutions no longer view new coal mines as smart investments, and indeed divestment is now a dominating trend. From a conservation perspective, an island cooperative research centre could promote a better understanding of the landscape-scale island–sea connections and ways of improving future resilience. A set of environmental accounts underpinned by comprehensive long-term monitoring data could track the health and condition of these assets, legislations and policy changes could streamline and strengthen regulatory process, and entrepreneur profit-for-purpose entities and community partnerships could help drive and implement the changes needed and contribute to the cost.

The potential to reinvent the future of the GBRWHA, and island conservation nationally, is not only an imperative but is an achievable goal, and Australians should be committed to make that process a success.

climate change (Secretariat of the Convention on Biological Diversity 2014). Climate change is already having a profound effect on Australia's island biota; for example, the now extinct Bramble Cay melomys, the Great Barrier Reef (Box 16.1), the sub-Antarctic fellfield plant *Azorella macquariensis* (Chapter 10), and bird taxa exposed to more frequent and intense threats leading to reduced productivity and a shift in geographic range (Garnett *et al.* 2014). Climate change needs to be accepted and managed like other potentially threatening processes (Chapter 15).

While these biological and ecosystem changes are being documented, the impacts associated with a changing climate also provide opportunities for innovation within the science and practice of conservation of biological diversity, particularly on islands. Indeed, it was climate change that created many of Australia's islands during the Holocene, and that event itself probably accelerated evolutionary process on the emergent islands that over time manifested in increased biodiversity through speciation and ecosystem-level shifts. Much of the debate about the impacts of climate change on islands has revolved around the physical loss of low-lying islands due to sea-level rise. There seems to be far less discussion that rising sea levels will also create new islands around the Australian coastline. These opportunities may remove pest species as new islands are formed, creating additional refuges for native species and ecosystems to balance the loss of other island refuges – and perhaps even increase their overall area.

The challenges presented by the impacts of climate change might be a sorely needed catalyst for the adoption of conservation interventions that many practitioners and risk-averse public institutions refuse to accept. These might include expanded translocation efforts (such as that which would have prevented extinction of the Bramble Cay melomys), ecosystem-level engineering of new seabird breeding areas as existing ones become submerged, and introduction of non-native herbivores to control weeds that benefit from changes in rainfall and temperature patterns.

Of critical importance is the consideration that while climate change is occurring, the magnitude and direction of change at smaller scales, like those that islands occupy, is far less understood. Most Australian islands are poorly surveyed for their biological diversity, and few are subject to regular monitoring programs. Development of long-term monitoring programs is an essential prerequisite to documenting, understanding and predicting future changes at the island scale. The results of climate change will in many ways be poorly understood until they occur. We need to invest more in monitoring, and include multi-branched decision pathways to guide ecosystem responses to unknown scenarios of change.

Emerging theme 6: the need for holistic management in island restoration

If there is a single common distinguishing feature of island conservation intervention in recent decades, it must surely be bigger and bolder thinking. In a little over 50 years, interventions have grown from eradications of single invasive species on small islands (e.g. Maria Island, Hauraki Gulf, New Zealand, <5 ha) in the 1960s (Russell and Broome 2016), to the current attempt to eradicate single species on large islands (feral cats on Dirk Hartog Island at 62 000 ha, which if successful will be the largest island globally from which cats have been removed), to the removal of multiple invasive species on islands or archipelagos to minimise ecosystem-wide trophic impacts from single-species eradications (Zavaleta *et al.* 2001; Glen *et al.* 2013). On Macquarie Island (12 865 ha), eradications commenced with weka (*Gallirallus australis*) then targeted feral cats (*Felis catus*), followed by a simultaneous multi-species eradication of rabbits (*Oryctolagus cuniculus*), black rats and house mice (*Mus domesticus*) (Springer 2011). By 2016 the results of that intervention were apparent across the entire island, with improved slope stability and population recovery of many endemic plants, vegetation communities, invertebrates and 14 seabird species of which eight now have a reduced threat status assigned by BirdLife Australia. This program culminated in the single biggest improvement in threatened species conservation in Australian history (Chapter 6).

Further afield, practitioners have targeted entire archipelagos such as the Mercury Islands in New Zealand with multi-species eradication programs rendering the island group free of invasive mammals (Towns *et al.* 2016). The emerging trend is our growing capacity to eradicate more species, on larger islands and even entire archipelagos, more cost-effectively. This presents opportunities to restore what has been lost and to use these islands as arks for species threatened by extirpation on the mainland where pests cannot yet be managed effectively. Since 1980 in Australia, 46 species of vertebrate animals and vascular plants have been translocated to islands (Chapter 7). For mammals the success rate has been 90%, which clearly illustrates the value of these conservation efforts.

Contemporary outcomes are extending beyond invasive species eradications and translocations as species conservation tools alone, to the restoration of entire ecosystems (Zavaleta *et al.* 2001; Chapter 14). Increasingly, island conservation ecologists have begun considering the trade-offs when planning control or eradication on islands. Ecosystem-level consequences of the loss of, or dramatic reduction in, some invasive species may result in unpredicted and undesirable consequences, specifically the ecological release of other invasive species on the island. For example, the eradication of invasive herbivores in the Galapagos released invasive plants such as *Rubus niveus* from grazing (Atkinson *et al.* 2011), and the removal of feral goats and pigs from Sarigan Island initially resulted in a population explosion of exotic vines (Caut *et al.* 2009).

The eradication or control of predators to secure and increase seabird populations is becoming a tool to re-establish previous soil conditions such as nutrient status and pH, with subsequent ecosystem- level changes in plant germination and survivorship, soil biota and invertebrate diversity and abundance (Towns *et al.* 2009; Mulder *et al.* 2011). This might be just one example of future whole-of-island restoration projects that would be underpinned by a better understanding of how island terrestrial ecosystems are interlinked with marine ecosystems, allowing simultaneous management of both to enhance island conservation goals.

Practitioners are becoming more versed in understanding the relationships between pest animals, pest and native plants, and other abiotic ecosystem features. For example, some invasive plants may perform a functional role as important habitat for endemic species on islands where they have been long-established. Invasive *Casuarina* trees provided nesting habitat for lesser noddies (*Anous tenuirostris*) on Cousine Island (Seychelles), so their removal was delayed until indigenous plantings were sufficiently established (Samways *et al.* 2010). On St Bees Island in the Great Barrier Reef, feral goat eradication was predicted to result in an increased biomass of native plants but also in the weed *Lantana camara*. An increase in shrubs would lead to higher fire risk due to the increase in fuel accumulation. To manage this, land practitioners decided to conduct local burns in parallel with the goat eradication exercise to mitigate fire hazards which could have been deleterious to the island's koala population. Planned burning was undertaken during the goat eradication using the increased fuel loads, to assist in control of the fire-sensitive lantana but also to reduce the potential for high-intensity fires in the future.

Concluding comments: future directions for the implementation of a public policy on islands

Australia has access to a large body of experience, knowledge and skills, and conservation successes are being achieved. However, it is clear that these are not sufficient to stem the loss of biodiversity on Australia's islands. In some cases, the conservation approaches used to date have been ineffective and need to change. Conservation of biodiversity on Australian islands needs to be underpinned not just by conservation science and practice but also by social and economic elements through sound collaborations with local communities, recognition and respect for their cultures and a realisation of their economic drivers. There is a need for sound public policy and appropriate governance structures to guide a renaissance in island conservation. We are not aware of any existing or proposed initiative, developed at a national level, to enact this.

We propose that the future direction of island management in Australia should incorporate four interdependent foundations (Table 16.1). This is a logical distillation of the 10 Commitments for islands proposed by Woinarski *et al.* (2014), and the emerging themes identified and discussed here and elsewhere in this book. We recommend that these four foundations become the remit of a central, multi-faceted organisation with a capacity to work across sectors while retaining integration.

Evidence-based management

There is need for a central repository of data and planning tools that can be used to support programs aiming to achieve conservation of biodiversity on Australia's islands. The repository need not be cus-

Table 16.1. A national strategic direction for Australia's island assets

10 Commitments (Woinarski *et al.* 2014)	Emerging themes	Proposed foundations (key policy elements required)
Implement a comprehensive stocktake of island life	1) A need to identify, better understand and monitor the values of Australia's islands	Evidenced-based management
Design and implement a comprehensive biosecurity program for islands, prioritised by risk and biodiversity significance		
Design and develop a national program that systematically prioritises, implements, monitors and improves island conservation management interventions	The need for explicit prioritisation of conservation interventions on islands The need to adapt island management within the context of climate change The need for holistic management of island restoration	Sound return on investment
Ensure that Australian islands are included in national conservation management programs	The need to review national and international programs in the context of islands Establish a Centre of Excellence for Island Studies	National coordination in partnership with states and territories
Act to conserve island threatened species, and amend the EPBC Act to better safeguard islands of biodiversity significance		
Develop strategic risk-based programs to translocate threatened species among islands and mainland areas		
Develop broad-based sustainability programs for island communities, incorporating social, economic and environmental factors	Many islands can be more effectively managed by incorporating socio-economic considerations within a risk framework	People are not simply stakeholders, they are fundamental to the long-term success of conservation initiatives
Improve collaboration among island managers, residents and visitors across Australia, including ongoing support for and coordination of the island arks program		
Increase resourcing to, and longer-term security of, island IPAs, and enhance their management capacity		
Integrate island-based ecotourism industries into holistic management arrangements		

todian of all available data, but there is a need for collection, collation and a capacity to analyse, not just biological and physical data but also socio-economic data to support foundations of island knowledge and subsequent conservation initiatives.

Sound return on investment

The scope of conservation of Australia's islands is large and it will be essential to maximise return on investment. The opportunity to offset costs through economic development initiatives as part of, or running in parallel to, conservation programs is of utmost importance. We need the capacity – and the willingness to use it – to nationally prioritise investments in island conservation and demonstrate accountability of the (environmental) returns. Entrepreneurial approaches are needed to more clearly identify opportunities at the junction between conservation and economic initiatives.

National coordination in partnership with states and territories

The multiplicity of jurisdictions over Australia's islands and adjacent marine systems is hampering ecologically appropriate management across realms, i.e. land to coast to sea. It is also hindering effective prioritisation of resources. Formation of a central island governance and coordination body, including representation and advice from each jurisdiction, would provide an opportunity to begin considering complementary management arrangements of island systems across jurisdictions and to identify key synergistic funding arrangements.

People are not simply stakeholders, they are fundamental to the long-term success of conservation initiatives

Conservation programs on Australia's islands cannot be singly focused on biodiversity outcomes. On inhabited islands or those with a human community of interest, community support will be essential. We strongly recommend that on these islands, conservation initiatives be developed and prioritised to recognise the risks and opportunities inherent to developing community support. Furthermore, we recommend that conservation initiatives be developed in conjunction with programs demonstrating not just the direct benefit of maintaining biodiversity, but also the subsequent benefits for the broader community, its culture(s) and its economic needs.

References

Atkinson R, Gardener M, Harper G, Carrion V (2011) 50 years of eradication as a conservation tool in Galapagos: what are the limits? In *The Role of Science for the Conservation of the Galapagos: A 50 Year Experience and Challenges for the Future* (Eds M Wolff and M Gardener) pp. 183–198. Routledge, Oxford.

Bassett IE, Cook J, Buchanan F, Russell JC (2016) Treasure Islands: biosecurity in the Hauraki Gulf Marine Park. *New Zealand Journal of Ecology* **40**, 250–266.

Buckley R (1994) A framework for tourism. *Annals of Tourism Research* **21**, 661–665. doi:10.1016/0160-7383(94)90126-0

Caut S, Angulo E, Courchamp F (2009) Avoiding surprise effects on Surprise Island: alien species control in a multitrophic level perspective. *Biological Invasions* **11**, 1689–1703. doi:10.1007/s10530-008-9397-9

Garnett ST, Pavey CR, Ehmke G, VanDerWal J, Hodgson L, Franklin DC (2014) Adaptation outlines for species that are both highly sensitive and highly exposed. In *Climate Change Adaptation Plan for Australian Birds* (Eds ST Garnett and DC Franklin) pp. 79–258. CSIRO Publishing, Melbourne.

Geosciences Australia (2004) *Geodata Coast 100K*. Commonwealth of Australia. <http://www.ga.gov.au/metadata-gateway/metadata/record/gcat_a05f7892-eae3-7506-e044-00144fdd4fa6/GEODATA+COAST+100K+2004>

Gynther I, Waller N, Leung LK-P (2016) Confirmation of the extinction of the Bramble Cay melomys *Melomys rubicola* on Bramble Cay, Torres Strait: results and conclusions from a comprehensive survey in August–September 2014. Unpublished report to the Department of Environment and Heritage Protection, Brisbane.

Glen AS, Atkinson R, Campbell KJ, Hagen E, Holmes ND, Keitt BS, Parkes JP, Saunders A, Sawyer J, Torres H (2013) Eradicating multiple invasive species on inhabited islands: the next big step in island restoration? *Biological Invasions* **15**, 2589–2603. doi:10.1007/s10530-013-0495-y

Lindenmayer D, Dovers S, Morton S (Eds) (2014) *Ten Commitments Revisited: Securing Australia's Future Environment*. CSIRO Publishing, Melbourne.

Mulder CPH, Anderson WB, Towns DR, Bellingham PJ (Eds) (2011) *Seabird Islands: Ecology, Invasion and Restoration*. Oxford University Press, New York.

Ogden J, Gilbert J (2009) Prospects for the eradication of rats from a large inhabited island: community-based ecosystem studies on Great Barrier Island, New Zealand. *Biological Invasions* **11**, 1705–1717. doi:10.1007/s10530-008-9398-8

Pressey RL (1998) Algorithms, politics and timber: an example of the role of science in a public, political negotiation process over new conservation areas in production forests. In *Ecology for Everyone: Communicating Ecology to Scientists, the Public and the Politicians* (Eds R Wills and R

Hobbs) pp. 73–87. Surrey Beatty and Sons, Sydney.

Pressey R, Watts M, Barrett T, Ridges M (2009) The C-plan conservation planning system: origins, applications, and possible futures. In *Spatial Conservation Prioritization: Quantitative Methods and Computational Tools.* (Eds A Moilanen, KA Wilson and H Possingham) pp. 211–234. Oxford University Press, Oxford.

Russell JC, Broome KG (2016) Fifty years of rodent eradications in New Zealand: another decade of advances. *New Zealand Journal of Ecology* **40**, 197–204. doi:10.20417/nzjecol.40.22

Samways MJ, Hitchins P, Bourquin O, Henwood J (2010) *Tropical Island Recovery: Cousine Island, Seychelles.* Wiley Blackwell, Oxford.

Secretariat of the Convention on Biological Diversity (2014) *Island Biodiversity: Island Bright Spots in Conservation and Sustainability.* Secretariat of the Convention on Biological Diversity, Montreal.

SoE (2016) *Australia State of the Environment 2016.* Commonwealth of Australia. <https://soe.environment.gov.au>

SPRAT (2017) *Species Profiles and Threats Database.* Commonwealth of Australia. <http://www.environment.gov.au/cgi-bin/sprat/public/sprat.pl>

Springer K (2011) Planning processes for eradication of multiple pest species on Macquarie Island: an Australian case study. In *Island Invasives: Eradication and Management. Proceedings of the International Conference on Island Invasives* (Eds CR Veitch, MN Clout and DR Towns) pp. 228–232. IUCN, Gland, Switzerland.

Stolton S, Dudley N (2012) *Managing Natural World Heritage.* United Nations Educational, Scientific and Cultural Organization, Paris.

Tourism Australia (2016) *Coastal and Aquatic Experiences.* Factsheet. <http://www.tourism.australia.com/statistics/consumer-demand-research.aspx>

Towns DR, Wardle DA, Mulder CPH, Yeates GW, Fitzgerald BM, Parrish GR, Bellingham PJ, Bonner KI (2009) Predation of seabirds by invasive rats: multiple indirect consequences for invertebrate communities. *Oikos* **118**, 420–430. doi:10.1111/j.1600-0706.2008.17186.x

Towns DR, Borrelle S, Thoresen J, Buxton R, Evans A (2016) Mercury Islands and their role in understanding seabird island restoration. *New Zealand Journal of Ecology* **40**, 235–249. doi:10.20417/nzjecol.40.27

Walter RK, Hamilton RJ (2014) A cultural landscape approach to community-based conservation in Solomon Islands. *Ecology and Society* **19**, 41–50. doi:10.5751/ES-06646-190441

Wentworth Group of Concerned Scientists (2014) *Blueprint for a Healthy Environment and a Productive Economy.* Wentworth Group of Concerned Scientists. <http://wentworthgroup.org/wp-content/uploads/2014/11/Blueprint-for-a-Healthy-Environment-and-a-Productive-Economy-November-2014.pdf>

Woinarski J, Ball D, Burbidge AA (2014) Islands. In *Ten Commitments Revisited: Securing Australia's Future Environment.* (Eds D Lindenmayer, S Dovers and S Morton) pp. 117–128. CSIRO Publishing, Melbourne.

Woinarski JCZ, Garnett ST, Legge SM, Lindenmayer DB (2017) The contribution of policy, law, management, research, and advocacy failings to the recent extinctions of three Australian vertebrate species. *Conservation Biology* **31**, 13–23.

Zavaleta ES, Hobbs RJ, Mooney HA (2001) Viewing invasive species removal in a whole-ecosystem context. *Trends in Ecology and Evolution* **16**, 454–459. doi:10.1016/S0169-5347(01)02194-2

INDEX

4Cs approach, to island management 64, 66–70

Aboriginal lands 14, 15, 117, 119
Aboriginal people 2, 4, 26, 115–26, 150, 185; *see also* Indigenous
abutilon, Norfolk Island 100, 103
Abutilon julianae 100, 103
A. listeria 20
Acacia bivenosa 75
A. mangium 25
Acanthaster planci 242
Acanthiza pusilla archibaldi 46, 228
Acanthornis magnus greenianus 46, 228
Accipiter hiogaster natalis 20
Achyranthes margaretarum 100, 103
Acridotheres tristis 30
Actitis hypoleucos 16
adaptive management 197
Adele Island 18, 31
Adirondack Park 76
Admiralty Islands 152
Aerva javanica 30
Agathis robusta 150
ahikā roa 215
Ahuahu *see* Great Mercury Island
Airlie Island 26
albatross, black-browed 16
light-mantled 16
wandering 16, 92
Albatross Island 32
Alces alces 76
alien species vectors 158
Allocasuarina verticillata 187
Alpha Island 33, 105
Ammophila arenaria 30
Analytical Hierarchy Process 76
Andropogon gayanus 28
Angel Island 75
Anindilyakwa Land Council 22
Anoplolepis gracilipes 27, 86
Anous minutus see noddy, black
A. stolidus see noddy, common
A. tenuirostris melanops 18, 227, 244
ant, African big-headed 91, 157, 197, 199
Argentine 28
black crazy 197
ginger 28
Guinea *see* ant, pennant
pennant 197, 199
yellow crazy 27, 86
Antarctic expeditions 153
antechinus, Carpentarian 23
Anthornis melanura see bellbird
Anthoxanthum odoratum 30
Aorangi Island, NZ 206
Apium insulare 20
Aplonis fusca fusca 47
A. fusca hulliana 47
Apteryx spp. *see* kiwi
Araucaria cunninhamii 150
A. heterophylla 21
Arctocephalus forsteri 19
A. gazelle 19
A. pusillus 19, 193, 195
A. tropicalis 19
Ardenna pacifica see shearwater, wedge-tailed
A. tenuirostris see shearwater, short-tailed
Arenaria interpres 16
Arenga listeria 20
Argusia argentea 196
artificial light emissions 3, 141–2
Ashmore Islands 16, 27–8, 49
Ashmore Reef 12, 13–14, 50, 195
Ashmore Reef National Nature Reserve 12
Asparagus asparagoides 31
Asplenium listeria 20
A. obtusatum ssp. *northlandicum* 23
assemblage gaps 213
Astel Island 32
Asystasia alba 20
Atrichornis clamosus 100, 228
A. rufescens 226, 228
Australia's Biodiversity Conservation Strategy 2010–2030 48, 50
Australian Bird Atlas Project 182
Australian Conservation Foundation 61
Australian External Territories 13–14, 49
Australian islands tenure 13–15
Australian islands, geography 11–13
Australian Wildlife Conservancy 15, 34, 106, 110, 183
Avicennia marina 75
Avoid Island 19, 183
Axis porcinus 30
Azorella macquariensis 23, 154, 155, 243
A. selago 22

Babel Island 26
Badger Island 26, 116
baiting 28, 31, 87, 88, 91, 94, 95–6, 199
Balaenoptera musculus 197
Ball, Henry Lidgbird 152
Ball's Pyramid 152

bandicoot, eastern barred 33, 100, 109
golden 23, 32, 33, 100, 105, 107
southern brown 24, 109
western barred 24, 100, 106, 183
Bandjin people 124, 125
Bare Sand Island 19, 25
barley-grass 29
Barra Base 25
Barranyi (North Island) National Park 14, 25
Barrow Island 2, 18, 21, 24, 26, 31, 34, 51, 52, 74, 105, 131–43, 239
biosecurity 135–8
environmental culture 136–40, 142
environmental monitoring 134–5
LNG plant 134
Barrow Island boodie 33
Barrow Island Environment Month 139–40
Bass Strait islands 17, 19, 180
Bathurst Island 12, 22, 23, 25, 28, 32
bats 23, 87, 206
coastal sheath-tail 24
Lord Howe long-eared 21, 47
bauxite mining 25
bear, black 76
beche de mer 3
Bedout Island 18, 74
bee, green carpenter 187
beetles, darkling 212, 213
Beilschmedia taraire 210
bellbird 211, 213
Bentick Island 24
Bernier Island 2, 24, 31, 32, 34, 49, 52, 154
bettong 109
Bettongia gaimardi 109
B. lesueur see boodie
B. penicillata see woylie
Beyeria subtecta 188
Bickerton Island 25
Big Dog Island 15, 26
Big Green Island 32, 87, 174, 183
Big South Cape Island 206
Bigge Island 122, 123
bilateral agreements 51
bilby 100
biodiversity, arks 4, 99–110
conservation, 4Cs approach 61–70
future 63–4
status in Australia 62–3
features 74–5, 76
financial value 69
hotspots 51
loss, island 46
monitoring 56
policy, improving 53–6
biological resilience 208–10, 212–13, 215
biosecurity planning, WA islands 52
biosecurity 3, 88, 157, 225, 240
Barrow Island 135–8
Fraser Island 157
Heard and McDonald Islands 157
Lord Howe Island 157
New Zealand 207–8, 213, 214, 217
Pilbara islands 34
sub-Antarctic islands 158, 159
Tasmania 32–3
Victoria 33
bird decline, New Zealand 205–6
BirdLife Australia 182, 192, 243
birds 15–18, 21, 30, 46, 75, 86, 87, 88, 91, 92–3, 149, 152, 178, 182, 192–3, 195, 196–7, 210–13, 223, 227
breeding sites 16–17, 18, 54–5, 65, 96, 223
extinctions 46, 47, 90, 92
and management of climate change 227–8
sanctuaries 15–18, 106, 205
bitou bush 28
blackbird, common 30
Bombus terrestris 30
Bonaparte Archipelago 121
boobook, Lord Howe southern 47
booby, Abbott's 16, 20
brown 15, 16, 17, 74
masked 16, 17, 18, 21, 65, 74
red-footed 15, 16, 18
boodie 24, 33, 100, 105, 106, 183
Boondelbah Island 20
border control 135, 136, 137, 138, 214
boronia, coast 20
Boronia anemonifolia subsp. 20
Boston Island 15, 29, 32
Boullanger Bay islands 17
Bountiful Islands 19
Bowen Island 14, 24, 49
boxthorn, African 26, 29, 30, 31, 180
Brachypeza archytus 20
Brachyurophis morrisi 23
Brampton Island 24, 175
Breaksea Islands 156, 174, 184
Bribie Island 13, 25, 178–9
Bribie Island Environmental Protection Association 178–9
bridal creeper 31
bridal flower 47
brodifacoum 95
Bromby Islands 18
Brook Islands 124
broom, canary 30
Broughton Island 24, 25, 87
Bruny Island 23, 26, 32, 51, 52, 117, 167, 170, 174, 178
Bruny Island Bird Festival 178
Bruny Island Environmental Network 178
Buccaneer Archipelago 13
Buckley's Hole 179
buffalo 28
buffel grass 30, 135

bumblebee 30
Burrup Peninsula 31, 32
Bush Heritage Australia 121, 122
bush wren 206
Butchulla people 150
Butler, Harry 136, 139
buttonquail, painted 227
Bynoe Harbour 25

cabbage tree, Norfolk Island 100, 103
Cabbage Tree Island 20, 87
Caladenia insularis 20
Calidris ruficollis 16
  C. tenuirostris 17
Callaeas wilsoni see kokako
Callitris columellaris 150
Calyptorhynchus lathami halmaturinus 187, 227–8
Canis familiaris see dingo
Cape Barren Island 26, 117
Cape York Bioregion 23
Capra hircus see goat, feral
Capricorn Eddy 196
Capricornia and Bunker group 196
Capricornia Cays 19, 196–7
Capricornia Cays, pisonia forests 198–9
Caranx ignobilis 194
Carcharhinus falciformis 14
Carduelis carduelis 30
Caretta caretta see turtle, loggerhead
Carnac Island 31
Carnac Island Nature Reserve 34
Carpobrotus rossii 29
Cartier Island 12, 49
Cartier Island Marine Reserve 12, 14
Casuarina sp. 244
cat, feral 5, 22, 27, 28, 29, 31–2, 33, 46, 52, 56, 72, 77, 86, 87, 90–4, 102, 105, 106, 109, 110, 154, 156–7, 174, 178, 184, 197, 206, 208, 212, 243
  Bruny Island cat management program 178
Catharacta lonnbergi 16
Catharanthus roseus 29
cattle, feral 28, 29, 157
celery, island 20
Cenchrus ciliaris 30, 135
  C. echinatus 72, 77
centipede, giant 210, 211, 212
Cereopsis novaehollandiae see goose, Cape Barren
Cervus unicolor see deer, sambar
chaff flower, Phillip Island 100, 103
Chalcophaps indica natalis 20
Charadrius mongolus 16
Chatham Island 23
Chelonia mydas see turtle, green
Chevron Australia 132, 134, 135, 137, 139, 140, 142
chicken 27, 197
Chionis minor 193
  C. minor nasicornis 16, 153
Christinus guentheri 21
Christmas Island 12, 13, 14, 15–16, 19–20, 25, 27, 49, 50, 86, 167, 224, 241
  phosphate 25, 131
Christmas Island National Park 241
Chrysanthemoides monilifera 28
Churchill Island 33, 186
Clarke Island 26, 119, 184
climate change 27, 30, 156, 221–9
  conservation options 224–6
  Fraser Island 156
  Great Barrier Reef 242
  impact on birds 223
  modelling 223–4
  past 222–3
  policy 54
  prediction tools 226–7
cloud forest 54
cockatoo, glossy black 187, 227–8
Cockatoo Island 26, 31, 132
coconut 27
Cocos (Keeling) Islands 11–12, 13, 24, 27–8, 49, 224
Cocos nucifera 27
Coelostomidia zealandica 211
Coenocorypha iredalei 206
Collaborative Australian Protected Area Database 123
collective impact 69, 240
Collocalia linchi natalis 20
Colobanthus kerguelensis 22
Colubrina peduncalata 20
Columba livia 91
  C. vitiensis godmanae 47
community
  conflict, with tourism 169
  consultation 238–40, 246
  delivery, of nature conservation 184–5
  driven action groups 178–9
  growth, and tourism 169
Conilurus penicillatus 22, 24, 28
conservation
  costs 76–7
  disputes 67
  funding 62
  goals, of island communities 67–8
  management tools 226–7
  partnerships 177–89
  planning algorithms 78
  planning tools 72–80
  programs, and Indigenous cultures 68–9
  tourism 170–1
continuous species response curves 77–8
Cook, Captain 102, 150
Coral Sea Commonwealth Marine Reserve 14, 16
Coral Sea Islands Territories 12, 14
Coral Sea Islands, inhabitants 14, 24–5, 50
Cordyline obtecta 100, 103
Coringa Island 12, 199

Coringa-Herald Nature Reserve 14, 16
Cormocephalus rubriceps see centipede, giant
cormorant, black-faced 17, 156
  Heard Island 16, 153
  pied 17
Corynocarpus laevigatus 211
Cossack 2
Coturnix chinensis 24
Cotyledon orbiculata 31
Cousine Island, Seychelles 244
crab, blue 20
  Christmas Island red 20
  ghost 194
  Jackson's 20
Crab Island 194
CRISPR–9 226
Crocidura trichura 20, 27, 46, 86
crocodile, estuarine 194
Crocodile Islands 12, 17, 18, 117
Crocodylus porosus 194
Croker Island 12, 14, 22, 25, 28, 117
cruise industry 122, 174
cryopreservation 226
Cryptoblepharus egeriae 20, 46
  C. gurrumul 23
Cryptolaemus montrouzieri 199
Cryptostegia grandiflora 28
Ctenotus lancelini 100
cultural impacts, on resilience 215
curlew, eastern 16
currawong, Lord Howe 21
Curtis Island 117, 132
cushion plant 23, 154, 155
Cyanoramphus cookii see parrot, green
  C. erythrotis 27, 92
  C. novaezelandiae 213
  C. novaezelandiae erythrotis 47
  C. novaezelandiae subflavescens 47
cyclones 26–7
Cyrtodactylus sadleiri 20

daisy, silver-leaf 93
  Singapore 29
daisy-bush, small flower 188
Dama dama see deer, fallow
Dampier Archipelago 2, 26, 32, 52
Dampier Archipelago nature reserves 34
Daption capense see petrel, cape
Dasyurus hallucatus see quoll, northern
Database of Island Invasive Species Eradications 87
Daydream Island 166
De Witt Island 30
Deen Maar Island 117
deer, fallow 30, 87
  hog 30
  rusa 28
  sambar 28, 30
deer hunting 30
Dendrocnide peltata 20
Department of Conservation (DOC), New Zealand 206–7, 213, 215, 216, 217
Dermochelys coriacea 193
dibbler 24, 100
Diclipera maclearii 20
dingo 24, 32, 99, 102, 150, 151, 173
Diomedea exulans see albatross, wandering
  D. melanophris 16
Dirk Hartog Island 2, 13, 15, 26, 31, 32, 52, 56, 87, 110, 243
Dirk Hartog Island National Park 33, 34
Discoplax hirtipes 20
Djiru people 124, 125
djoongari 101, 104, 105; *see also* mouse, Shark Bay
DNA fingerprinting 215–16
dog, feral 32, 33, 92
  Polynesian 207, 215
Doole Island 104, 107
Dorre Island 2, 24, 34, 49, 52
dotterels 17
Double Island, NZ 208, 211, 214
Double Island, WA 18
dove, Christmas Island emerald 20
Dream Island, NZ 151
Dromaius ater 27, 47
  D. baudinianus 27, 47
  D. novaehollandiae diemenensis 27
Dryococelus australis 90, 110, 152
duck beak, Christmas Island 20
Ducula bicolor 46, 195
  D. whartoni 20
dugong 155
Dunk Island 124, 125
dunnart, Butler's 23
Dysoxylum bijugum 100, 103
Dyxoxylem spectable 210

EarthCheck 172, 174
East Australian Current 196
East Island 12, 16, 28
Easter group 19
Eastern Kuku Yalanji IPA 120
ecological restoration, Mercury Islands 208–15
  New Zealand islands 205–18
economic growth, by tourism 169–70
ecosystem function, Mercury Islands 210–11
eco-tourism 25, 69, 159, 165, 168, 169, 173, 174, 238–9
Egernia coventryi 24
egret, eastern reef 16
  little 16
Egretta garzetta 16
  E. sacra 16
Elcho Island 12, 17, 22, 25, 118
Elizabeth and Middleton Reefs Marine National Nature Reserve 14

Elizabeth Island 15
Emoia nativitatis 20, 46, 47, 110, 241
emu, Kangaroo Island 27, 47
 King Island 27, 47
 Tasmanian 27
endemism 19–21, 50, 56, 85
English Co. Islands 12, 18
environmental
 culture, Barrow Island 136–40, 142
 legislation 4, 48–57
 monitoring, Barrow Island 134–5
 policies 48–57
eradication, priority threats 56
Eremiornis carteri 100, 105
Eretmochelys imbricate see turtle, hawksbill
Erica lusitanica 30
Erskine Island 197
Eucalyptus cneorifolia 50
 E. globulus 23
 E. viminalis subsp. *cygnetensis* 108
Eudyptes chrysolophus 16
 E. schlegeli 193
Eudyptula minor see penguin, little
euphorbia, Norfolk Island 100, 103
Euphorbia norfolkiana 100, 103
 E. paralias 30
European settlers, animal transportation 102
 early island use 2–3
evidence-based management 244–5
Exmouth Gulf Islands 18, 26, 74
expeditioners, as alien species vectors 158
 sub-Antarctic 159
extinction 3, 28, 33, 46–7, 54, 62, 85, 109–10, 177, 222, 241
 Christmas Island 27, 62, 86
 Lord Howe Island 85–6, 90–2
 Macquarie Island 27, 92–3
 New Zealand 206

fairy-wren, white-winged 33, 100, 105
Family Islands 124
fantail, Lord Howe grey 47
Faure Island 15, 26, 31, 32, 34, 87, 104, 106, 110, 154, 183
Felis catus see cat, feral
Feral Animals on Offshore Islands Database 102
fern, native 23
 prickly shield 93
Field Island 19, 49
Fischer Island 32, 87
Fish Habitat Protection Areas 34
fishing 15, 25, 65, 153, 156, 158
Flat Witch Island 23
flea, European rabbit 93
Flickingeria nativitatis 20
Flinders Island 2, 13, 15, 17, 26, 29, 32, 50, 184
Flinders, Matthew 150
Flying Foam Islands 3
flying-fox, Christmas Island 20, 27
founder animals 107
fox, red 32, 33, 52, 72, 87, 88–90, 102, 104
Franklin Islands 24
Fraser Island 13, 24, 25, 29, 67, 117, 147, 148, 149, 150–2, 159, 173, 239
 biosecurity 157
 climate change 156
 logging 150
 pests 156–7
 sand mining 150
 tourism 150, 173
Fregata andrewsi 16, 20
 F. ariel see frigatebird, lesser
 F. minor see frigatebird, great
French Island 13, 15, 24, 26, 30, 33, 102
Freycinet Special Purpose Zone 34
Friends of Bass Strait islands 180
Friends of Churchill Island 186
Friends of Glossy Black Cockatoo 187
Friends of Islands partnerships 179–81
Friends of Lord Howe Island 180
Friends of Maatsuyker Island 180
frigatebird, Christmas 16, 20
 great 15, 16, 18
 lesser 3, 15, 16, 17, 18, 74
frog, green and golden bell 24
funding, IPAs 126
funding, island conservation 56, 241
fur-seal, Antarctic 19
 Australian 19, 193, 195
 New Zealand 19
 Sub-Antarctic 19

Gabo Island 26, 33
Galapagos Islands 244
Galeocerdo cuvier 191
Gallicolumba norfolciensis 47
gallinule, white 47
Gallirallus australis see weka
 G. philippensis andrewsi 16, 100, 104
 G. philippensis macquariensis 27, 92
 G. sylvestris 21, 90, 152
Gallus gallus domesticus 27, 197
 G. varius 27
gamba grass 28
Garden Island 15, 26, 31, 52
gas exploration, and island conservation 131–43
Gavis immer 76
Gecarcoidea natalis 20
geckos 208, 211
 Asian house 27, 28
 Duvaucel's 211, 215
 four-clawed 27
 giant 20
 Lister's 20, 46
 mourning 23, 27

Gehyra mutilata 27
genetic algorithms 78, 79
genetic analysis 226
genetic status of source population 107
Genista monspessulana 30
Geograpsus stormi 20
gerygone, Lord Howe 47
  Norfolk Island 21
Gerygone insularis 47
  G. modesta 21
Gidley Island 75
Gigis alba 16, 65
Girramay people 124, 125
Girringun Aboriginal Corporation 125
Girringun Aboriginal Rangers 125
Girringun IPA 118, 120, 124–5
Girringun National Park 125
Girugarr 125
glacial maximum 222
glacial retreat, Heard and McDonald Islands 27, 153, 156
glider, squirrel 24
  sugar 102, 108
Global Strategy for Plant Conservation 226
glory pea, Phillip Island 47
Gloucester Island 24
goat, feral 28, 29, 52, 72, 90–1, 103, 106, 110, 152, 157, 197, 206, 207, 208, 212, 244
godwit, bar-tailed 16, 17
goldfinch, European 30
Goold Island 124
goose, Cape Barren 24, 108, 109
Gorgon Project 34, 132–4, 135, 136, 137, 138, 140
gorse 30
goshawk, Christmas Island 20
Goulburn Islands 12, 14, 25, 117
Granite Island 29, 32, 108
grass tree, Kangaroo Island 187
grazing 26, 29, 93, 94, 96
Great Barrier Reef 1, 12, 13, 17, 19, 23–4, 75, 154, 157, 175, 182, 183, 192, 193, 196, 223, 242
  IPAs 120, 124–5
  tourism 1, 25, 166, 167
Great Barrier Reef Marine Park 118, 124–5
Great King Island, NZ 206
Great Mercury Island, NZ 208, 209, 212, 213–14, 215
Great Sandy Marine Park 152
Green Island, NZ 208, 215
Green Island, Qld 117, 166, 167
Green Island, SA 15
Greenpeace 61
greenshank 17
green-washing 168
Grewia insularis 20
Griffiths Island 15
Groote Eylandt 12, 18, 22, 25, 28, 132
ground-dove, Norfolk Island 47
guano 195, 196, 197, 198
guardianship, Māori *iwi* 216
guava, cherry 90
Gugu Badhun people 124
Guidelines for Establishing the National Representative System of Marine Protected Areas 123
guild species 211
gull, kelp 16
Gulnay people 124, 125
Guluwuru Island 32
Gur A Baradharaw Kod Torres Strait Sea and Land Council 118

habitat enhancement 224–5
Habitat Tuateawa 214
Haematopus fuliginosus 17
  H. longirostris 17
halacrocorax fuscescens 17, 156
Haliaeetus leucogaster see sea eagle, white-bellied
Hamish Saunders Memorial Trust Island Survey Program 184
hare, brown 30
hare-wallaby, banded 24, 101, 106, 183
  rufous 24, 101, 183; *see also* mala
  spectacled 33, 101, 105
Hartog, Dirk 2
Hasselborough, Frederick 153
Hatch, Joseph 153
Haul Round Island 17
hawk-owl, Christmas Island 20
Hayman Island 24, 166, 167
Healthy Country plan 51, 52, 118, 120, 121, 122
Heard, John 152
Heard and McDonald Islands 12, 14, 16, 19, 49, 147, 148, 149, 151, 152–3, 156, 157, 159, 193
  glacial retreat 27, 156
Heard Island 12, 22, 151, 152, 153, 157, 159, 193, 224
heath, Spanish 30
Hemidactylus frenatus see gecko, Asian house
Hemidiena thoracica 212
Hemiphaga novaeseelandiae see kērēru
  H. novaeseelandiae spadicea 47
Henri Freycinet Harbour 34
Henry, Richard 205
Hermite Island 31, 33, 87, 105, 107
heron, night 194
Heron Island 2, 166, 167, 195
hibiscus, Phillip Island 100, 103
Hibiscus insularis 100, 103
high-visitation islands 185–7
hihi 207
Hinchinbrook Island 124, 125
Hinchinbrook Island National park 51
HMS *Supply* 152
Home Island 24
honeyeater, helmeted 226
  yellow-tinted 228
Hook Island 166

Hoplodactylus duvaucelii see gecko, Duvaucel's
Horsburgh Island 16, 27
horses, feral 28, 106, 157
Houtman Abrolhos 15, 18, 19, 26, 34, 52, 192, 223, 227
Howea foresteriana 152
Hoya aldrichii 20
Hunter Island 184
hunting, by Indigenous owners 25
  recreational 30
Huperzia australiana 96
Hydromys chrysogaster 105
Hydrurga leptonyx 19
Hypotaenidia philippensis macquariensis 47

ice-plant, common 26, 29, 30
Ile de Phoques Island 32
Ile du Golfe 23
Indigenous
  cultures, and conservation programs 68–9
  owners, conservation by 185
  sea rights 126
  wildfire 123
Indigenous Land Corporation 118
Indigenous Land Use Agreements 116, 118, 122
Indigenous Protected Areas 4, 14, 22, 51, 52, 115–26, 239
Indigenous Rangers 14, 22, 25, 32, 116, 121, 123, 125, 126, 185
Inner Sister Island 184
INPEX 122
interconnectedness, island 195
introduced species *see* invasive species
invasion risk assessment 215–16
invasive plants, as habitat 244
invasive species 23, 27–32, 33, 52, 56, 71, 72, 74, 75, 87, 90, 91, 92, 94–5, 149, 156–7, 158, 182, 197, 206, 208, 211–12, 213, 214, 215, 217, 225, 243–4
Invasive Species Council 182
invertebrates 20, 21, 28, 29–30, 50, 56, 86, 90, 92, 96, 132, 149, 153, 157, 158, 208, 209, 210, 211, 212, 213
Investigator Group 14
Ischaemum nativitatis 20
island communities, conservation goals 67–8
  planning 237
island conservation 4–6, 182
  improving 45–57
  interventions 241
  management priorities 71–80
  and oil/gas exploration 131–43
  partnerships 177–89
island-endemic species extinctions 46
Island Eradication Advisory Group 217
island management 32–4
  4Cs approach 61–70
  and climate change 241–3
  collaboration 240–1, 245–6
  future 235–46
  holistic 243–4
  and risk framework 238–40
  to support marine species 195, 197
Island Point 15
islands, community values 24–6
  ecological links 237
  ecological monitoring 236–8
  importance to marine life 191–9
  inhabited 3, 24
  interconnectedness 195
  intrinsic values 3–4
  philanthropy 183–4
  resorts 166
  and resource development 131–2
  threats 26–32
  use, historical 2–3
  wide recovery plans 51–2
Islands Decision Support System 71, 73–80
Isle of Wight (hotel) 166
Isoodon auratus 23, 32, 100, 105, 107
  I. obesulus 109

Judgement Rocks 32
junglefowl, green 27
Jurien Marine Park 34

kākā 213
kaka, Norfolk Island 47
kakapo 205, 216
kākāriki 213
Kalumburu 121
kangaroo, eastern grey 109
Kangaroo Island 1–2, 13, 14, 15, 19, 20, 24, 29, 32, 51, 52, 102, 108, 167, 186–7, 195, 223
  birds 227–8
  threatened plants 188
Kapiti Island, NZ 205, 206
kapok 30
karaka 211
Karioi-Maunga ki to moana project 216
Keeling Island, introduced reptiles 27
kelp forests 23
Keppel Island Group 117
kerakenga 101
kērēru 210, 211, 213
Kerguelen cabbage 22
K'gari-Fraser Island *see* Fraser Island
Kikuyu grass 28
Kimberley islands 2, 13, 15, 18, 21, 26, 30, 32, 33, 52, 56, 122
Kimberley Science and Conservation Strategy 122
King Island 13, 17, 21, 23, 26, 50, 51, 223, 228
Kingfisher Bay resort 173
kingfishers 212, 213
Kingston and Arthur's Vale Historic Area 25
kitchen design, Barrow Island 136
kiwi 205, 216
knot, great 17

koala 24, 29, 33, 102, 108, 244
kohekohe 210
kokako 207, 217
Koolan Island 26, 31, 132
Kooragang Island 24
Korapuki Island, NZ 208, 212, 213, 214
kuri *see* dog, Polynesian

labour costs 76–7
Lacepede Islands 18
Lady Elliot Island 167, 196
Lady Julia Percy Island 13, 33
ladybird, Australian 199
Lagorchestes conspicillatus see hare-wallaby, spectacled
L. hirsutus see hare-wallaby, rufous; mala
Lagostrophus fasciatus 24, 101, 106, 183
Lalage leucopyga leucopyga 47
Lampropholis delicata 28
lantana 28–9, 244
Lantana camara see lantana
Larus dominicanus 16
Lasiorhinus latifrons 107
Lathamus discolour 108
Leeuwin Current 227
Leggadina lakedownensis 101
Leionema equestre 188
Lepidochelys olivacea see turtle, olive ridley
Lepidodactylus listeria 20, 46
L. lugubris 23, 27
Leporillus conditor see rat, greater stick-nest
Leptonychotes weddelli 19
Lepus apensis 30
Lerista carpentariae 23
Leucocarbo atriceps nivalis 193
Leunig, Michael 178
Lichenostomus melanops cassidix 226
light emissions, impact on turtles 141–2
lighthouse stations 3, 180
Lihou Cays 12
Lihou Reef Nature Reserve 14, 16
Limosa limosa see godwit, bar-tailed
Lindeman Island 166
Linepithema humile 28
Litoria aurea 24
L. dentata 28
Little Barrier Island, NZ 205
Little Dog Island 15
Living Artists (SALA) Festival 187
Lizard Island 167
Lobodon carcinophagus 19
logging, Fraser Island 150
Long Island 18, 166
loon, common 76
Lord Howe Island 12, 14, 21, 25, 51, 54, 147, 148, 149, 151, 152, 158, 166, 167, 180, 223, 224
ant control 152
biosecurity 157
birds 21, 152, 193
extinctions 85–6, 90–2
introduced species 28, 90, 157
ocean warming 156
pest control 90–2
tourism 158, 169–70, 173
translocations 110
weed control 91, 152, 180
Louth Island 15
Low Rock 17
Lowendal Islands 18, 74
lungtalanana 26, 119, 184
Lycium ferocissimum see boxthorn, African
Lycodon aulicus capucinus 86

Maatsuyker Island 23, 155, 180
Macquarie Harbour islets 184
Macquarie Island 2, 13, 18, 19, 23, 26, 33, 51, 56, 87, 92–7, 110, 131, 147, 148, 149, 151, 153–5, 158, 159, 224, 243
Antarctic expeditions 153
birds 157, 192–3
climate change 156
cushion plant decline 155
eradication programs 157, 243
extinction 27, 92–3
feral cat control 93
introduced species 30, 157
invertebrate recovery 96
invertebrate vectors 158
multi-vertebrate eradication program 154
pest control 4, 92–7
rabbit control 93, 94–5
rodent, impact of 92, 94–5
vegetation recovery 96
weeds 30
Macquarie Island cabbage 93
Macronectes giganteus 16
Macropus eugenii see wallaby, tammar
M. giganteus 109
M. rufogriseus 109
Macrotis lagotis 100
Magnetic Island 117, 167, 170
'mainland islands', restoration of 216–17
mala 24, 33, 101, 105; *see also* hare-wallaby, rufous
mallard × Pacific black duck hybrids 91
mallee, Kangaroo Island narrow-leaved 50
mallow, tree 29
Malurus leucopterus edouardi 100, 105
Malva arborea 29
Mandingalbay Yidinji IPA 120
mangroves 75
Māori *iwi* 206–7, 215, 216
Marchinbar Island 32
Maret Islands 122
Maria Island, NZ 243
Maria Island, Tas 23, 26, 29, 87, 108, 109, 183
marine
barriers 213
environment, value of islands 191–9

protected areas 120, 123
species, management 195, 197
turtles 18–19, 22, 193
Marion Island, South Africa 110
marooning 33
marram grass 30
marten, American 76
Martes americana 76
MARXAN 72–3, 74
Maslow's Hierarchy of Needs 66
Matters of National Environmental Significance 45, 48, 49, 54, 241
Mawson, Douglas 153
McDonald, William 153
McDonald Island 12, 16, 19, 22, 152–3, 193, *see also* Heard and McDonald Islands
Melaleuca armillaris 23
Melanodryas cucullata melvillensis 23, 46
Melomys rubicola see melomys, Bramble Cay
melomys, Bramble Cay 3, 27, 46, 47, 54, 62, 110, 156, 221, 224, 241, 242, 243
Melville Island 12, 22, 23, 25, 28, 32
Menura superba 102
Mercury Islands 208–15, 243
Merri Island 15
Meryta latifolia 100, 103
meryta, broad-leafed 100, 103
Mesembryanthemum crystallinum see ice-plant, common
meteorological events 26–7
Mewstone Island 30
Middle Island, Ashmore Reef 12, 16, 28, 31
Middle Island, NZ 208, 209, 215
Middle Island, Vic 15, 33
Milingimbi Island 17, 25
military activity, impact of 27
Milman Island 19
Mimopeus opaculus 212
Mimopeus spp. *see* beetles, darkling
Mimosa pigra 28
mimosa 28
mining 22, 25, 27, 131, 150
Mirounga leonine 19, 193
Montague Island 19, 25, 28, 87
Montebello Islands 3, 18, 26, 31, 33, 52, 74, 87, 105
Montebello Renewal project 87
moose 76
morepork 211, 213
Moreton Bay islands 25
Moreton Island 2, 13, 50
Mornington Island 13, 25
Mossman river grass 72, 77
Motuweta isolate see wēta, tusked
Moulter Cay 193
mouse, house 27–8, 29, 31, 32, 52, 86, 87, 91, 92, 94, 96, 102, 106, 109, 154, 197, 215, 243
northern hopping 22, 23
northern short-tailed 101
Shark Bay 24, 101, 183; *see also* djoongari
Mt Chappell Island 26, 116, 119
Mt Gower 54, 86, 91
Mt Lidgbird 54
Muhr-Amalee Restricted Access Area 125
Muiron Islands 34, 49, 155
multi-vertebrate eradication program, Macquarie Island 154
Murrayfield Station 26, 117
Mus domesticus see mice, house
mussels 158
Mustela ermine see stoat
Mutiny on the Bounty 65
Mutton Bird Islands 152
mutton birds *see* shearwater, short-tailed
myna, common 30
Myoporum laetum 211
Mystacina robusta 206
Mytilus galloprovincialis 158
myxoma virus 93, 94

Natator depressus see turtle, flatback
National Indigenous Sea Country Statement 126
national parks, Australia 61
National Representative System of Marine Protected Areas 120, 123
National Reserve System 50, 120
national risk assessment 56
native title 11, 15, 52, 115, 116, 118, 121, 122, 123, 125, 150
Natural Heritage Trust 62, 123
natural resources, custodianship 61–3
Nature Conservancy 61, 122
nature conservation, community delivery 184–5
nature protectionists 63, 64
nature-based tourism 1, 167, 169, 174, 178, 238–9
Nematoceras dienema 96
N. sulcatum 96
Neophema chrysogaster 223, 228
Neophoca cinerea 19, 186, 195
Nestor meridionalis 213
N. productus 47
new conservation 63, 66, 67–70
New South Wales islands 12, 14
birds 16–17
fauna refuges 24
inhabitants 25
invasive species 28
threats 28
weeds 28
Ngari Capes Marine Park 34
Ngati Hei 215
ngilkat 101
Ningaloo Coast 18, 155, 156, 159
Ninox natalis 20
N. novaeseelandiae 211
N. novaeseelandiae albaria 47
N. novaeseelandiae undulata 28, 227
noddy, Australian lesser 18, 227, 244

black 16, 17, 195, 196, 199
common 16, 17, 55, 74
non-government environmental organisations 181–3
NGO-owned islands 183
non-native species management 156–7, 237
Norfolk Island 2, 12, 14, 25, 28, 49, 65–6, 87, 152, 166, 167, 175, 223, 224, 225
birds 65–6, 227
conservation initiatives 65–6
endemic species 21
plant exchange 102–3
Norfolk Island Birdweek 175
Norfolk Island owl 28, 227
North East Herald Island 199
North East Isle 32
North Keeling Island 12, 50
North Peron Island 17
North Stradbroke Island 2, 13, 117, 132, 167, 169
North West Island 33, 105, 196
Northern Territory islands 12, 14, 18–19
birds 17
flora 22–3
inhabitants 25
vertebrates 23, 28
weeds 28
Notiomystis cincta 207
Notomys aquilo see mouse, northern hopping
nuclear tests 3, 105
Numenius madagascariensis 16
N. phaeopus 16
nutrient transfer, by animals 195, 210
Nuyts Archipelago 14
Nycticorax caledonicus 194
Nyctophilus howensis 21, 47
Nywaigi people 124, 125

ocean-derived nutrients 195
Oceanites oceanicus 16
octopus bush 196
Ocypode sp. 194
Ohinau Island, NZ 212, 215
oil exploration, and island conservation 131–43
oil spills 159
Olearia microdisca 188
Oligosoma alani 209
O. lichenigera 21
O. whitakeri see skink, Whitaker's
Ommatophoca rossi 19
Onychoprion anaethetus see tern, bridled
O. fuscatus 16, 55, 65
orchid, Ridley's 20
Oriental (ship) 152
Orpheus Island 167
Oryctolagus cuniculus see rabbits
osprey 24, 74
Pilbara 24
Outstanding Universal Value (OUV) properties 147, 148, 154, 156–9
Ovis aries see sheep
owl, masked 91, 157
Norfolk Island 28, 227
Tiwi Islands masked 23, 228
oystercatcher, pied 17
sooty 17

Pachycephala pectoralis contempt 21
P. pectoralis xanthoprocta 65
Pachyptila crassirostris 16
P. desolata see prion, Antarctic
P. turtur 156, 174
pademelon, Tasmanian 109
palm, Christmas Island 20
kentia 152
Pambasso Island 32
Pamona Island, NZ 179–80
Pandanus christmatensis 20
P. elatus 20
Pandion cristatus see osprey
P. haliaetus cristatus 24
Papasula abbotti 16, 20
Papilio memnon 20
parakeet, Lord Howe Tasman 47
Macquarie Island 27, 92
Macquarie Island red-fronted 47
Norfolk Island green 65
Parantechinus apicalis see dibbler
Paratrechina longicornis 197
pardalote, forty-spotted 184
Pardalotus quadragintus 184
Parkinsonia aculeata 28
parrot, eastern ground 150
forest 213
green 21, 175
orange-bellied 223, 228
swift 108
Passer domesitcus 197
Passiflora foetida 30
passion vine 30
Pavo cristatus 30
peafowl, Indian 30
Peak Island 19
Pedra Branca Island 156, 184
Peel Island 169
Pelagodroma marina 17, 214
Pelecanoides georgicus 16
P. urinatrix 16, 214
Pelecanus conspicillatus 17
pelican, Australian 17
Pelsaert Island 18, 55
penal colonies 2, 166
penguin, gentoo 16
little 3, 17, 24, 26, 33, 88, 108, 214

macaroni 16
royal 193
Penguin Island 31
Penguin Parade 88, 185–6
Pennicott Wilderness Journeys 174, 183–4
Pennisetum clandestinum 28
Peperomia rossi 20
Perameles bougainville 24, 100, 106, 183
P. gunnii 33, 100, 109
periwinkle, pink 29
Persoonia prostrata 47, 48
pest management 3–4, 32, 56, 85–97, 182, 215, 216, 243–4
Lord Howe Island 90–2
Macquarie Island 4, 92–7
New Zealand 206–18
Petaurus breviceps see glider, sugar
P. norfolcensis 24
petrel, cape 16, 93, 96
common diving 16, 214
Gould's 20
grey 93–4, 96
grey-faced 212, 214
providence 21, 152
Pycroft's 212, 214
soft-plumaged 96, 155
South Georgian diving 16
southern giant 16
white-faced storm 17, 214
white-headed 94, 96
Wilson's storm 16
Petrogale lateralis 107
P. persephone 24, 101
P. rothschildi 101
Petroica multicolor multicolor 21, 65
Pezoporus wallicus 150
Phaethon lepturus fulvus see tropicbird, white-tailed
P. rubricauda see tropicbird, red-tailed
Phalacrocorax fuscescens 17, 156
P. atriceps nivalis 16, 153
P. varius 17
Phascolarctos cinereus see koala
phasmid, Lord Howe 90, 110, 152
phebalium, Kangaroo Island 188
Pheidole megacephala see ant, African big-headed
Philesturnus carunculatus 206
Phillip Island, Vic 1, 13, 15, 24, 26, 30, 33, 88–90, 166, 167, 185–6
Phillip Island, Norfolk Island 87, 103
Phillip Island Nature Parks 1, 90, 185–6
Phoebetria palpebrata 16
phosphate mining 25, 27, 131
Phreatia listeria 20
pig's ear 31
pigeon, Christmas imperial 20
feral 91
Lord Howe white-throated 47
Norfolk Island New Zealand 47
pied imperial 46, 195
pigface 29
pigs 28, 29, 32, 90, 91, 102, 103, 152, 156, 157, 194, 206, 207, 244
Pilbara islands 30, 34, 73, 74, 75, 78
pine, cypress 150
hoop 150
kauri 150
Norfolk Island 21
pinnipeds 19, 191, 193, 195, 237
pipistrelle, Christmas Island 3, 20, 27, 46, 47, 62, 78, 86, 109, 241
Pipistrellus murrayi see pipistrelle, Christmas Island
pisonia 195, 196, 197, 198–9
Pisonia brunoniana 215
P. grandis 195, 196, 197, 198–9
Pisonia Island 19
Pitcairner families 65
planning units 74
plant
dispersal, by seabirds 196–7
extinctions 46
preservation 226
translocation 102–3
Pleurophyllum hookeri 93
plover, hooded 186
lesser sand 16
Plumage League 61
Poa annua 156, 157, 158
P. cookii 22, 96
P. foliosa see tussock
Polynesian settlement, New Zealand 205–6, 215
Polystichum vestitum 93
pomaderris, Kangaroo Island 188
Pomaderris halmaturina subsp. *halmaturina* 188
population preservation 226
Porphyrio albus 47
P. hochstetteri 207
possum, brushtail 24, 109, 217
ringtail 109
potoroo, Gilbert's 101
long-nosed 109
Potorous gilbertii 101
P. tridactylus 109
Pouteria costata 210
Predator Free 2050 217
Prime Seal Island 23, 184
Princess Melikoff Trust 183
Pringlea antiscorbutica 22
prion, Antarctic 16, 94
Antarctic 16, 94
fairy 156, 174
fulmar 16
prioritisation process 77, 78–9

Procellaria cinerea see petrel, grey
Procelsterna albivitta 65
Pseudantechinus mimulus 23
Pseudocheirus peregrinus 109
Pseudomys fieldi see mouse, Shark Bay; djoongari
Pseudopanax crassifolium 211
Psidium cattleianum 90
Pterodroma lessonii 94
 P. leucoptera leucoptera 20
 P. macroptera see petrel, grey-faced 212, 214
 P. mollis see petrel, soft-plumaged
 P. pycrofti see petrel, Pycroft's 212, 214
 P. solandri 21, 152
Pteropus melanotus natalis 20, 27
Ptilotula flavescens melvillensis 228
Puffinus assimilis 214
 P. carneipes see shearwater, flesh-footed
 P. gavial see shearwater, fluttering
 P. griseus 214
Pulu Keeling 50
Pulu Keeling National Park 12, 16
Pulvinaria urbicola 199
Pygoscelis papua 16
python, carpet 24

quail, king 24
quarantine areas 2
Quarantine Expert Panel 135
Quarantine Management System 34, 135
Queensland islands 13, 14, 19
 birds 17
 flora 23
 inhabitants 25
 introduced species 28–9
 weeds 29
Queensland Trust for Nature 183
quokka 24, 172, 186
quoll, northern 22, 32, 100, 107

rabbits 5, 28, 29, 32, 52, 87, 92–6, 103, 154, 157, 174, 207, 208, 212, 215, 243
 rabbit control, Macquarie Island 93, 94–5
 rabbit haemorrhagic disease virus (RHDV) 95
rail, banded 92
 Cocos buff-banded 16, 100, 104
 Macquarie Island 27
 Macquarie Island buff-banded 47
Raine Island 19, 191, 193, 197, 199
Ramsar sites 16, 17, 50
Rangitoto-Motutapu Islands, NZ 215, 216
Raoul Island, NZ 91
Raragala Island 32
rat, black 27, 28, 29, 31, 32, 33, 46, 52, 72, 86, 87, 90, 91, 92, 93, 94, 96, 102, 105, 109, 110, 152, 154, 174, 197, 206, 208, 212, 217, 238, 243
 brown *see* rat, Norway
 brush-tailed rabbit 22
 brush-tailed tree 24
 bulldog 20, 27, 47
 bush 24
 canefield 23
 greater stick-nest 23, 101, 104, 106
 Maclear's 20, 27, 47
 Norway 27, 91, 94
 Pacific 31, 91, 207, 208, 211, 212, 215
 pale field 106
 rabbit-eared 28
 ship *see* rat, black
 swamp 109
 water 105
Rattus exulans see rat, Pacific
 R. lutreolus 109
 R. macleari 20, 27, 47
 R. nativitatis 20, 27, 47
 R. norvegicus see rat, Norway
 R. rattus *see* rat, black
 R. sordidus 23
 R. tunneyi 106
Recherche Archipelago 18, 34, 52
recolonisation, Mercury Islands 212–13
recreational hunting 30
Red Mercury Island 208, 211, 214, 215
Reef Guardians 183
refuges 19–24, 225
 endemic species 19–21
 fauna 23–4
 flora 21–3
Regional Forest Agreements 51
reinvasion, island 215–16
reptiles 20, 23, 27, 50, 149, 208
residents, and mammal control 216
resilience 208–10, 212–13, 215, 224
 social determinants 213–15
Resolution Island, NZ 205
resource development, and island conservation 131–43
restoration, whole-island 4
Rhinella marina see toad, cane
Rhipidura fuliginosa cervina 47
Rhysophora stylosa 75
rights of guardianship, Māori *iwi* 215
robin, Norfolk Island scarlet 21, 65
 Tiwi Island hooded 23, 46
rock art sites 122
rock-wallaby, black-footed 107
 Pearson Island 24
 Proserpine 24, 101
 Rothschild's 101
rodent control 31, 86, 91–2, 93, 94–6, 206, 208, 217, 238
rodents 23, 27, 28, 29, 90, 156, 157
Rodondo Island 13, 23, 32
Roebourne 2
Rosemary Island 32

Rotamah Island 24
Rottnest Island 2, 15, 18, 26, 31, 34, 52, 87, 166, 167, 172, 186, 195
Rottnest Island Authority 34, 172, 186
Rowley Shoals 13
Royal National Park 61
Royal Society for the Protection of Birds 61
rubber vine 28
Rubus niveus 244
Rumex crispus 30
runoff 157, 242
rye-grass 29

saddleback, South Island 206
Salisbury Island 19
Saltwater Traditional Owners 125
sand mining, Fraser Island 150
  Melville Island 25
sandpiper, common 16
Sandstone Island 15
Sandy Island 17
Sarcophilus harrisii 100, 108, 109
Sarigan Island 244
Satellite Island 167
satinay 150
Saunders, Hamish 184
scale, honeydew 211, 212
scale insects 199, 211, 213
Schouten Island 184
Scott and Seringapatam Reefs 13
scrub-bird, noisy 100, 228
  rufous 226, 228
scrubtit, King Island 46, 228
sea country IPAs 117, 118, 121, 123
Sea Country Workshop 126
sea eagle, white-bellied 24, 74
sea level rise 27, 54, 156, 222, 224, 227, 243
sea rights, Indigenous 126
sea spurge 30
seabirds 208
Seagull Island 17
seal, Crabeater 19
  Leopard 19
  Macquarie Island 27
  Ross 19
  southern elephant 2, 19, 193
  Weddell 19
Seal rocks 193
sealing 26, 153
sea-lion, Australian 19, 186, 195
seawater temperatures 156
seed dispersal 157, 158, 197
Setonix brachyurus see quokka
shag, Heard 193
shark, silky 194
  tiger 191
Shark Bay 34, 49, 51, 154–5, 156, 183
sharkwood 100, 103
shearwater, flesh-footed 208, 214, 215
  fluttering 212–13, 214
  little 214
  short-tailed 2, 3, 17, 33, 156, 180, 192
  sooty 214
  wedge-tailed 17, 55, 65, 195, 196
sheathbill, black-faced 193
  Heard Island 16, 153
sheep 29, 32, 106, 110
she-oak, drooping 187
Shoalwater Bay Islands nature reserves 34
shrew, Christmas Island 20, 27, 46, 86
silvereye, Lord Howe 21
Sir Edward Pellew group 12, 17, 28
skink, Arafura snake-eyed 23
  blue-tailed 20, 46
  Carpentarian 23
  Christmas Island forest 20, 46, 47, 110, 241
  garden 28
  Lancelin Island 100
  Lord Howe Island 21
  robust 209
  swamp 24
  Whitaker's 209, 213
skua, subantarctic 16
Sloping Island 184
Sminthopsis butleri 23
snails 21, 122, 208, 213
Snake Island 26
snake, Arnhem shovel-nosed 23
  black tiger 24
  blind 27
  wolf 86
snipe 206
social conservationists 63, 64
socio-economic linkages 238–40, 246
Solanum bauerianum 47
Solenopsis geminata 28
Sonchus oleraceus 31
South Australian islands 13, 14–15
  birds 17
  fauna refuges 24
  inhabitants 25–6
  pinnipeds 19
  weeds 29
South Coast Region Regional Management Plan 34
South Percy Island 29
South West Isle 32
South-west Island 28
sowthistle, common 30–1
sparrow 197
species management, intensive 225–6
SPEXAN 72
Sphagneticola trilobata 29

Sphenodon punctatus see tuatara
spider orchid, French Island 20
Spilopsyllus cuniculi 93
Spilsby Island 15, 29
spinifexbird 33, 100, 105
spleenwort, Christmas Island 20
Spyridium eriocephalum var. *glabrisepalum* 188
spyridium, MacGillivray 188
SS Makambo 90
St Bees Island 29, 244
St Francis Island 24
St Peter's Island 2, 29
Stanley Island, NZ 208, 214
starfish, crown-of-thorns 242
starling, common 30, 91
  Lord Howe Tasman 47
  Norfolk Island Tasman 47
Steep Island 26, 117
Sterna dougalli see tern, roseate
  S. sumatrana 17
  S. vittata see tern, Antarctic
Sternula albifrons 17, 18
Stewart Island, NZ 151, 216
Stilbocarpa polaris 93
stint, red-necked 16
stoat 205, 216, 217
stone-curlew, southern bush 24
Stradbroke group 13
Streblorrhiza speciosa 47
Strepera graculina crissalis 21
Strigops habroptilus see kakapo
stromatolites 154
Sturnus vulgaris 30, 91
sub-Antarctic islands, biosecurity 158, 159
Sula dactylactra see booby, masked
  S. leucogaster see booby, brown
  S. sula see booby, red-footed
Summerland Peninsula 26
Sunday Island 15, 30, 31, 74
superb lyrebird 102
Supporters of Tiritiri Matangi 207
Sus scrofa see pigs
Swainson Island 156
swallowtail, Christmas 20
Swan Island 15
Sweers Island 167
swiftlet, Christmas Island 20
Syncarpia hilli 150
systematic conservation planning 55–6, 72–9, 224, 241

takahe 207
Takapourewa, NZ 206
Taphozous australis 24
taraire 210
Tasman Island 32, 87, 172, 183, 184
  cat eradication program 183
Tasmanian devil 100, 108, 109
Tasmanian islands 13, 15
  Aboriginal occupation 26
  birds 17–18, 193, 228
  endemic species 21
  introduced species 26, 29–30
  vegetation communities 23
  weeds 30
tattler, grey-tailed 16, 17
  wandering 16, 17
Taukihepa, NZ 206
tawapou 210
Taylor Island 15
Te Hauturu-o-Toi, NZ 205
temporal optimisation models 78–9
tern, Antarctic 16, 96
  black-naped 17
  bridled 16, 17
  Caspian 29
  crested 16, 17, 29, 76
  fairy 29
  lesser crested 17
  little 17, 18
  roseate 17, 22
  sooty 16, 55, 65
  white 16, 65
ternlet, grey 65
Tetramorium bicarinatum 197, 199
Thalasseus bengalensis 17
  T. bergii see tern, crested
Thevenard Island 31, 32, 132
Thistle Island 15, 29, 32
thornbill, King Island brown 46, 228
Threatened Ecological Communities 63
threatened species, trends 62–3
Three Hummock Island 184
thrush, Christmas Island 20
  Lord Howe 47
  Norfolk Island 47
thylacine 27, 109
Thylacinus cynocephalus 27, 109
Thylogale billardierii 109
Tiritiri Matangi Island 207, 216
Tiwi Islands 12, 14, 17, 18, 21, 25, 223, 225, 228
toad, cane 22, 28, 32, 123, 156
tools, for climate change predictions 226–7
  for managing conservation 226–7
Torres Strait islanders 2, 4, 6, 25, 115–20, 126, 185
tourism 1–2, 159, 185–7
  Bruny Island 178
  and communities 169
  conservation 170–1
  cruises 122, 174
  and economic growth 169–70
  economic value 237
  Great Barrier Reef 1, 25, 166, 167
  Heard Island 153
  impacts of 168–70

and island conservation 165–75
history on Australia's islands 166–7
K'gari-Fraser Island 150, 173
Lord Howe Island 152, 158, 169–70
nature-based 167, 238–9
Norfolk Island 65
overseas departures 166
Rottnest Island 172
trade deficit 166–7
types 168
tourists, alien species vectors 158
traditional knowledge 68–9
Traditional Owners 2, 4, 25, 26, 33–4, 52, 115–26
Traditional Use of Marine Resources Agreement 118, 125
transformative species 210
Translocation Proposal 107–8
translocations 32, 33, 52, 99–110, 206, 207, 209, 237, 243
effect on ecosystems 108–9
island selection 107–8
Lord Howe Island 110
Maria Island 109
monitoring 104, 106–7
'Treasure Islands' initiative 214, 240
tree frog, bleating 28
Trefoil Island 26, 117
trevally, giant 194
Trichosurus vulpecula see possum, brushtail
triller, Norfolk Island long-tailed 47
Trimouille Island 33, 105
Tringa brevipes see tattler, grey-tailed
T. incana see tattler, wandering
tropicbird, red-tailed 15, 16, 17, 21, 65, 152
white-tailed 15, 16, 20
Trumpeter Islets 156
Tryon Island 198, 199
tuatara 206, 207, 208, 210, 211, 212, 213, 215
Tullaberga Island 33
Turdus merula 30
T. poliocephalus erythropleurus 20
T. poliocephalus poliocephalus 47
T. poliocephalus vinitinctus 47
Turnix varius scintillans 227
turnstone, ruddy 16
turpentine bush, Kangaroo Island 188
Turquoise Coast Island nature reserves 34
turtle, flatback 18, 19, 132, 133–5, 141–2, 183, 193, 194, 197
green 2, 3, 18, 19, 22, 132, 166, 191, 193, 195, 197, 199
hawksbill 18, 19, 22, 193
leatherback 193
loggerhead 193, 195, 196
olive ridley 18, 193, 194, 197
turtle cannery 2–3, 166
turtle rookeries 19, 193
turtles, impact of light emissions 141–2
tussock 93, 96
tussock grass, short 22, 96

Twelve Apostles 33
Typhlops braminus 27
Tyto novaehollandiae castanops 91
T. novaehollandiae melvillensis 23, 228

Ukerabagh Island 117
Ulex europaeus 30
Unwins Island 32
Ursus americanus 76
Uunguu IPA 121–3

Vanderlin Island 28
Varanus Island 26, 31, 132
vectors, alien species 158
vehicles, impact on mammals 134
Vespula germanica 30
Victorian islands 13, 15
birds 18
endemic species 20
fauna refuges 24
inhabitants 26
introduced species 30, 33
weeds 30
Vissher Island 32
volcanism 12, 153
Vombatus ursinus 109
Vulpes vulpes see fox, red

wallabies 77, 227
wallaby, Bennett's 109
tammar 24, 108
Wardang Island 15, 29
Warrgamay people 124, 125
Warul Kawa IPA 116
Warungnu people 124
wasp, European 30
parasitic 199
wastewater treatment, Rottnest Island 172
Wedge Island 15, 29, 87, 107
weed control 91, 135, 152, 180
weeds 28–31
Weetapoona Aboriginal Corporation 26
weka 32, 92–3, 94, 154, 243
Wellesley Islands 13, 14, 17, 25, 29
Wessel Islands 12, 18, 25
West Island, Ashmore Reef 12, 14, 28
West Island, Cocos (Keeling) Islands 12, 24, 27
West Island, NT 28
West Island, SA 29
West Moncoeur Island 13
West Wallabi Island 18
Western Australia, conservation policy 52
Western Australian islands, birds 18
endemic species 21
fauna refuges 24
indigenous use 26
introduced species 30–2

pinnipeds 19
translocations 33
uses 26
weeds 30–1
wēta, tree 212
tusked 208, 209, 210, 212, 215
wetlands 14, 15, 50, 172
whale, blue 197
whaling 2, 26, 65, 152
whimbrel 16
whistler, Lord Howe golden 21
Norfolk Island golden 65
white-eye, Capricorn 196
Christmas Island 20
robust 47
slender-billed 21
white-chested 21, 47
Whitsunday Islands 1, 28, 166, 167
whole-island restoration 4
Wigram Island 167
Wild Duck Island 19
Wildcare 180–1
wildfire, Indigenous 123
wildlife breeding aggregations 54–5
WildMob 65–6, 175, 182–3
Willis Islets 12, 25
Wilson Island 199
Wilsons Promontory islands 33
wombat 109
southern hairy-nosed 107
Wooded Island 18
woodhen, Lord Howe 21, 90, 152
Woody Island 31, 34, 87
workforce, environmental culture 136–40, 142
Working on Country program 123
World Heritage Convention 147–8
World Heritage Islands, Australian 49, 147–60
World Wide Fund for Nature 61, 182
woylie 101, 104, 106
Wunambal Gaambera Healthy Country Plan 2010–2020 122
Wunambal Gaamberaj people 121–3

Xanthorrhoea semiplana tateana 187
Xenicus longipes 206
Xylocopa aerates 187

Yellowstone National Park 61

Zehneria alba 20
Zeitz Foundation 66, 69
Zeuxine exilis 20
Zosterops albogularis 21, 47
Z. lateralis chlorocephalus 196
Z. lateralis tephropleura 21
Z. natalis 20
Z. strenuus 47
Z. tenuirostris 21

www.ingramcontent.com/pod-product-compliance
Ingram Content Group UK Ltd.
Pitfield, Milton Keynes, MK11 3LW, UK
UKHW062002041026
15591UKWH00016B/1127

9 781486 306602